Springer Transactions in Civil and Environmental Engineering

More information about this series at http://www.springer.com/series/13593

Andrea Benedetto · Lara Pajewski
Editors

Civil Engineering Applications of Ground Penetrating Radar

Springer

Editors
Andrea Benedetto
Department of Engineering
University Roma Tre
Rome
Italy

Lara Pajewski
Department of Engineering
University Roma Tre
Rome
Italy

ISSN 2363-7633 ISSN 2363-7641 (electronic)
Springer Transactions in Civil and Environmental Engineering
ISBN 978-3-319-04812-3 ISBN 978-3-319-04813-0 (eBook)
DOI 10.1007/978-3-319-04813-0

Library of Congress Control Number: 2015933624

Springer Cham Heidelberg New York Dordrecht London

Printed on acid-free paper

Springer International Publishing AG Switzerland is part of Springer Science+Business Media
(www.springer.com)

Introduction

Ground Penetrating Radar (GPR) is a safe, advanced, and non-invasive sensing technique that has several traditional and novel applications, sometimes standardised by national or international regulations. It can be effectively used for subsurface investigation, three-dimensional imaging of composite structures, and diagnostics affecting the whole life-cycle of civil engineering works.

The major GPR strengths, on which its success in the civil engineering field is based, are related to the non-destructiveness and non-intrusiveness of the surveys, notably lower costs compared to traditional methods, high-speed data acquisition, reliability and representativeness of measurements. The time needed for a survey is typically one order of magnitude shorter than with possible alternative technology; this results in obvious business benefits of reduced costs, together with limited or eliminated charges associated to restricting access of other activities to the investigated area. GPR provides significant, dense and accurate data; the resolution is higher, compared to competing geophysical technologies as seismic, transient electromagnetic, electrical and magnetic approaches. The main performance limitations occur in the presence of high-conductivity materials, such as clay or salt-contaminated soils, and in heterogeneous conditions causing complicated electromagnetic-scattering phenomena. Considerable expertise is necessary to effectively design and conduct a survey; moreover, the interpretation of radargrams is generally non-intuitive, thus specific competences are needed to enable measurements to be transformed into clear pictures and engineering decision-making data. We are confident that, thanks to the improvement and evolution of both hardware and software technologies, ground-penetrating radar will become an even more efficient, effective, extensively used and less-invasive technique in the near future.

Looking at the current interest of the scientific community, technicians and professionals all over the world, towards GPR and its civil engineering applications, it could seem that the history of this electromagnetic technique is very long. Then, it may sound odd that its origin is assumed in the first applications of the radio-wave propagation above and along the surface of the Earth that were developed about 60 years ago. The first use in the field of civil engineering is commonly considered to have taken place in Egypt and it was oriented to identify the water table depth: in 1956, El Said implemented a research programme, funded by the Egyptian National Research Council, for the geophysical prospection of underground water in the deserts. It is interesting to underline that the methodology adopted by El Said was essentially the same as is currently used to estimate the thickness of layers or the burial depth of targets. In particular, he used a continuous-wave transmitter to diffuse electromagnetic energy in the ground through an antenna laid on the soil surface. A radio receiver measured the wave reflected by the water table. The distance between receiver and transmitter was known. Following procedures nowadays still used, he calculated the depth of the table by measuring the time delay of the received wave.

The whole history of ground penetrating radar is intertwined with its various applications. A significant activity in the field of civil engineering started up in the 1960s and has become mature in the 1970s. Mines and underground deposit inspections were very frequent. Moreover, in the line of the lunar science missions, strong efforts were spent to improve new technologies that seemed very promising for the subsurface examination. Additional and promising uses were observed in archaeology and geology. The electrical characterisation of geological materials, as well as the relationships between electrical conductivity and dielectric polarisation, were topics of great interest in the research community. In the 1980s, the GPR borehole configuration was successfully proposed for the assessment of nuclear waste disposal sites. Starting from these experiences, the borehole configuration has become a relevant standard for hydrological studies of porous media. Another application, that is now likely the most financed, is mine detection for security and humanitarian purposes. In recent years, the GPR was proposed and successfully used for the localisation of people buried or trapped under snow or debris, aiding rescue activities in disaster scenarios such as avalanches, collapsed buildings and earthquakes.

In the civil engineering field, GPR is currently used for inspection, monitoring and design purposes. The detection of utilities and buried objects, as well as the surveying of road pavements, bridge decks, tunnels, and the measurement of moisture content in natural soils and manmade materials, are the main applications. In addition, interesting examples concerning the use of the ground penetrating radar in structural, geotechnical and railway engineering have to be mentioned.

This book is a deliverable of the COST (European COoperation in Science and Technology) Action TU1208 "Civil Engineering Applications of Ground Penetrating Radar."

COST is the longest-running European (EU) framework supporting cooperation among scientists and researchers across Europe; founded in 1971, it has been confirmed in Horizon 2020. It contributes to reducing the fragmentation in EU research investments, building the European Research Area (ERA) and opening it to cooperation worldwide. It also aims at constituting a "bridge" towards the scientific communities of emerging countries, increasing the mobility of researchers across Europe, and fostering the establishment of excellence in various key scientific domains. Gender balance and early-stage researchers are strategic priorities of this programme.

COST does not fund research itself, but provides support for networking activities carried out within Actions: these are bottom-up science and technology networks, centred around nationally funded research projects, with a 4-year duration and a minimum participation of five COST Countries. The Actions are active through a range of networking tools, such as meetings, workshops, conferences, training schools, short-term scientific missions, and dissemination activities; they are open to researchers and experts from universities, public and private research institutions, non-governmental organisations, industry, and small and medium-sized enterprises.

The Action TU1208 is running in the "Transport and Urban Development" COST domain; it started in April 2013 and focuses on the exchange of scientific-technical knowledge and experience of GPR techniques in civil engineering, aiming at promoting throughout Europe a more effective use of this inspection method. The ambitious and interdisciplinary project of the COST Action TU1208 is being developed within the frame of a unique approach based on the integrated contribution of university researchers, software developers, geophysicists, civil and electronic engineers, archaeologists, non-destructive testing equipment designers and producers, end users from private companies and stakeholders from public agencies. About 300 participants from 130 institutions in 28 COST Countries (Austria, Belgium, Croatia, Czech Republic, Denmark, Estonia, Finland, France, Germany, Greece, Ireland, Italy, Latvia, Malta, Macedonia, The Netherlands, Norway, Poland, Portugal, Romania, Serbia, Slovakia, Slovenia, Spain, Sweden, Switzerland, Turkey, UK) and a COST Cooperating State (Israel) have joined the Action. Partners from COST Near-Neighbour Countries (Albania, Armenia, Egypt, Jordan, Russia, Ukraine) and International Partner Countries (Australia, Philippines, Rwanda, United States of America) are participating, too. Applications from further Countries are currently under examination.

During the first year of activity, the partners worked on highlighting the advantages and limitations of the currently available equipment, surveying procedures, and electromagnetic/numerical methods for the interpretation of experimental data. Such studies led to a comprehensive assessment of the state of the art in the field of the civil engineering applications of GPR, and to the identification of open issues and gaps in knowledge and technology. The results of this wide and in-depth review activity were fruitfully discussed during Action's meetings and are presented in this book.

The organisation of the book reflects the scientific structure of the Action, which includes four Working Groups (WGs). In particular, the WG 1 of the COST Action TU1208 focuses on the design of innovative GPR equipment, the building of prototypes, the development of testing and calibration procedures, and the optimisation of new systems. The WG 2 deals with surveying of transport infrastructures and buildings, sensing of underground utilities and voids, testing of construction materials and estimation of soil water content. The WG 3 is developing accurate and fast electromagnetic scattering approaches for the characterisation of complex scenarios, inversion and imaging techniques, and data processing algorithms for the elaboration of GPR data collected during civil-engineering surveys. Finally, the WG 4 focuses on the applications of GPR outside from the civil engineering field, as well as on the combination of GPR with other non-destructive testing techniques.

The book is opened with the first chapter of Part I—"GPR Instrumentation"—authored by G. Manacorda et al. and entitled "Design of Advanced GPR Equipment for Civil Engineering Applications," where the main issues in designing ground penetrating radar equipment dedicated to civil engineering applications are described. A comprehensive review on the commonly available system architectures along with the main design challenges to build an effective tool are herein provided. Overall, the work mostly focuses on three major areas where solutions to technical challenges are nowadays more than ever needed, namely, radio-frequency system design, antenna design and data analysis.

The transmitting and receiving antennas are among the most critical parts of a ground penetrating radar, performing the essential functions of transferring electromagnetic energy to the surveyed scenario with the required pattern, bandwidth and efficiency, and receiving the energy scattered-reflected by the environment. For this reason, Part I is complemented with the chapter by L. Pajewski et al., entitled "Antennas for GPR Systems." This contribution offers a review on the antennas currently used in GPRs, suggesting ideas for their improvement, and resuming the numerical and experimental methods for their electromagnetic characterisation.

Part II of the book is entitled "GPR Surveying of Pavements, Bridges, Tunnels and Buildings; Underground Utility and Void Sensing." It includes five contributions on this topic.

The chapter authored by J. Stryk et al. is entitled "Innovative Inspection Procedures for Effective GPR Surveying of Critical Transport Infrastructures (Pavements, Bridges and Tunnels)." This work thoroughly reviews individual applications, which are currently in use, and outlines those that are still in the phase of research and verification. An overview on issues that need to be dealt with GPR is also addressed, to enable the larger applicability of this non-destructive method in critical transport infrastructures.

The following chapter is entitled "Inspection Procedures for Effective GPR Surveying of Buildings," by V. Pérez-Gracia and M. Solla. It focuses on the main achievements in surveying different types of buildings, on the software development for enhancing data interpretation and on laboratory studies that can be overall relevant for the analyses of complex scenarios. Open issues are also defined as a final conclusion, based on the revision of different works.

In their chapter "Inspection Procedures for Effective GPR Sensing and Mapping of Underground Utilities and Voids, with a Focus to Urban Areas," C. Plati and X. Dérobert present some studies showing the ground penetrating radar performances and limitations in locating and mapping objects such as pipes, drums, tanks, cables and underground features or in detecting subsurface voids related to subsidence and erosion of ground materials, from single-channel systems to the potential of multi-channel three-dimensional imaging and integrating systems. The Authors also discuss the importance of achieving cost-effective installations from the deployment of GPR prior to directional drilling for the prevention of damage to existing utilities.

L. Krysinski and J. Hugenschmidt authored the chapter "Effective GPR Inspection Procedures for Construction Materials and Structures," in which a review of approaches related to the assessment of construction details and material properties by using ground penetrating radar is presented. The analysis of the authors relies on the assessment of electromagnetic properties as a fundamental mean for understanding both materials physical properties and as an inherent part of any GPR structural study necessary for correcting uncalibrated electromagnetic parameters. Major directions of research along with some benefits and limits of different approaches are herein described.

To complete Part II of the book, the chapter by F. Tosti and E. Slob, entitled "Determination, by Using GPR, of the Volumetric Water Content in Structures, Substructures, Foundations and Soil," describes the use of several instruments and processing techniques for the evaluation of volumetric water content in concrete structures and unsaturated soils, at different investigation scales. Strength points and main drawbacks of the commonly used approaches for moisture sensing are discussed, relative to the most recent research studies on this issue. In addition, recently developed methods on this field of application are introduced.

Part III of the book is entitled "Electromagnetic Methods for Near-Field Scattering Problems by Buried Structures; Data Processing Techniques;" it includes overall four contributions on these issues

This part of the book is opened by the chapter "Methods for the Electromagnetic Forward Scattering by Buried Objects," written by C. Ponti, in which the usefulness in using dedicated tools for the solution of forward electromagnetic scattering by buried objects is outlined, with the main purpose of interpreting the GPR responses. A review on the most established approaches in the modelling of impulse radar systems, such as Finite-Difference Time Domain or space-time integral equations, is developed. Furthermore, the issue of implementing novel approaches to approximate the integral equations via series expansions with lower computational complexity, when adopting a Method of Moments discretisation, is addressed. The spectral-domain Cylindrical Wave Approach is presented.

In the following chapter, entitled "Development of Intrinsic Models for Describing Near-field Antenna Effects, Including Antenna-Medium Coupling, for Improved Radar Data Processing Using Full-Wave Inversion," A.P. Tran and S. Lambot deal with the proper description of antenna effects on GPR data and resume the methods that have been developed for this purpose. Traditional numerical methods are computationally expansive and often not able to provide an accurate reproduction of real measurements. The Authors thoroughly describe how intrinsic modelling approaches, through which radar antennas can be effectively described taking into account their fundamental properties, have demonstrated great promise for fast and accurate near-field radar antenna modelling in order to reliably estimate medium electrical properties.

In the chapter "GPR Imaging via Qualitative and Quantitative Approaches," I. Catapano et al. resume the issue of solving an inverse scattering problem, where a set of parameters describing the underground scenario must be retrieved starting from samples of the measured electromagnetic field. The authors provide an overview of different approaches and algorithms for both quantitative and qualitative buried scatterer reconstruction.

N. Economou et al. complete Part III of the book with a chapter on "GPR Data Processing Techniques," wherein the difficulties in automating data analysis are mainly addressed. In this regard, after providing the reader with a deep understanding of the state of the art and open issues in the field of GPR data processing techniques, the authors present an overview on noise suppression, deconvolution, migration, attribute analysis and classification techniques with a particular focus on data collected during civil engineering surveys.

This book is concluded with Part IV "Different Applications of GPR and Other Non-Destructive Testing Technologies in Civil Engineering," which includes four chapters.

The first contribution is entitled "Applications of GPR for Humanitarian Assistance and Security," by X. Núñez-Nieto et al. This chapter reviews a series of published works in the frame of the ground penetrating radar applications for humanitarian assistance and security, with a special reference to the detection of mines and unexploded ordnances. The location of underground spaces and the GPR use in rescue operations is also addressed, wherein its contribution in locating human remains or living victims in disaster areas is always more demanded. The authors analyse specific systems, methodologies and processing algorithms specifically developed for these applications.

The following chapter, entitled "Applications of GPR in Association with Other Non-Destructive Testing Methods in Surveying of Transport Infrastructures," written by M. Solla et al., reviews a compilation of works in the frame of the applications of GPR combined to other non-invasive methods in the evaluation of transport infrastructures. The authors demonstrate that these integrated approaches have significantly benefited the procedures for inspection and they successfully solved some of the limitations of traditional methods in monitoring roads and pavements, concrete and masonry structures, and tunnels.

The next chapter, entitled "Advanced Electric and Electromagnetic Methods for the Characterisation of Soil," by M. Van Meirvenne, deals with the detailed spatial characterisation of soil properties with different electric and electromagnetic methods, which is essential for the management of soil to provide all its functions and essential services to the environment. Electrical resistivity sensors, ground penetrating radar systems and electromagnetic induction sensors are herein thoroughly compared by outlining potential targets of each measurement technique, along with advantages and limitations. Despite the strengths of every type of sensing system, it is suggested by the author an increased integration of soil sensors into multi-sensor systems enabling their fused processing as a future challenge for enhancing the reliability of soil analyses.

The last chapter, entitled "Applications of radar systems in planetary sciences: an overview," by F. Tosti and L. Pajewski, focuses on the remarkable results and sophistication of radar systems achieved over the history in several planetary explorations, by dividing the treatment according to different planets and celestial bodies investigated.

We would like to thank very much the Authors of all the Chapters, for contributing to this book. We are also sincerely grateful to Springer, to Dr. Pierpaolo Riva, Springer Engineering and Applied Sciences Editor, and to the Springer Editorial Staff, for giving us the opportunity to publish this book, for their patience, suggestions and support, and for the efficient handling of the editorial process. Finally, we would like to thank COST for funding the Action TU1208 "Civil Engineering Applications of Ground Penetrating Radar."

Andrea Benedetto
Lara Pajewski

Contents

Part I
GPR Instrumentation

Design of Advanced GPR Equipment for Civil Engineering Applications

Guido Manacorda, Raffaele Persico and Howard F. Scott

Abstract This chapter describes the issues to be addressed in the design of Ground Penetrating Radar equipment dedicated to civil engineering applications. Radar is well known for its ability to detect aircraft, ships, vehicles, birds, rainstorms and other above-ground objects. It relies for its operation on the transmission of electro-magnetic energy, usually in the form of a pulse, and the detection of the small amount of energy that is reflected from the target. The round-trip transit time of the pulse and its reflection provide range information on the target. The application of radar in the detection of buried objects is quite old; there are details of such work dating back to 1910, with the first pulsed experiments reported in 1926 when the depths of rock strata were determined by time-of-flight methods. The design of effective Ground Penetrating Radars requires solutions to technical challenges in three major areas:

- Radio Frequency system design.
- Antenna design.
- Data analysis.

Hence, this chapter reviews the commonly available GPR system architectures and summarises main design challenges to build an effective tool.

G. Manacorda (✉)
Georadar Division, IDS Ingegneria Dei Sistemi S.p.A., via Enrica Calabresi, 24-Loc. Montacchiello, 56121 Pisa, Italy
e-mail: g.manacorda@idscorporation.com

R. Persico
Institute for Archaeological and Monumental Heritage IBAM-CNR via Monteroni, Campus Universitario, 73100 Lecce, Italy
e-mail: r.persico@ibam.cnr.it

H.F. Scott
OSYS Technology Ltd., Ouseburn Building, Albion Row, Newcastle upon Tyne NE6 1LL, UK
e-mail: howard.scott@osys.co.uk

A. Benedetto and L. Pajewski (eds.), *Civil Engineering Applications of Ground Penetrating Radar*, Springer Transactions in Civil and Environmental Engineering,
DOI 10.1007/978-3-319-04813-0_1

1 Introduction

The use of Ground Penetrating Radars (GPRs) in civil engineering is well established and there are several applications where it is currently utilised; they include the location of buried services, the detection voids or cavities, locating steel reinforcement in concrete, geotechnical foundation investigations, as well as archaeological, environmental and hydro-geological surveys.

The main applications of GPR to transportation infrastructures generally include measuring the thickness of pavement layers, detecting voids beneath layers, detecting and locating reinforcing bars, inspecting pavement structure, and mapping of the underground utilities.

Ground Penetrating Radars are designed to probe up to a few metres into the ground through material that is, usually, non-homogenous and, unlike free-space, strongly absorbs radar signals. The frequency range that has been found to be useful for such an application lies within the limits of 100 MHz to 2 GHz.

Requirements for civil engineering applications differ, depending upon the application, and each one imposes a particular set of constraints on the design of an effective GPR. For example, the majority of buried plant is within 1.5–2 m of the ground surface, but it may have a wide variation in its size, may be metallic or non-metallic, may be in close proximity to other plant and may be buried in any one of a wide range of soil types, with implications for large differences in both the absorption and the velocity of propagation of electro-magnetic waves, and consequent effects upon GPR performance. For this application, the most important performance criterion is depth of penetration, with resolution (the ability to distinguish between closely space objects), whilst being important, is a secondary consideration.

On the other hand, surveys of concrete or asphalt pavements requires very high resolution for accurately measuring the thickness of layers composing roadways or the runways; the same applies to the assessment of bridge decks where GPR signals can be analysed to detect potentially corroded areas.

Performance characteristics of GPRs are also often affected by ground conditions that may vary rapidly within the area of a radar survey where, for example, variations in water content can be crucial and, particularly in urban areas, where there could be imported backfill of inconsistent quality. Consequently, it can sometimes be problematic to achieve both adequate penetration of the radar energy and good resolution, and some design compromises may have to be accepted.

In addition, a further issue concerns the interpretation of GPR data, which is not trivial in many situations; in this respect, the latest developments in GPR are oriented towards the design of equipment featuring real-time 3D high resolution images of surveyed areas.

Images, such as that shown in Fig. 1 can easily be understood even by an unskilled operator; however, this visualisation improvement can be effective only if the GPR performs well in terms of signal quality and detection range; in fact, if the received signal is too weak, as would be the case in wet, muddy ground, enhanced graphics software will solve neither the basic signal problem nor the detection performance.

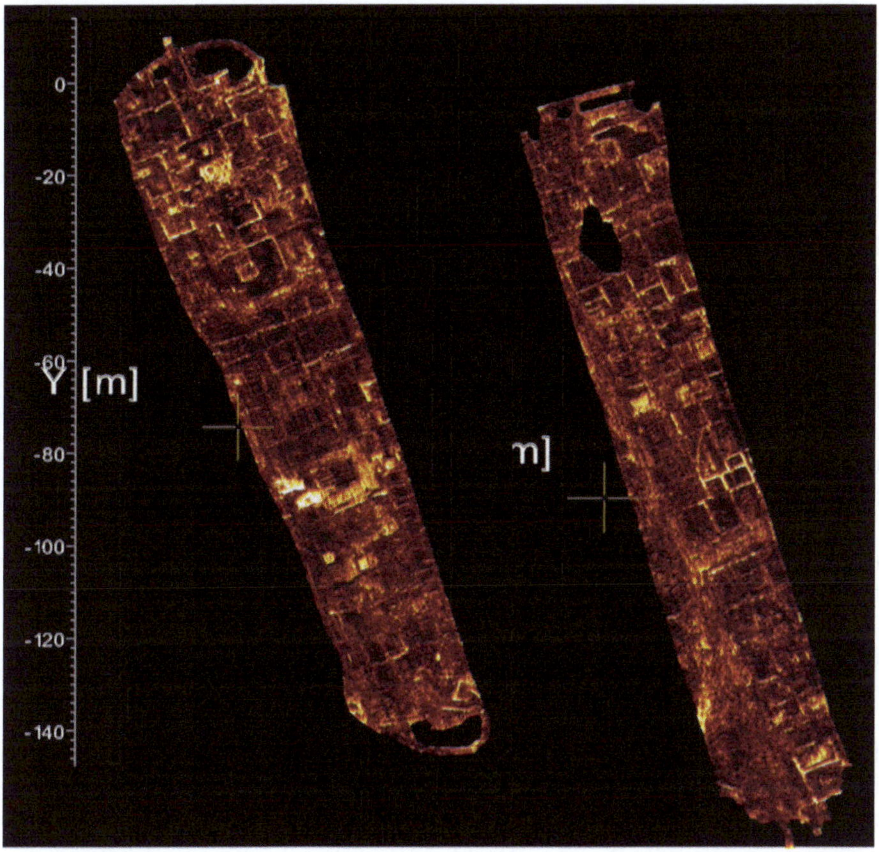

Fig. 1 High resolution GPR image of the archaeological site of Empúries (Spain) (Courtesy of Geostudi Astier—Italy)

Consequently, the design of high performance equipment is a complex but fascinating task for engineers and researchers as it involves a wide range of expertise such as electromagnetic wave propagation in media, antenna technology, radar design and electronics as well as advanced signal processing techniques and computer graphics.

2 The Radio Frequency System

2.1 Introduction

The purpose of the Radio Frequency (RF) system is:

- To generate an electrical signal of appropriate power level, frequency range and spectral characteristics, and to apply it to the transmit antenna.
- To process energy collected by the receive antenna into a form suitable for data analysis.

The RF energy is usually in the form of a short pulse, a frequency modulated burst of electromagnetic energy or discrete frequencies transmitted in a known sequence. RF system design and costs are significantly affected by the choice of modulation technique. Pulse modulation is, at present, cheaper to implement and is the method used in most commercial systems. Performance benefits, particularly increased dynamic range, are available from Frequency Modulated Continuous Wave (FMCW) and stepped frequency systems, but practical limitations imposed by the physics of the process make the benefits difficult, but not impossible, to realise in practice. Other modulation techniques are possible, such as noise and pseudorandom coding, but these are very seldom used.

The critical performance parameter of the RF system is dynamic range. Pulse modulated systems use sampling receiver techniques where, typically, the dynamic range will be 70 dB. Time varying gain is usually applied which can increase the dynamic range to 90 dB or more. In addition, averaging may be used (if acquisition times permit) which can provide a further increase. Frequency modulated systems, either continuous wave or stepped, can increase dynamic range.

However, when the RF system is connected to practical antennas, internal system reflections, ground returns and transmit to receive antenna leakage generate time dependent clutter (clutter is the term used for returns identified by the system as targets that do not correspond to intended targets or to noise). Unless special measures are taken, clutter limits dynamic range regardless of whether time or frequency domain systems are used.

The final stage of the RF system transforms the analogue signal into digital form for data analysis and display purposes; this conversion must be executed with a suitable sampler resolution that prevents the limiting effects on dynamic range of quantization noise.

2.2 Time Domain GPR

Usually, Time Domain GPRs produce transmit signals with the required frequency range by using an impulse generator based upon an avalanche transistor. A typical pulse obtained from such a device is shown below, together with its spectrum (note that the signal consists of just a single cycle with a period of, approximately, 10 ns, as shown in the diagram Fig. 2). Although this is a cost-effective means of producing a signal with usable characteristics, the physical mechanism is a random process that may produce noise and jitter, which limits the inherent dynamic range of the system.

The receivers for such systems are based upon the methods used in high frequency time domain sampling oscilloscopes which also have fundamental limits on their dynamic range that rarely exceeds 70 dB.

The top diagram in Fig. 3 depicts a typical GPR system, consisting of an impulse source and receiver connected by transmission lines to transmit and receive antennas. The system is deployed close to the ground surface, and interactions

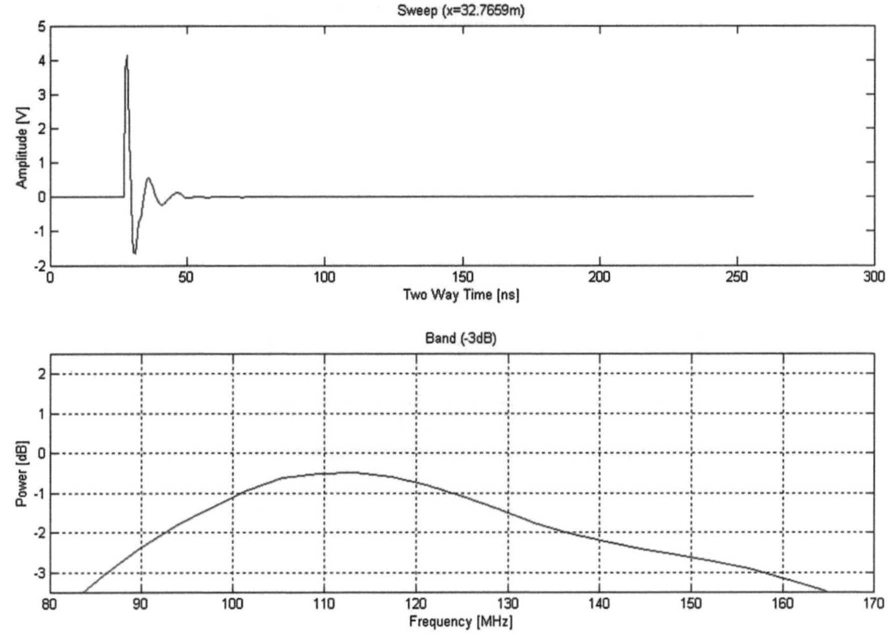

Fig. 2 Impulse radar pulse and spectrum envelope

occur between the radar and the ground and between its internal components. The major interaction paths are marked.

These interactions are extended in time and define what is known as the 'Impulse Response' of the system. It is also known as the 'Clutter Profile'. This is shown in the Fig. 3 bottom diagram as a decaying received signal—with respect to time and hence distance from the radar. Reflections from targets buried in the ground must be large enough for their peaks to be visible above the clutter profile. The system clutter profile is a critical system performance parameter and the radar must be designed to minimise its decay time and to make it, as far as possible, independent of the electrical properties of the ground.

2.3 Frequency Domain GPR

Frequency domain radar systems have as long a history as their time domain counterparts. For some applications, the advantages of simple Continuous Wave (CW) systems is that they avoid the complication of modulation circuitry, have no minimum or maximum range as well as maximising power on the target.

However, because they depend upon the Doppler shift principle they also have the disadvantage of only being able to detect moving targets. The main use of such systems has been military, where they provide a means of determining the point of

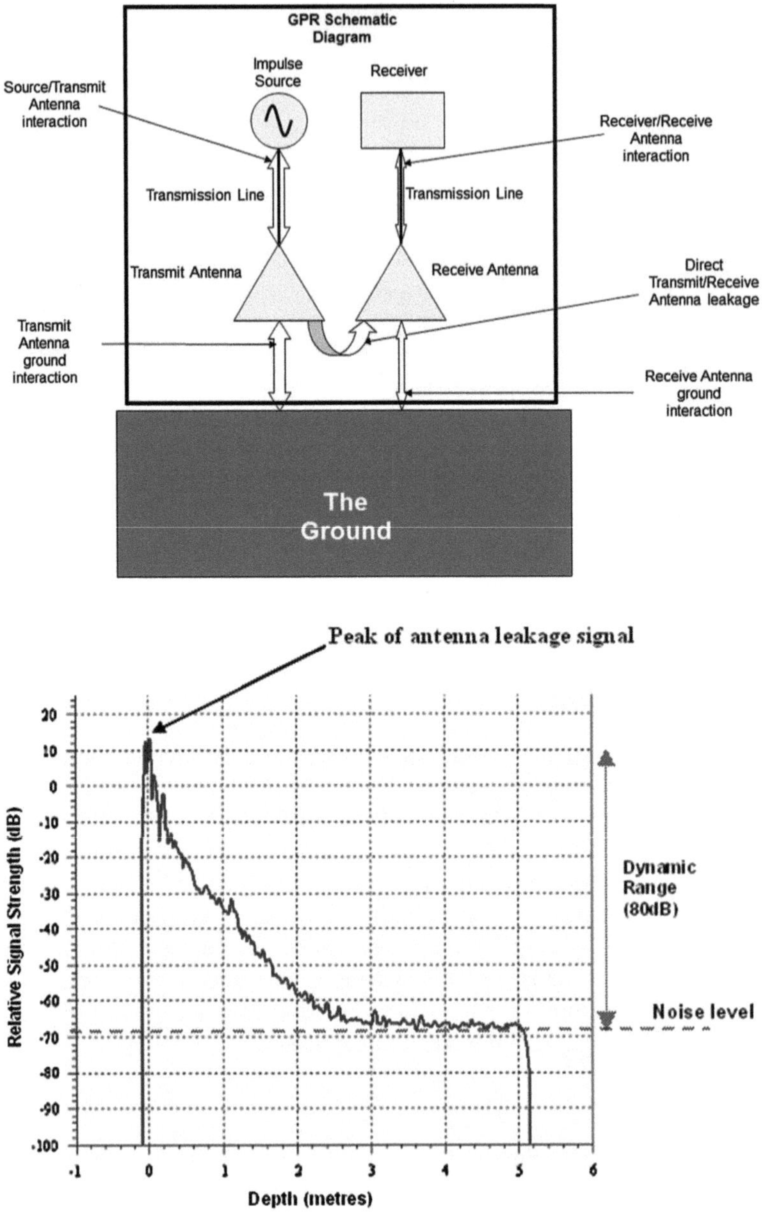

Fig. 3 Impulse radar GPR Scheme with major signal interaction paths (*above*) and consequent GPR clutter response (*below*)

closest approach by guided weapons to their targets so that the warheads may be detonated at the correct time.

Being unable to detect targets unless they are moving and producing a Doppler shift clearly makes CW radars unsuitable for GPRs. If, however, the source is able to produce a range of frequencies continuously varying with time, then it is possible to detect targets that do not move. Such radars are known as Swept Frequency Continuous Wave (SFCW). Usually the signal is generated by a source whose frequency can be controlled by the application of an external DC voltage.

In both the generation and reception of frequency domain signals, the technology is very different from that their time domain counterparts, and some aspects of the performance of such systems, particularly noise and dynamic range, are superior.

2.3.1 Principles of Frequency Modulated Continuous Wave (FMCW) Receivers

FMCW radar systems are able to detect motionless targets because of the time delay between the energy originally transmitted and the reception of energy reflected from the remote target. Because the frequency of the transmitted signal is constantly varying with time, the signal from the target will be different to that currently being transmitted. If the transmitted frequency changes linearly with time, then the frequency difference is a direct measure of the range of the target (Fig. 4).

By choosing the rate of change of frequency appropriate to the target range, the difference frequency can be set so that it lies within a range that can be processed by audio frequency devices. A typical difference frequency might be 10 kHz, so that a narrow-band filter can be used to minimise the noise bandwidth and, hence, maximise the dynamic range.

By applying a sample of the transmitted signal and the received signal to a non-linear receiver, sum and difference frequencies are created. A low pass filter easily separates the required difference and sum frequencies.

The diagram in Fig. 5 depicts a simple FMCW radar system. In the GPR application illustrated, the same signal interactions are present as in the impulse radar.

Detection is achieved by taking a sample of the transmitted waveform, and using it as a Local Oscillator (LO) input to a diode. The much weaker received signal is also fed into the diode, which acts as a 'phase coherent' detector. This

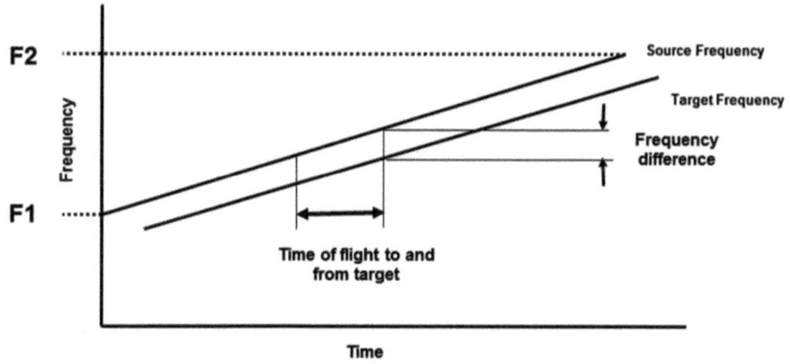

Fig. 4 FMCW difference frequency generation

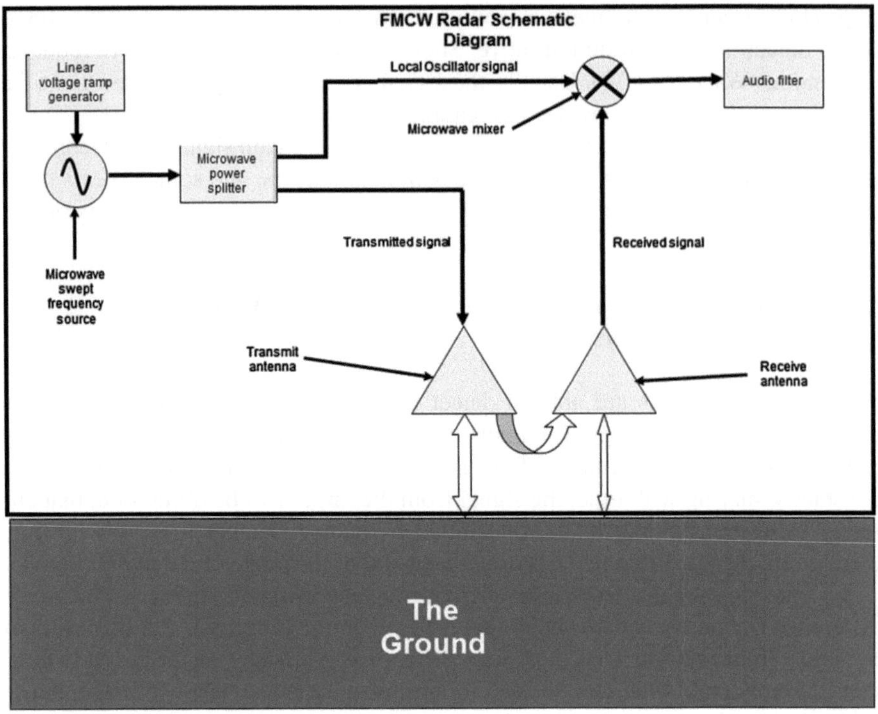

Fig. 5 FMCW radar schematic diagram

signal mixing (heterodyning) in the diode creates sum and difference frequencies from the LO and received signals. The sum frequency is outside the frequency range of the receiver and is thus filtered out, whilst the difference frequency is processed further to yield the required target information. This processing takes the form of a transformation of the difference frequency data, by means of a Fourier Transform, into the time domain so that range information may be extracted.

FMCW radars offer wider dynamic range, lower noise figures and can radiate higher mean powers than the time domain counterpart. In addition, the frequency range used is easily tailored to suit the characteristics of the material and targets under investigation. If, however, degradation of the system resolution by spectral widening of the Intermediate Frequency (IF) is to be avoided, then a high degree of the linearity in the variation in frequency as a function of time is required.

2.3.2 Time Domain to Frequency Domain Transformation

Fourier, in his *Théorie analytique de la chaleur* (1822), stated that any periodic signal can be represented by the sum of an infinite series of sine waves separated by the repetition frequency of the time domain signal. The process has become known as the Fourier Transform. Further, an inverse process may be applied to the series of sine waves to transform them from the frequency domain into the time domain. This is known as an Inverse Fourier Transform.

All time domain signals are represented by a single value that describes their amplitude at any instant in time. On the other hand, the frequency domain description of a sinusoid requires both amplitude and phase. The output, therefore, of the Fourier Transform requires two channels of information to convey a full description. This can be amplitude and phase, but it can also be the real and imaginary parts of a complex number. In order to carry out the Inverse Fourier Transform correctly, both the real and imaginary parts of the frequency domain signal must be known. The simple schematic of a FMCW system as shown above, cannot supply the complete information required.

As the repetition frequency decreases, then the frequency difference between the sinusoids of the Fourier Transform becomes smaller until, in the limit when there is only a single occurrence of the signal, the frequency spectrum becomes continuous. The diagram below shows a single occurrence of a real time domain signal, together with its Fourier Transform. Note that the time domain signal illustrated has a DC component, and the spectrum of the output of the Fourier Transform is distributed about zero frequency. With radar pulses, there would be no DC component and the spectrum of the Fourier Transform would be distributed about the centre frequency of the radar system (Figs. 6 and 7).

To obtain the correct Inverse Fourier Transform, both the real and imaginary components of the frequency domain version must be used in the calculation, as shown below. The general case is that, in principle, the time domain signal may also have real and imaginary parts.

The above has implications for the design of a FMCW radar system, because an extra channel of information is needed so that both real and imaginary parts (or amplitude and phase) of the frequency domain signal can be captured. To achieve this, the Local Oscillator output is divided into two signals which differ in phase by 90°, and each is applied to its own mixer. The signal from the receive antenna is also divided, with no relative phase shift. These are mixed with the two LO signals to generate the real and imaginary parts. They are more usually known as the 'In-Phase' (I) and 'Quadrature' (Q) components.

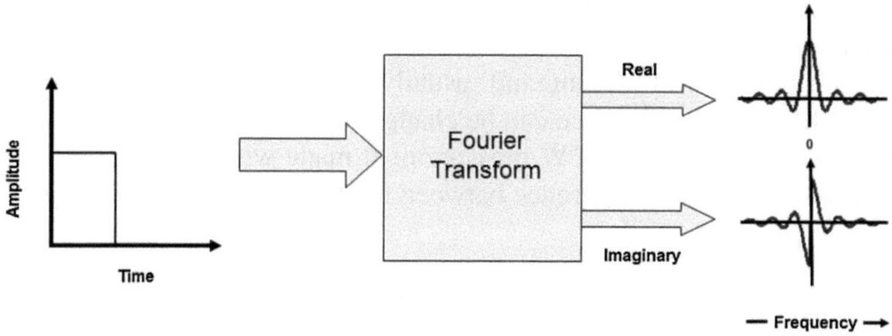

Fig. 6 Fourier transform concept

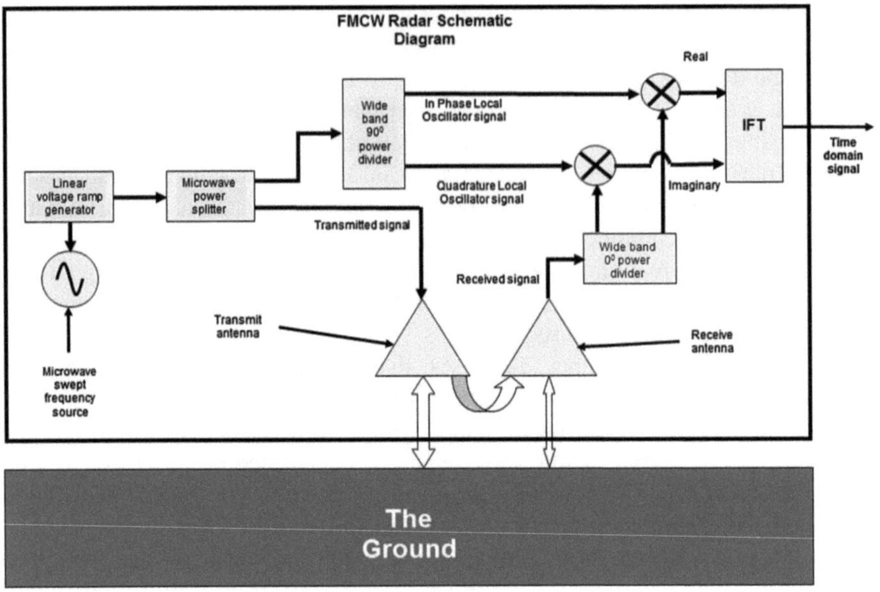

Fig. 7 FMCW radar with I and Q outputs schematic diagram

2.3.3 Stepped Frequency Continuous Wave Systems

Thus far, the microwave source has been described as a voltage controlled swept frequency generator where the output frequency is continuously variable; the main advantages and disadvantages of such a system have been outlined above.

Another class of microwave sources is available where the frequency can be changed in discrete, highly repeatable and stable, steps. In this case, each measurement is made at a constant frequency and, hence, the output of the both the In-Phase and Quadrature mixers is a constant voltage. Because both Local Oscillator signals are coherent with the received signal, the output of the mixers is proportional to the difference in phase between the transmitted and received signals, but the difference frequency, as described for the continuous wave source, cannot exist because there is no change in frequency. In this sense, it is a CW radar and, technically, is termed a homodyne system.

By the application of a command, usually issued by a digital system, the frequency of the microwave source can be changed to a new value and, after allowing for the settling time, another CW measurement made where the mixer outputs are proportional to the phase difference between the transmit and receive signals at the new frequency.

As the microwave source is systematically 'stepped', in equal increments, through its complete frequency range, the DC voltages from both mixers are digitised and stored. At the completion of the process, the record of the voltages of the mixers (which are proportional to change in phase of the received signals, with respect to the transmitted signal) may be displayed as a function of frequency of the transmitted signal. The result is indistinguishable from the difference frequency that would have been obtained from the continuously swept source. However, it is not possible to apply a filter to the data, so the Nyquist requirement that the sampling rate provides a range of twice that to the most distant target to be detected (whether that target is below or above ground) must be strictly observed.

The advantage gained from the extra complication of a stepped frequency source is that the transmitted CW signals are extremely stable and spectrally pure to a degree that cannot be obtained from a continuously swept source. Also, the accuracy of the frequencies provides an extremely linear sweep, which eliminates one of the principal weaknesses of FMCW sources.

3 Ground Penetrating Radar Antennas

The purpose of the antenna system is to illuminate the target and to collect the resulting scattered energy from the environment. Two main types of system are used:

- Monostatic.
- Bistatic.

Monostatic systems use a common radiating element for transmitting and receiving the signal; most GPRs use bistatic antennas with the transmitter and receiver hosted in the same enclosure.

The key requirements for the antennas are:

- An ability to radiate a wide range of frequencies, typically over the range 100–1000 MHz or more.
- Linear phase characteristics over the operating frequency range.
- Predictable (and, preferably, constant) polarisation characteristics over the operating frequency range.

The key design objectives for a GPR antenna are to achieve

- a high degree of isolation between transmit and receive antennas;
- a low return loss from the feed point to minimise ringing and thus clutter generation;
- a good immunity from radio frequency interference.

Two general types of antenna have been used so far; dispersive and non-dispersive. In dispersive antennas different frequencies are radiated at different times, usually with high frequencies being radiated first followed by the low frequencies. This type of waveform is sometimes known as a "chirp". Although some antennas are classed as non-dispersive, in fact all antennas are dispersive to some degree. Examples of these classes of antennas are:

- Dispersive—Spirals (logarithmic, exponential and Archimedean), exponential slot, Vivaldi.
- Non-dispersive—TEM horn, biconical, bow tie, resistive lumped element loaded, resistive continuously loaded.

The design of the antenna system is influenced by the nature of the required targets. All targets tend to depolarise electromagnetic waves incident upon them, the most extreme example of this being long, thin (in terms of wavelength) objects. If the long thin object is metallic then the reflected wave's polarisation is parallel to its axis; if it is non-metallic, then the reflected polarisation is orthogonal to its axis (Roberts and Daniels 1996).

In the case of radar systems designed to locate pipes and cables, this is a very useful property that can, in principle, be used to improve the response to the required target. A complete knowledge of the depolarisation characteristics of targets in general, particularly how they behave over a range of frequencies, can be a powerful aid in target detection and classification. There are many descriptions in the literature of antenna configurations specifically designed to extract polarisation information.

3.1 Array of Antennas

Optimum GPR system performance, particularly for high resolution detection of shallow objects, is obtained when the whole of the system is designed around a specific target type or geometry.

For example, whenever a high density of utilities is expected in several orientations (e.g. in an inner city road junction), a single antenna GPR must be scanned over a dense orthogonal grid (i.e. along two directions at right angles to each other) with a step no larger than say 0.5 m; using a larger step might be adequate when trying to trace a single pipeline across a field where the approximate entry and exit points are known, but is totally inadequate when mapping a complex layout of underground assets.

Thus, when the collection of data requires the execution of a large number of profiles, the use of an array of antennas lined up in the transversal direction with respect to the direction of movement, and operating simultaneously, will provide the most efficient method.

This architecture enables another advantage in respect to the analysis of collected data; in a certain sense, an array GPR implements a scheme very similar to the one used by "double threshold detection" radar.

Non-array GPRs are examples of, "single threshold detection radars", where the operator decides whether a target is present on the basis of one "peak detection decision"; in other words, the output of the receiver is compared to a threshold or bias level.

This bias level, which is dependent upon the sensitivity of the display and the human operator's visual perception, affects the probability of generating a false alarm and of missing a genuine target; in other words, when a received signal component that has been generated by noise or clutter exceeds the bias level, it can be mistaken for a return from a genuine target and there is a "false alarm". On the other hand, when a signal received from a genuine target is interpreted as a noise pulse or clutter (which occurs when the signal return is below the bias level), there is a "missed detection".

Missed detection and false alarm rates are subject to trade-off; it means that the number of false alarms (missed detection) may be decreased (increased) as the bias level is raised, and vice versa; therefore bias level is a primary parameter in the radar's design.

For solving problems related to the human operator, "double threshold detection" radars have been introduced. Such systems impose detection criteria whereby there must be several occurrences of the threshold of detection being exceeded in a defined period of time before a target is confirmed; therefore, the performance of these equipment are less vulnerable to the sensitivity of human operators.

Analogously, in GPR operations, the use of antenna arrays improves the detection of buried utilities because elongated targets (e.g. pipes or reinforcement bars in concrete) produce echoes (in the form of hyperbolae) in the same position in most (ideally in all) of the data windows; therefore, the operator can easily distinguish this category of target from those that are concentrated (e.g. a stone), and performance, in terms of probability of detection, is improved.

A further benefit of this array based architecture is the possibility of using advanced signal processing techniques; data collected with the array are geometrically coordinated and can be stored as a three-dimensional data set that may be displayed as slices in vertical or horizontal planes. These are effectively microwave images, and image-processing techniques may be applied to enhance wanted features; in this respect, as the target sought has identifiable properties (a pipe is long and thin), then this may be taken into account in the processing by the application of appropriate filters, such as line finding algorithms.

Finally, the latest developments in GPR are tending towards providing the capability of generating real-time 3D high resolution images of surveyed areas. This objective is achievable by having a complete, very dense, coverage of the surveyed area, in the form of a 3D data set.

This could be achieved, albeit less efficiently, by performing hundreds of 2D profiles, very close to each other, with a single antenna GPR. However, since this data collection procedure is very time consuming, several dense arrays have been developed with the capability of producing 3D data volumes with a single scan.

4 Data Processing and Analysis

The processing of GPR data, with its several aspects both theoretical and practical, could form a "chapter" in its own right. At first sight, it appears that the mathematical ill-posedness of the problem (Colton and Kress 1992) makes it impossible to retrieve any details of the buried scenario. This should not make us unnecessarily pessimistic. However, this means that, whether the exploited processing algorithm, there will be a finite resolution and a finite quantity of information extractable from the data. Consequently, it is illusory to think that gathering an indefinitely growing number of data will achieve an increasingly precise image, and it is even harmful to think of prolonging indefinitely the processing in order to improve the achieved results. This over-processing is the equivalent of squeezing a lemon in an attempt to extract more juice than it contains.

That being said, several categories of processing can be identified, depending upon whether a 1D, 2D or (more rarely) a 3D approach is taken and depending upon whether the problem is treated by a linear or (more rarely) nonlinear method. Within each approach, the configuration of the antennas has also to be taken into account (Persico et al. 2005), as well as the height of the measurement line (Persico 2006), the frequency band (Sala and Linford 2012) and the a priori information available in relationship with the case history at hand. In particular, if reliable a priori information is available, that is if the nature of underground targets are already presumed, and only some of their geometrical features are looked for, then a forward modelling technique may be exploited (Daniels 2004; Utsi 2012).

4.1 One-Dimensional Processing

Proceeding in order of increasing complexity, and taking for granted the preliminary step of the zero timing, 1-D processing is, in general, a technique where GPR traces (gathered at any fixed measurement point and presented either in time or frequency domains), are processed independently from each other. This can be done if a one-dimensional scenario is scanned so that a 1-D model of the propagation can be exploited (Persico and Soldovieri 2004; Pieraccini et al. 2006), or the interest is in equalizing the deformation of the signal when it is reflected by the targets of interest.

In particular, it can be the case that incident waves may characteristically be scattered by some targets, allowing them to be recognised and, possibly, distinguished from other objects that may be geometrically similar, but of a different nature. This procedure is called deconvolution (Daniels 2004; Jol 2009), and is sometimes exploited where such an identification is particularly important, as e.g. in de-mining operations (Daniels 2004).

The predominant characteristic of GPR data is that it diminishes in amplitude as a function of time. This is caused by the geometrical spreading of the radiated (and scattered) energy plus the electrical losses usually present in the soil. These effects combine to make the echoes from the deepest targets possibly much weaker than those from shallower targets. By applying an increase in gain versus time along the received signal, it is possible to compensate for this attenuation and make those targets visible.

It should be noted that varying the gain as a function of depth is a non-stationary processing step, which can cause a spurious enlargement of the spectrum of the traces, causing a deterioration of the image. This enlargement of the band is theoretically easily explained by the fact that a variable gain is equivalent to a multiplication of the time domain trace by a monotonically increasing function.

In the frequency domain, this is equivalent to the convolution product of the two spectra that, in general, leads to an enlargement of the band.

This effect can, usually, satisfactorily be mitigated by means of a further 1-D processing procedure consisting of filtering the traces to limit their spectra to be within the original band. This can be done via software making use of an ideal filter (the causality requirement is not essential because the GPR post-processing is not a real time operation) or an algorithm imitating a physically realisable filter, as e.g. a Butterworth filter (Di Lorenzo 2013). Any choice has its pros and cons. In particular, an ideal filtering process can, theoretically, anticipate the depth of the target whereas a "real filter" might add some further distortion in the band (in any case, both of these effects are usually negligible).

Variable gain and 1-D filtering procedures are usually contained in the routines available in commercial codes for GPR data processing, as e.g. the Reflexw (Sandmeier 2003) or the GPRslice (Goodman and Piro 2013). It is worth noting that variable gain and 1-D filtering are also usually exploited within the processing chain performed in 2-D or 3-D contexts. In other words, 1-D processing steps can be (and in practice are) mixed with 2-D or 3-D processing steps. To summarise with a final formula the 1-D filtering, let $\hat{T}(\omega)$ represent the spectrum of a trace $T(t)$.[1]

A 1-D filter is the multiplication of $\hat{T}(\omega)$ by an established filtering function $H(\omega)$, so to achieve a filtered spectrum $\hat{T}_F(\omega)$ given by:

$$\hat{T}_F(\omega) = H(\omega)\hat{T}(\omega). \tag{4.1}$$

4.2 Two-Dimensional Processing

The most common forms of 2-D processing are spatial filtering and migration. The difference between 1-D and 2-D filtering is that, in the second case, several traces are treated and combined together in some manner to produce a comprehensive representation known as a B-scan. The most common 2-D spatial filters are implemented by multiplying the spectrum of the data by a 2-D filtering function.

More precisely, labelling the 2-D Fourier transform of the datum as $d(x, t)$ with respect to the measurement abscissa and to the time as $\hat{\hat{D}}(k, \omega)$, the most common 2-D filter is implemented by multiplying this spectrum by an established filtering function $H(k, \omega)$ to produce the filtered spectrum of the data $\hat{\hat{D}}_F(k, \omega)$, given by

$$\hat{\hat{D}}_F(k, \omega) = H(k, \omega)\hat{\hat{D}}(k, \omega) \tag{4.2}$$

Included in this category is the FK filter, which aims to reject the effect of possible reflection from targets in air (Chan and Stewart 1994) and the Background

[1] The data may be gathered either with a pulsed or a swept frequency GPR, in the latter case $T(t)$ is meant as an equivalent trace in the "synthetic" time domain.

Removal (BKG), that is aimed to remove background clutter caused by quasi-horizontal reflectors that may mask targets of interest (Persico and Soldovieri 2008).

In particular, the quasi-horizontal "disturbing" reflections can be generated from either actual physical discontinuities (such as the air-soil interface, and also, possibly, any layered structure of the underground scenario) or from an apparent constant target, physically generated (e.g.) by a vehicle that moves the antennas or by ringing (Daniels 2004) of the antennas. The background removal can be performed on all of the traces or on a limited subset centred on the current trace (which is called a moving average). In theoretical terms, it can be shown that the first case is a particular instance of the second.

It is important to outline that any "therapy" on the signal has disadvantages. In particular, it is impossible to filter out completely the whole of the undesired parts of the image and leave intact the useful parts. Therefore, the nature and the spatial band of the filtering have to be determined by the data, and by the expertise of the human operator. The background removal on all of the traces may generate "new" horizontal artefacts (paradoxically, in a sense) when, in particular, the top of a strong reflector makes the level of the average trace significantly different from the level of most traces at some points along the depth axis.

Indeed, more complex 2-D filtering processes have been proposed, as e.g. the eigenimages (Kim et al. 2007), based on the singular value decomposition (SVD) (Bertero and Boccacci 1998) of the matrix of the data, that is the function $d(x, t)$ (we can gather only a finite number of data, which means that, in the end, there is a sampled version of $d(x, t)$ representing a continuous function). In particular, the matrix $d(x, t)$ can be decomposed along component matrixes (called eigenimages) by means of its SVD, and the subsequent eigenimages are weighted by the associated singular values, forming a decreasing sequence, which makes them more affected by the noise. The filtering, in this case, consists of retaining the first eigenimage or a few eigenimages corresponding to the highest singular values.

Beyond 2-D filtering, there are 2-D focusing procedures, which constrain, within certain limits, the "signature" of the target so that it is within its actual geometric size (more precisely the radar cross-section of the target below the scan trajectory. The most exploited focusing algorithm is 2-D migration, operating either in the frequency domain (Stolt 1978) or the time domain (Schneider 1978). The migration is an approximated closed (integral) form solution for the shape of the buried targets. Several approaches can be followed to describe it, and the object function associated with the buried target can be given by the contrast of dielectric permittivity (Persico 2014) or by some equivalent electric field in the medium of propagation (Stolt 1978; Schneider 1978). Whatever the object function, there are several forms of the 2-D migration formulas. Reported here are two of the most common, labelling as $O(x', z')$ the object function.

In particular, denoting again with $\hat{\hat{D}}_F(k, \omega)$ the filtered (and possibly "enhanced" by some variable gain) data, a 2-D migration formula in frequency domain is given by

$$O(x', z') = \int_{-\infty}^{+\infty} \int_{-\infty}^{+\infty} \hat{\hat{D}}_F(k, \omega) \exp(jkx') \exp\left(jz' \sqrt{\frac{4\omega^2}{v^2} - k^2.} \right) dk d\omega \quad (4.3)$$

and a 2-D migration formula in time domain is given by

$$O(x', z') = \frac{\partial}{\partial z'} \int_{-\infty}^{+\infty} dx \int_{\frac{2\sqrt{(x-x')^2+z'^2}}{v}}^{+\infty} \frac{d_F(x,t)}{\sqrt{t^2 - \frac{4\left[(x-x')^2+z'^2\right]}{v^2}}} dt \tag{4.4}$$

where $d_F(x, t)$ is the space-time domain representation of the filtered datum corresponding to $\hat{\hat{D}}_F(k, \omega)$ in the frequency and wave number domain.

In Eqs. (4.3) and (4.4) v is the propagation velocity of the electromagnetic waves in the soil and, to enhance clarity, some inessential constants have been omitted before the integral. In both cases, measurements at the air soil interface have been implicitly assumed. Equation (4.3) is also known as the 2-D Stolt's migration formula, whereas Eq. (4.4) is also known as 2-D Kirchoff's migration formula. Details on their derivation, starting from the Maxwell's equations, are available in (Persico 2014). What is important to emphasise here is that the migration is, in any case, an approximate procedure based on a linear approximation of the scattering phenomenon (Chew 1995) [which is actually intrinsically nonlinear (Colton and Kress 1992)] and on the assumption that targets' depths are large in terms of wavelength (Persico 2014). Another important point is that the migration (at least in its basic forms) is a procedure that neglects the losses of the soil (Lesselier and Duchene 1996).

However, the theory underlying the migration algorithms [that are also collectively known as diffraction tomography (DT) (Lesselier and Duchene 1996; Meincke 2001)] provides not only a solution of the problem, but also allows an analysis that can estimate the resolution limits and the required sampling rate in the spatial and frequency (or alternatively time) domain. In particular, the essence of DT is the identification of an algebraic relationship between the spectrum of the data and that of the object function, which allows estimations to be made of the available resolution and the required rate of sampling data in the spatial and frequency domains. The available resolution for shallow targets is of the order of $\frac{\lambda_s}{2}$ (λ_s being the centre frequency wavelength in the soil), and degrades as the depth increases. Several more detailed formulas describe, in different ways, this decay (Jol 2009; Persico 2014; Sheriff 1980). The vertical resolution for shallow targets is expected to be of the order of $\frac{v}{B}$, B being the available frequency band and v is the propagation velocity of the electromagnetic waves in the soil. Usually, the Ultra Wide Band (UWB) antennas exploited for GPR prospecting have a bandwidth of the same order as the centre frequency f_c, so that $\frac{v}{B} \approx \frac{v}{f_c} = \lambda_s$.

Also, with regard to the vertical resolution, because the soil attenuates the higher frequencies more than the lower frequencies (Daniels 2004; Jol 2009), the "received bandwidth" reduces as the depth of the target increases (Sala and Linford 2012), and so some degradation vs. the depth of the vertical resolution is expected too.

Based on the DT, moreover, the expected required spatial step for the data is, of the order of, $\frac{\lambda_{s\,min}}{4}$ ($\lambda_{s\,min}$ being the minimum expected wavelength in the soil).

Invariably, the GPR signal never consists of a single harmonic, and the spatial steps may be redundantly narrow with respect to the lowest harmonic components of the signal. With a pulsed system, the spatial stepping of the signal cannot be varied to make it larger for the lower harmonics, and for a stepped frequency system it is impractical. If, however, it is necessary, for computational efficiency in particular cases, the redundancy of the spatial stepping necessary for gathering the data can be reduced in the processing phase through some suitable "harmless" decimation of the data.

The reasoning outlined above demonstrates that spatial sampling is not independent of the time variability of the signal. This is predicted by Maxwell's equations, where the spatial and time (or frequency) variabilities are coupled. The necessary frequency step (for swept frequency systems) is of the order of $\frac{v}{2D}$, where D is the maximum expected penetration depth of the signal.

Sometimes, D is implicitly understood as the maximum depth of interest, but caution should be exercised. For example, if the range of depth interest is, say, one metre, as it may be in the case of asphalt monitoring, it should be borne in mind that radiation penetrates to greater depths. In some situations, deeper targets may cause disturbances within the depth range of interest if the frequency step is too large.

Finally, for a pulsed system, a time step of the order of $\frac{1}{B}$ is necessary. All of the required sampling rates, either in space, frequency or time, are essentially determined by the anti-aliasing requirement. More details are available, e.g., in (Persico 2014).

It is a common practise suitably to combine (possibly with some interpolation) several processed B-scans together to obtain a "parallelepiped" containing information representing a volume of the sub-surface. This allows visualisation of the buried scenario relative to a series of planes parallel to the three coordinate Cartesian planes (i.e. the planes with equations, $x = 0$, $y = 0$ and $z = 0$, respectively). With some additional effort, if needed in particular cases, an image of the underground scenario along an oblique plane or even a curve surface can also be produced. The images at fixed depth (i.e. the images on the planes at constant z, also called depth-slices or time-slices) are often particularly useful, because they provide a map of the buried infrastructure in planes parallel to the air soil interface. This can be important for the mapping of the buried services or for archaeological investigations (Conyers 2004). Three dimensional perspective representations can also be produced from the reconstructed parallelepiped of the sub-surface volume. However, these involve the choice of some on-off energy thresholds in order establish the inner and the outer regions of the targets.

The choice of threshold is intrinsically linked to the mechanism of human vision. Indeed, when we observe the scenario of any room, what we usually distinguish are the external surfaces of the objects present that, in most but not all cases,[2] appear to be "sharp". With the reconstruction of buried targets there are two problems, namely: (i) even more so than in air, buried targets may not be intrinsically sharp, but diffuse because of smoothing mechanism, such as in the case of ingress of water, the spreading of polluting substances from leaking pipes,

[2] E.g. let us think of an unequally dense, or 'patchy' fog.

or targets broken into discrete pieces of varying sizes. (ii) even if the target is sharp, its reconstruction, in general, may not be because of the spatial filtering properties of the scattering operator.

In these circumstances it is, in most cases, unavoidable that the appearance of the reconstruction is dependent on the chosen threshold level, which introduces some degree of arbitrariness, which must be assessed by human expertise. An examination of the perspective reconstruction at several threshold levels, in order to test heuristically the sensitivity to this parameter, is well advised (Leucci et al. 2011).

These three dimensional representations, as well as horizontal slices, are sometimes improperly referred to as "3-D" processing. In fact, it is more exact to say that it is pseudo-3-D processing based on an amalgamation of several 2-D solutions.

4.3 Three-Dimensional Processing

Authentic 3-D processing must be based Maxwell's equations without imposing any invariance direction. This is computationally expensive and is, therefore, carried out more rarely. Adopting a linear approximation of the scattering phenomenon is commonly done also in 3-D processing, and also in 3-D it is possible to determine DT relationships (Persico 2014) and migration formulas (Stolt 1978; Schneider 1978; Persico 2014). In particular, expressions for 3-D migration formulas, in frequency and time domains, are provided in Eqs. (4.5) and (4.6), respectively.

$$O(x',y',z') = \int_{-\infty}^{+\infty}\int_{-\infty}^{+\infty}\int_{-\infty}^{+\infty} \hat{\hat{\hat{D}}}_F(k_1,k_2,\omega) \times \exp(jk_1x')\exp(jk_2y')\exp\left(jz'\sqrt{\frac{4\omega^2}{v^2}-k^2.}\right) dk_1\,dk_2\,d\omega \tag{4.5}$$

$$O(x',y',z') = \frac{\partial}{\partial z'}\int_{-\infty}^{+\infty}\int_{-\infty}^{+\infty} \frac{d_F\left(x,y,\frac{2\sqrt{(x-x')^2+(y-y')^2+z'^2}}{v}\right)}{\sqrt{(x-x')^2+(y-y')^2+z'^2}} dx\,dy. \tag{4.6}$$

Symbols in Eqs. (4.5) and (4.6) are direct extensions of the homologous symbols in Eqs. (4.3) and (4.4). In general, there is a plane for the measurements instead of a line, so that the spectrum of the data is a triple Fourier transform with respect to two spatial variables and time. Similarly to Eqs. (4.3)–(4.6), in Eqs. (4.5)–(4.6) it is implicit that the measurements are taken from the air-soil interface. The data sampling and resolution limits available in this framework are of the same order as the homologous 2-D version (Persico 2014). However, in comparison to a 2-D model, a 3-D version is able to provide a visualisation with respect to the required transect, i.e. the theoretical maximum allowed spacing between adjacent B-scans.

This is expected to be of the order of $\frac{\lambda_{s\,min}}{4}$, i.e. the same as the maximum allowed spatial step of the data along any single B-scan. In other words, in order not to lose information, and avoid spatial aliasing phenomena, the measurement points should form a square grid on the air soil interface. From a practical point of view, the theoretically required transect is an important result. In fact, while there is usually no problem in guaranteeing a spatial step of $\frac{\lambda_{s\,min}}{4}$ along a single B-scan (the GPR is in most cases equipped with an odometer that allows, in most cases, an even narrower spatial sampling rate), to gather data with a transect of $\frac{\lambda_{s\,min}}{4}$ is impractical with a single bi-static antenna system. To quantify, if the relative dielectric permittivity of the soil is 4 and the maximum usable frequency is 500 MHz (which are, on average, two "favourable" hypotheses that relax the requirement) then the transect should not be larger than 7.5 cm. For practical reasons, transects used in the "real world" are normally of the order of 50 cm.

As already explained in par. 3.1, the development of GPR systems equipped with antenna arrays offer significant advantages because they allow the automatic gathering of several parallel B-scans at reciprocal distances of few centimetres, which allows the saving of time, and better spatial sampling of the datum.

4.4 Final Remarks on GPR Processing

In this overview, the most commonly used processing techniques have been described. There are, however, other focusing processes that complement the migration algorithm, namely tomographic inversions, both linear (Persico et al. 2005; Persico 2006; Meincke 2001) and nonlinear (Qin and Cakoni 2011). Linear inversion algorithms are usually based on the SVD of the linearized scattering operator. The linearization can be achieved in several ways, depending of the specific assumption adopted for the case at hand. The most applied linearization's are the Born (Chew 1995), Rytov (Devaney 1981) and Kirchoff approximations (Liseno et al. 2004).

Whatever the adopted method, in general, a linear inversion is theoretically more refined than a migration and, in particular, it does not require it to be assumed that soils are lossless and that targets are electrically deep. On the other hand, it is computationally much more cumbersome than a migration because the SVD of the relevant linear operator is, in general, not known in a close form and has to be implemented numerically. In general, with a "medium power" computer, it is possible to invert an investigation domain that is a few wavelengths (computed at the centre frequency) in extent, both in the abscissa and the depth directions, whereas the length of a B-scan can be of the order of hundreds of wavelengths. This problem can be addressed by combining a sequence of separately reconstructed investigation domains. However, optimising the combination is in itself a problem that, if not correctly treated, can affect the quality of the achievable result (Persico and Sala 2014).

A further niche aspect of processing GPR data is the possible application of tomographic inversion using nonlinear processing. This method is theoretically more realistic than a linear inversion, because the physical nature of the scattering process is also nonlinear. However, the computational effort is considerable, and nonlinear inversions are, usually, applied only in particular cases (and most often for the purpose of studying the algorithm rather than solving a practical problem). Computation times usually restrict investigations to domain sizes of the order of a very few square wavelengths (Qin and Cakoni 2011). Finally, any nonlinear inversion has to consider the problem of false solutions caused by the local minima generated by, possibly, minimising a non-convex cost-function (Persico et al. 2002).

This section closes with a reference to forward modelling, which consists of postulating a scenario that provides artificial data similar to those recorded. By adjusting the parameters of the simulation, the result is driven to be progressively more similar to the real measured data. This approach can be useful in particular cases when the data and/or the available a priori information allow a reasonably precise hypothesis on the nature, size and disposition of any buried anomalies to be constructed to provide the basis for the forward modelling process. There are several available codes for this, the best known of which is, probably, GPRmax (Giannopoulos 2003). However, forward modelling cannot be exploited as a general method, because (as stated at the beginning of this section), the problem at hand is ill-posed, which means that it is possible for two different buried scenarios to provide similar data.

5 Advanced Equipment for Civil Engineering Applications

5.1 Introduction

Ground-penetrating radar is one of a number of non-destructive inspection methods that is gaining acceptance in a wide range of applications in the Civil Engineering sector.

It has become a viable means of detecting and mapping buried infrastructure, locating reinforcement bars (rebars), voids and cracks in concrete, inspecting foundations, measuring asphalt layers' thickness in roadways and airport runways, evaluating the fouling of ballast in rail-tracks, etc.

Given the diverse nature of all these applications, requirements for building effective equipment are very different from case to case and a detailed, exhaustive description for all of these is not within the scope of this chapter; thus, the following paragraphs describe three examples of advanced GPRs

1. a dense array system designed for mapping underground assets;
2. a sparse array, high resolution system for the evaluation of bridge decks;
3. a continuous wave, reconfigurable GPR.

5.2 Dense Array GPR

One of the main objectives in developing any new GPR system is to reduce the time needed to collect data while maintaining the performance in terms of targets' detection; this is a key point in the design and is the origin of the 3-D GPR concept, i.e. a system which is able to detects all targets within the surveyed swath, irrespective of their layout.

In other words, such system has to collect a 3-D volume of GPR data, in a matrix dense enough to not require moving the GPR along transects as is necessary with a GPR collecting one profile at a time.

Then, the next question is: "how dense does the matrix need to for a practical 3-D GPR survey?" (Grasmueck et al. 2003).

Obviously surveying a large number of 2-D GPR profiles, close to each other and accurately positioned in Cartesian space, could accomplish the task of collecting a dense 3-D data set over the required volume, but this procedure is time consuming; thus, efforts have been focused on the design of GPR arrays that can produce a 3-D data volume with a single scan, being "de facto" equivalent to the volume collected by multiple surveys consisting of several parallel scans with a single GPR sensor (Fig. 8).

It is important to minimise the number of the number of profiles to be collected by the array, and the Nyquist sampling theorem establishes the lower bound of the minimum number, according the following equation (Grasmueck et al. 2005):

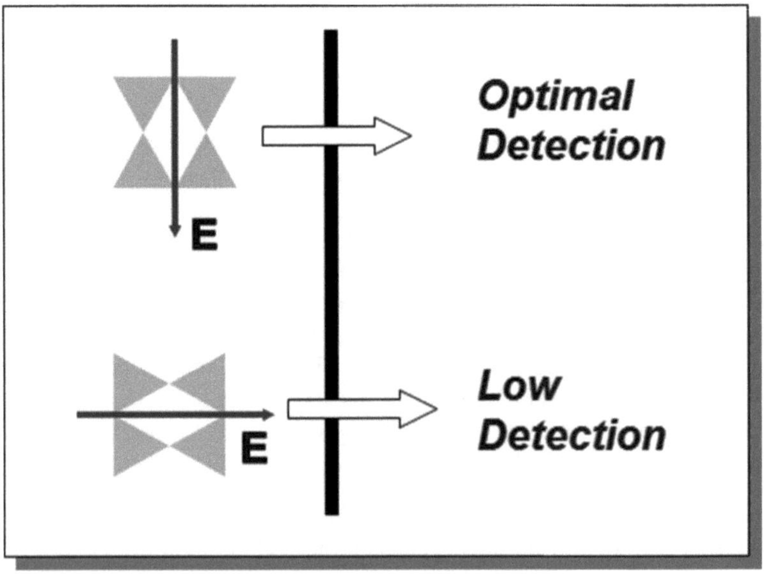

Fig. 8 3-D GPR data volume collected with a 200 MHz GPR system

$$\Delta x \leq \frac{\lambda_{\min}}{4\sin(\theta)} = \frac{c}{4\sin(\theta) f \sqrt{\varepsilon_r}} \tag{5.1}$$

where Δx is the spatial separation between two GPR profiles, c is the speed of light in vacuum, θ is the half beam of the GPR antenna (that can be assumed to be 60°), f is the main working frequency of the GPR, ε_r is the relative dielectric constant of the soil (maximum expected value is 15 while common value is 9).

This Eq. (5.1) defines the strictest constraint for the GPR design; since GPR working frequency is within 200 and 1000 MHz in the applications considered in this chapter, this constraint imposes a maximum separation between parallel scans of less than 14 cm for a 200 MHz GPR and 2.8 cm for the 1000 MHz GPR, assuming a standard soil with relative dielectric constant of 9.

However, to be fully conservative, the spacing between the parallel scans has to be well below the limit described above when the system is used in the presence of higher dielectric permittivity (e.g. wet clay with a permittivity up to, say, 20).

Thus, these issues affect the design of antennas, as typical bow-ties GPR are planar dipoles which cannot be squeezed together to comply with the requirement; moreover, as the spacing between transmitting and receiving antennas is reduced, so the mutual coupling between them increases, which is a very undesirable consequence leading to performance difficulties.

Another important issue for dense array GPR systems is the antenna deployment; as seen in par. 3 (Fig. 9), shallower, thin metallic objects (e.g. cables) are almost invisible to the GPR when the polarisation of the radiating waves are perpendicular to the objects' longitudinal axis (Fig. 10).

As these targets are an important class of object users want to locate, it is mandatory that the GPR collects data using more than one polarisation; this imposes the requirement to integrate (at least) some antennas orientated perpendicularly to the majority that form the dense array.

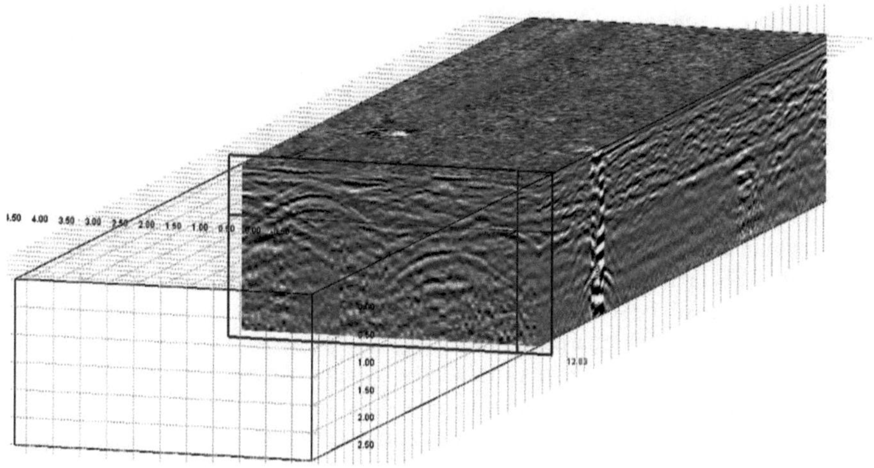

Fig. 9 Influence of the wave's polarisation for the detection of long, thin, metallic objects

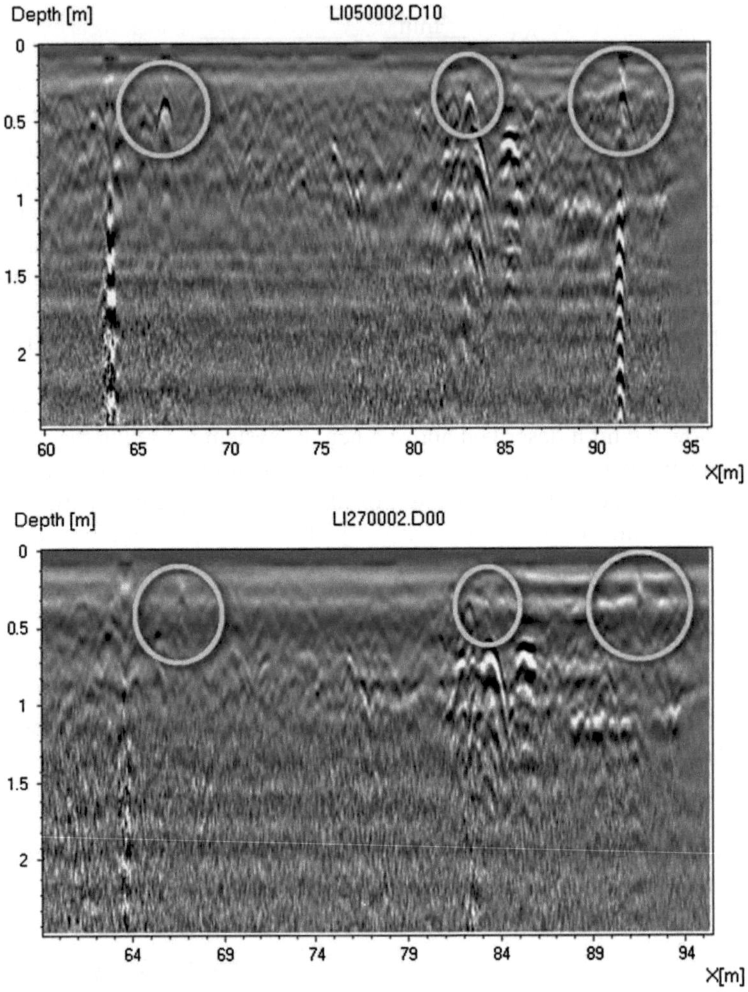

Fig. 10 Benefits from collecting data with 2 polarisations. Objects circled in white are detected just in the map above (collected with wave's the polarisation parallel to the objects' direction)

A further important practical consideration is that of the rate of data acquisition, which has to be high enough to permit productive use of the equipment; moreover, GPR operators work on roads, in live traffic, so the system should operate as fast as possible, so that it can be towed by a vehicle at a reasonable speed and not to require traffic guards in place.

GPR data positioning is also an issue; the system can move everywhere on the surveyed site and a GPS or total station have to be interfaced with acquisition software to mark system trajectory in order to correctly position data on the XY plane during the subsequent analysis stage. In this respect, requirement on position accuracy is a function of the GPR working frequency (thus it can be as small as 1 cm).

Fig. 11 A dense array GPR for the detection of underground assets (IDS Stream EM)

Moreover, adjacent swaths have to partially overlap in order not to leave any "gaps" in the surveyed zone and to assure that features of interest that could be on the edge of one swath won't be on the edge of the next one (Fig. 11).

Last but not least, the collected 3-D data volume has to be processed with a proper software tool that helps operators to analyse the large amounts of information (tens of Gigabytes) as quickly and efficiently as possible. In this respect, the requirement is to provide a quick and easy method of locating the position of all the targets by interacting with a 3-D representation of the collected data collected and then concluding the whole process by exporting all the features of interest directly onto a Computer Aided Design system (CAD).

In summary, the design of a 3-D dense array GPR is not trivial; it requires a solid understanding of system and antenna design as well as the development of innovative solutions (hardware and software) to overcome all the issues inherent to the practical implementation.

The benefits that arise from true 3-D GPR data collection are evident when looking at Fig. 12 or Fig. 1 at the beginning of this chapter; data interpretation, which is to date the 'holy grail' of GPR, can become much easier and hence can reduce the level of skill required to produce the underground assets map.

5.3 Bridge Deck Survey with High Resolution Ground Penetrating Radar

In the preceding decade, an increasing number of bridges all over the world have been classified as structurally deficient. In addition, these structural deficiencies can cause severe damage to the top pavement layers, often having a detrimental

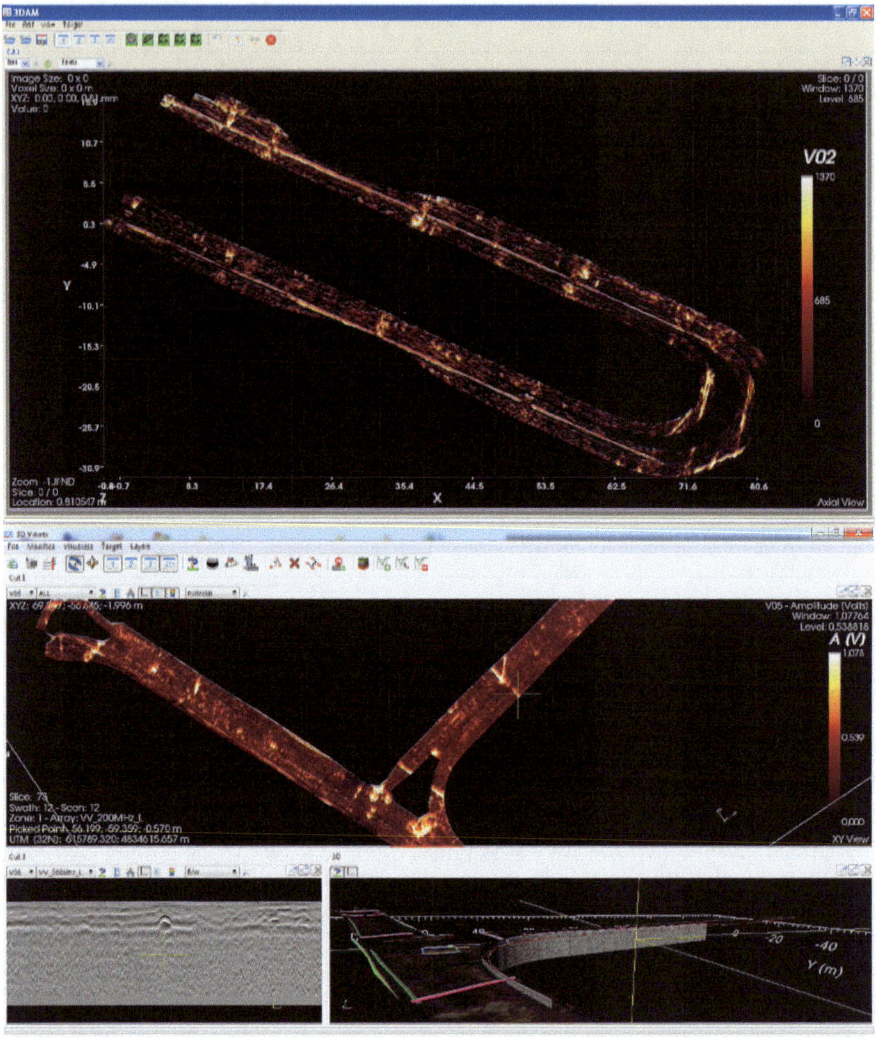

Fig. 12 Processing and display of dense array GPR data (IDS GRED HD)

effect on driving safety. The estimated financial cost of repair is very high and the social costs, in terms of safety reduction, are also becoming significant.

Basically, corrosion associated with rebars is the most significant contributor to deficiencies both in bridges and other concrete infrastructures.

Corrosion is usually caused by the presence of excessive moisture and chloride ions in the concrete adjacent to the reinforcing steel. The acid solution of moisture and chloride, which penetrates from the top of the concrete deck down to the reinforcement bar, depassivates the normally alkaline environment in concrete, initiating corrosion.

The solution reacts with the steel, generating oxidation processes. The early stage of corrosion is generally known as the initial passivity stage. The reinforcing bars are rapidly attacked by corrosion and yield expansive corrosion products, which can in turn accelerate the formation of cracks in the surrounding concrete. These oxidation products occupy a larger volume than the original intact steel, and internal expansive stresses lead to cracking and de-bonding. This is commonly known as delamination of the concrete cover material from the reinforcing layer.

Essentially, early detection of reinforcing bar corrosion, anticipating the delamination cracks, significantly reduces maintenance costs and extends the life of the reinforced structures.

There are some conventional methods for the detection of reinforcing bar corrosion: these are based on visual, mechanical, electrical and chemical techniques. However they are often inadequate for the acquisition of data in the field for a variety of reasons: they can be invasive and destructive, the data acquisition is often laborious, the procedures require lane closures or some traffic limitations for a significant time, with a reduction in safety in the work zones. In addition, any quantification of corrosion is difficult or unreliable. Obviously gross underestimates of the actual repairs lead to significant cost overruns.

GPR can yield data with very high spatial resolution and they can be acquired quickly, thus minimising disruption to traffic. It makes this non-destructive technique advantageous as a field tool.

The GPR inspection of a bridge to detect corrosion and cracking problems is based on the concept that as the moisture and chloride concentration increase, the relative dielectric constant and conductivity of the concrete increase. This leads to certain features of the radar waveform that can be reasonably associated with the presence of deterioration. In addition, an increase in the dielectric constant increases the amplitude of the reflections.

With regards to bridge inspection, theoretical studies and experimental investigations have typically been carried out to understand the effects on phase, attenuation and frequency of the GPR signal, of the temperature, of the water and chloride content and of the concrete. These studies and investigations aim to develop GPR wave form inversion techniques and to implement algorithms for solving the influence of delamination cracks on the GPR signal (Hugenschmidt and Loser 2008; Parrillo and Roberts 2006; Roqueta et al. 2011).

Moreover there is an associated growth in the conductivity that also increases the level of the signal attenuation through the contaminated concrete. This effect can be attributed to moisture and chloride ingress. However the signal characteristics that can typically be attributed to delamination, are not exclusively due to it, nor they are even due exclusively to moisture and chloride. There are other phenomena and conditions that can be responsible for them. These include properties and condition of the asphalt concrete overlay, environmental effects pertaining to moisture, deck structure, extent of deterioration, and the method of detecting and delineating repair areas on the exposed deck surface.

Consequently, a GPR tailored for this application must be designed to detect not only the increase of the signal attenuation, but a complete set of parameters whose combination can lead to the localisation of corroded areas; in principle, pavement and concrete thickness, moisture and damaged areas on the surface are all potential causes of corrosion.

This GPR should also feature a very dense data collection with a high working frequency (around 2 GHz with a fractional bandwidth of 100 %) and with two polarisations; in fact, a very good resolution is necessary in order to get all the feature of the surveyed structure.

Resolution is related to bandwidth of the radiated pulses and to propagation velocity of the electromagnetic waves; in concrete (which is the material which bridge decks are mostly composed of), propagation velocity ranges from 10 to 12 cm/ns, so that a 2 GHz system can guarantee a resolution (in range) that theoretically reaches 0.0125 m (quarter of a wavelength criteria).

Likewise the mapping of underground utilities, the use of an array of antennas give a clear benefit in this application; moreover, if waveforms are transmitted with two polarisations (by using 2 arrays of antennas), the detection of all kind of targets is enabled. As already explained, metallic targets such as rebars or cables are best detected when generating an electric field parallel to their axis, whereas non-metallic objects such as empty PVC pipes or cavities are best seen with a perpendicular field.

Another benefit of collecting data in both polarizations is that echoes generated by metallic targets perpendicular to the scan direction are weakly seen in data collected by dipoles radiating a field perpendicular to them (thus parallel to the antenna route), so that other objects below them can be more easily detected (for instance concrete beams) (Fig. 13).

All the data collected by the system are then processed with the overall objective of enabling the identification of corroded or potentially corroded areas; the

Fig. 13 Bridge deck survey with high resolution Ground Penetrating Radar (IDS HiBrighT)

GPR is not able to provide a quantitative information (e.g. the percentage of humidity in the concrete slab), but can still produce useful data enabling the analysis software to deliver geometric information on:

- the thickness of asphalt overlay;
- the depth of the interface between asphalt and concrete slab;
- the depth of the first layer of rebars;
- the thickness of the concrete slab over the rebars;
- the depth of the bottom layer of rebars (if within the penetration range).

Slab thickness values can be mapped over the surveyed area so that they can be compared with design values, yielding valuable information on zones where rebars could be more easily exposed to moisture ingress due to the protective slab not being sufficiently thick.

A 2-D image (time slice view) can be also produced at a variable depth and used for checking the presence and the depth of the reinforcement rebars (coloured in white in the following picture).

Then, dedicated software tools can extract other features from the data in order to produce the output shown in the following Fig. 14; essentially, by measuring the

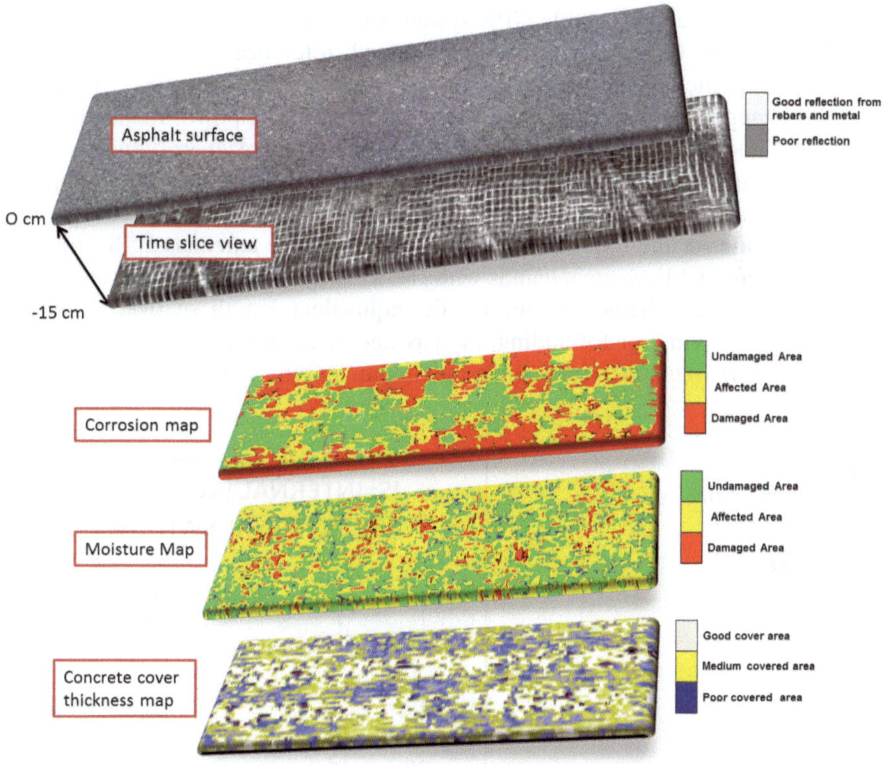

Fig. 14 Automatic output from the data processing software (IDS GRED HD Bridge Module)

amplitude of the rebars' echo and the propagation velocity of the electromagnetic wave, it is possible correlate and map them on the surveyed area, thus identifying those that can be affected by moisture and rebar corrosion.

It is worth remembering that these outputs can only be produced by high resolution GPR array systems that collect very densely sampled data.

In fact, corrosion is a phenomenon that mainly affects extended areas, thus output from the processing software have to be averaged over several GPR traces several evaluation tests have assessed the reliability of this procedure and of the relevant method (Manacorda et al. 2012).

5.4 The Reconfigurable GPR

In a reconfigurable GPR, some systems parameters may be varied, in a controlled manner, as a function of frequency or (theoretically) of time.

The concept of reconfigurable electromagnetic systems was initially introduced in communication systems, where it is sometimes an advantage to vary the direction of the antenna beam, or the range of frequencies, in order to, for example, counter undesirable fading in the transmission path.

The idea of a reconfigurable GPR system was first suggested in 2008 (Persico and Prisco 2008; Prisco and Persico 2008), with reference to stepped frequency systems. In particular, an ability to reconfigure the system can be exploited in order to change the integration time of the harmonics, the power radiated at each frequency, and the input impedance of the generator and receiver [which amounts to a reconfiguration of the electronics (Parrini et al. 2011)] or the equivalent length (more generally the size and the shape) of the antennas (Persico et al. 2011).

The scheme for the implementation of reconfigurable antennas is illustrated in Fig. 15, where two bow-tie antennas with two switches along each arm are shown. According to the scheme of Fig. 15, the equivalent length of the antennas is changed by allowing, or forbidding, the passage of current by means of switches. In Fig. 15 the antennas reach their maximum length if both the switches are

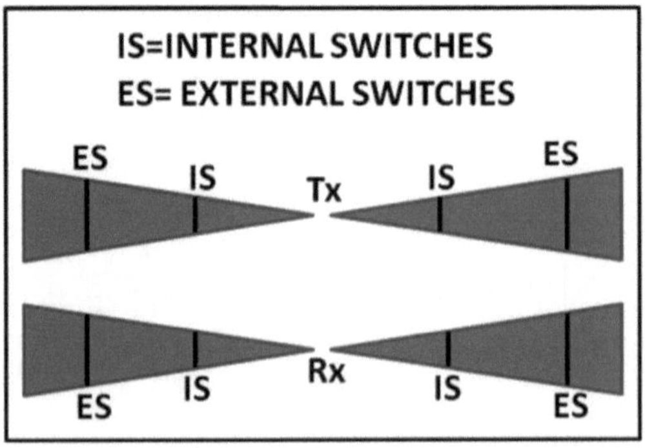

Fig. 15 Antenna configuration scheme

Fig. 16 Prototypal stepped frequency reconfigurable GPR, built within the AITECH project

closed; they assume an intermediate size if only the internal switches are closed and the external ones are open and they are at their shortest when both switches along each arm are open. Of course, when the switches are open the detached part of the arm still has some role as a passive element. However, this effect is not expected to be significant.

By means of the reconfiguration, the total available bandwidth of the system (with some mitigation of the relative increase of the size of the system) can be enlarged and some narrow band interference signals can be rejected.

An early prototype of stepped frequency system has recently been implemented, within the research project AITECH (http://www.aitechnet.com/ibam.html), by means of a collaboration between the Institute for Archaeological and Monumental Heritage IBAM-CNR, the University of Florence and the Ingegneria dei Sistemi IDS Corporation. The prototype is shown in Fig. 16.

Some first results have recently been made available (Persico et al. 2013): the system has a large equivalent bandwidth that ranges from 50 to 1000 MHz, provided by three pairs of equivalent antennas, connected to the system by means of switching on and off two series of switches.

The switches allow an equivalent shortening of the antennas, allowing them to operate at higher frequencies. Moreover, the system can prolong, in a selective way, the integration time of the harmonics, which helps to reject narrow band interference and increases the dynamic range. Finally, the power radiated at each frequency can be modulated, which enlarges the equivalent bandwidth of the radiated pulse.

The system has been successfully exploited in archaeological sites and historical monuments. On the basis of the first experimental results, it has been possible to identify some important development possibilities related to the pros and the cons of the reconfigurable technology. The main pros are those listed [they have been experimentally documented (Persico et al. 2014)]. The main cons that have been identified are some ringing (probably due to the electronic switches that disturb the matching of the end of the equivalent arms) and, most significantly, the present inability to visualise the data in real time. The work is in progress and new results, and insights, are being acquired on a daily basis.

6 GPR Calibration Procedures

There are several meanings for the term "calibration" relatively to GPR equipment: the first is the procedure of checking the proper operation of the apparatus and that the expected performance can be achieved.

In this respect, a test should be executed to ensure that the GPR is achieving the expected dynamic range, i.e. it transmits signals with the correct amplitude and that the internal noise level is within the expected limit. Deeper targets may become undetectable if the noise level is too high or if the transmitted signal is too weak.

Modern GPRs usually execute automatically these verifications at the system start-up and the operator is warned if any of the internal parameters do not match the required specifications.

There are, however, other variables, whose evaluation may not be trivial, that can affect the performance of a GPR; these concern the stability of the electronic circuits and the linearity of the time base.

If the scale of the time window becomes nonlinear, detected targets depth estimation is degraded and its reliability, therefore, becomes suspect; this problem may arise from electronic component fault and as well aging.

Degradation in GPR electronics' stability can produce severe effects in the quality of collected data; first, the efficacy of the background removal filter can be affected and the rejection of the strong signal due to the direct coupling between antenna transmitter and receiver becomes impossible. Also, because of aging, jitter can increase; this is defined as a short-term variation of the actual instants of a digital signal from their ideal position in time.

Jitter is equivalent to a phase modulation of the clock signal used to sample the data and it, therefore, produces an uncertainty in the sampling instant (this is also called aperture jitter).

Consider the effect of the aperture jitter on the analogue-to-digital (A/D) signal conversion performed in any pulse or CW GPR for saving the collected data in a digital media; the digital GPR signal is composed by N samples after an ideal A/D conversion performed with a period T_s

$$s_l = a(t_l)e^{j\omega t_l}, t_l = lT_s, l = 0, 1, 2, \ldots \tag{6.1}$$

The uncertainty of the sampling instant can be accounted for by replacing t_l with

$$\tilde{t}_l = lT_s + \delta_l \tag{6.2}$$

where δ is a random variable, representing the aperture jitter, which has a mean value of 0.

Thus

$$\tilde{s}_l = a(t_l + \delta_l)e^{j\omega(t_l+\delta_l)}. \qquad (6.3)$$

If we consider the Taylor expansion of $a(t_l + \delta_l)$ and $e^{j\omega(t_l+\delta_l)}$ about t_l, previous expression can be simplified as

$$\tilde{s}_l \cong \{a(t_l)[1 + j\omega\delta_l] + a'(t_l)\delta_l\}e^{j\omega(t_l)} \qquad (6.4)$$

where $a'(t_l)$ denotes the derivative of $a(t_l)$.

Thus

$$\tilde{s}_l = s_l + \{j\omega a(t_l) + a'(t_l)\}\delta_l e^{j\omega(t_l)} \qquad (6.5)$$

and the error generated on each sample is

$$\tilde{s}_l - s_l = \{j\omega a(t_l) + a'(t_l)\}\delta_l e^{j\omega(t_l)}. \qquad (6.6)$$

Equation (6.6) shows that the errors in the sampling time produce errors in the digitised sample amplitude that can be significant; therefore, the stability of the sampling circuitry is a key parameter, and aperture jitter values should not exceed 10 ps. This should be monitored to verify compliance throughout the life of the equipment (Manacorda and Miniati 2000).

Further requirements for "calibration" concern the relationship between physical properties of the surveyed medium and the quantities measured by the GPR; for example, to establish the real depth of a target, an accurate time to depth conversion factor is necessary.

Accurate measurement of target depth depends upon knowing the speed of propagation of electromagnetic waves in the ground; this is very variable and can cause errors in the estimate of the depth. The errors are usually resolved by obtaining the true depth of at least one target by an independent method, usually excavation and direct measurement. Depths of other targets are then expressed relative to the known depth on the assumption that, over the survey area, the speed of propagation does not vary.

There are, however, other techniques that have been used and proven capable of supplying reliable estimate of the velocity; amongst these, the CMP (common mid-point), WARR (wide angle reflection and refraction), the hyperbolic fitting to a local target and diffraction tail matching. A detailed description of all these methods can be found in the literature [e.g. (Daniels 2004)] (Fig. 17).

Finally, recent research [see for instance (Benedetto 2004)] has addressed the possibility of correlating the GPR response to geophysical characteristics of the surveyed material by means of a semi-empirical approach. Experimental validations of such methods are still in progress, but results are interesting and seem to open new possibilities primarily in those applications that are of interest for the Civil Engineering (for instance the evaluation of asphalt pavement and railway ballast).

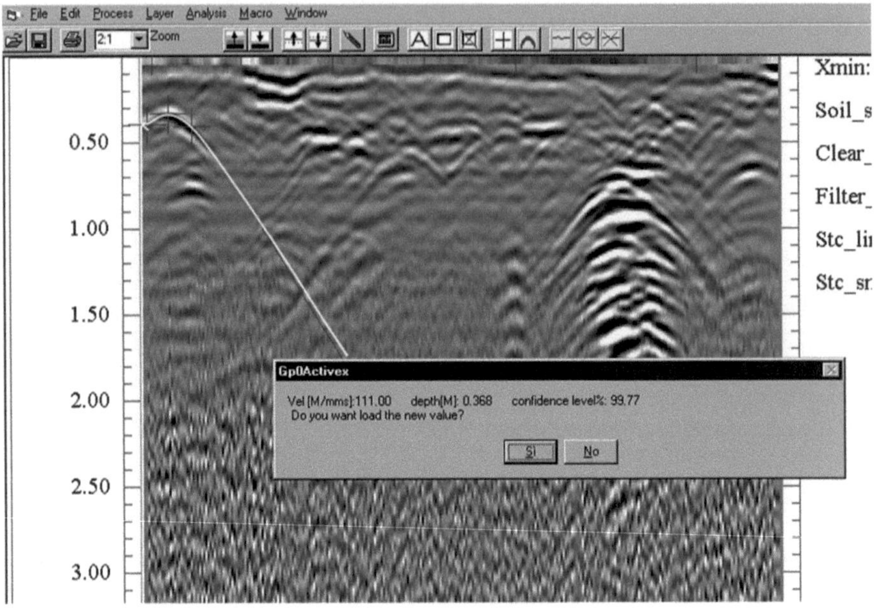

Fig. 17 Hyperbolic fitting and velocity estimation

7 Electromagnetic Compatibility and Regulations

The global use of the radio frequency spectrum is subjected to regulation drafted by relevant Authorities; the general principle is that if essential operation requires the movement of electrons then the equipment is regulated.

For equipment potentially radiating electromagnetic signals, radio licensing and electromagnetic compatibility (EMC) are important.

By definition, GPR looks downward into the earth, water, ice and man-made materials to non-destructively detect anomalies. Moreover, GPR is not an apparatus designed to radiate electromagnetic energy for radio communication purposes as it is not intended for air transmission.

However, GPR is governed by the radio regulations because of the pulsed nature of the signals employed, which does not fit into any established structure for radio licensing.

Even though GPR represents a class of Ultra Wide Band devices that radiate the major part of their emissions into the ground, a small part of that energy 'leaks' in all directions above the ground and, consequently, they may interfere with existing services.

The issue was first raised in the USA, where GPR technology came to the notice of the Federal Communications Commission (FCC), and it was decided to amend Part 15 of the FFC Rules specifically to acknowledge and control the use of GPR technology (2002).

The FCC imposed a limit on the frequency range available to GPR systems to below 960 MHz and offered special protection to the band of frequencies used by the Global Positioning System (GPS) extending from 960 to 1.61 GHz.

In the European Union (EU) the use of GPR is controlled by directives issued by the European Commission (EC) and the Government/National radio administrations, who control the use of GPR by setting conditions of use through their licensing regulations; it means that in some European countries a national licence is often required before the equipment can be used.

Compliance with global regulations has become a further issue in the design of GPR equipment because it is an advantage to minimise the energy received from backscattered signals from above ground objects that may hide weaker reflections from buried targets.

It is the case, however, that permitted airborne power levels are very small and this has adversely hindered the development of systems that use a large number of antennas, in arrays, by dictating the use of lower transmitter repetition frequencies, which has reduced possible GPR survey speeds to an impracticably low level that has made the survey of motorways and railways a dangerous undertaking.

Solving this problem is one of the most urgent tasks for the GPR designers in the coming years.

8 Conclusion

This chapter addresses the design of effective GPR systems suited to the Civil Engineering applications; this involves subjects ranging from radio frequency system and antenna design to data processing methodologies.

With respect to the generation of waveforms, two architectures are possible, impulsive (the most common) and frequency modulated. Whatever the architecture used, the most important requirements for a successful design are high sensitivity and wide dynamic range in order to maximise the usability of GPR so that it can cope with as wide a range of conditions as possible.

The recent development of antenna arrays increases the amount and quality of data collected on-site, and enhances the productivity of the process. Using a GPR array allows the acquisition of a large number of radar profiles, densely spaced, to reconstruct the entire 3-D volume surveyed.

To assure quick and reliable data collection, respecting Nyquist criteria, large and very dense GPR arrays are needed. Some further developments in this technology, which can be identified as research topics, are the study of innovative GPR antenna technology and GPR sensor multiplexing.

Issues that remain to be addressed are the selection of the optimal working frequencies and bandwidths, according to the characteristics of the targets of interest, as well as identifying the most effective waveform to be radiated by the antennas.

Other important objectives in developing new GPR systems are increases in productivity, and the provision of high performance in terms of detection capability and ease of data interpretation.

Some of these objectives have been accomplished in the three systems presented in paragraph 5, which represent the leading edge of the technology.

References

Benedetto, A.: Theoretical approach to electromagnetic monitoring of road pavement. In Proceedings of X International Conference on Ground Penetrating Radar, Delft, The Netherlands (2004)

Bertero, M., Boccacci, P.: Introduction to Inverse Problems in Imaging. Institute of Physics Publishing, Bristol (1998)

Chan, W.C., Stewart, R.R.: 3-D f-k filtering. CREWES Res. Rep. **6**, 15/1–15/7 (1994)

Chew, W.C.: Waves and Fields in Inhomogeneous Media. Institute of Electrical and Electronics Engineers, Piscataway, NJ (1995)

Colton, D., Kress, R.: Inverse Acoustic and Electromagnetic Scattering Theory. Springer, Berlin (1992)

Conyers, L.B.: Ground Penetrating Radar for Archaeology. AltaMira Press, Lanham (2004)

Daniels, D.J.: Ground Penetrating Radar, 2nd edn. IEEE press, New Jersey (2004)

Devaney, A.J.: Inverse-scattering theory within the Rytov approximation. Opt. Lett. **6**(8), 374–376 (1981)

Di Lorenzo, R.: Trading Systems: Theory and Immediate Practice. Springer, Berlin (2013). ISBN 978-88-470-2706-0

Giannopoulos, A.: GprMax2D V 1.5 (Electromagnetic simulator for Ground Probing Radar, the software is available at www.gprmax.org) (2003)

Goodman, D., Piro, S.: GPR Remote Sensing in Archaeology. Springer, New York (2013)

Grasmueck, M., Weger, R., Horstmeyer, H.: How dense is dense enough for a "real" 3D GPR survey?, society of exploration geophysicists. In: 73nd Annual International Meeting, Expanded Abstracts, pp. 1180–1183 (2003)

Grasmueck, M., Weger, R., Horstmeyer, H.: Full-resolution 3D GPR imaging. Geophysics **70**(1), K12–K19 (2005)

Hugenschmidt, J., Loser, R.: Detection of chlorides and moisture in concrete structures with ground penetrating radar. Mater. Struct. **41**(4), 785–792 (2008). doi:10.1617/s11527-007-9282-5

Jol, H.: Ground Penetrating Radar: Theory and Applications. Elsevier, Amsterdam (2009)

Kim, J.K., Cho, S.J., Yi, M.J.: Removal of ringing noise in GPR data by signal processing. Geosci. J. **11**(1), 75–81 (2007)

Lesselier, D., Duchene, B.: Wavefield inversion of objects in stratified environments: from back-propagation schemes to full solutions. In: Stone, R. (ed.) Review of Radio Science 1993–1996. Oxford University Press, Oxford (1996)

Leucci, G., Masini, N., Persico, R., Soldovieri, F.: GPR and sonic tomography for structural restoration: the case of the Cathedral of Tricarico. J. Geophys. Eng. **8**, S76–S92 (2011)

Liseno, A., Tartaglione, F., Soldovieri, F.: Shape reconstruction of 2D buried objects under a Kirchhoff approximation. IEEE Geosci. Remote Sens. Lett. **1**(2), 118–121 (2004)

Manacorda, G., Miniati, M.: An easy way of checking impulsive GPR performance. In: Proceedings of VIII International Conference on Ground Penetrating Radar, Gold Coast, Australia (2000)

Manacorda, G., Simi, A., Benedetto, A.: Bridge deck survey with high resolution ground penetrating radar. In: 14th International Conference on Ground Penetrating Radar, GPR2012, Shanghai, China (2012)

Meincke, P.: Linear GPR inversion for lossy soil and a planar air-soil interface. IEEE Trans. Geosci. Remote Sens. **39**(12), 2713–2721 (2001)

Parrillo, R., Roberts, R. Haggan, A.: Bridge deck condition assessment using Ground Penetrating Radar. ECNDT T.4.2.5 (2006)

Parrini, F., Persico, R., Pieraccini, M., Spinetti, A., Macaluso, G., Fratini, M., Dei, D., Manacorda, G.: A reconfigurable stepped frequency GPR (GPR-R). In: Proceedings of IEEE International Geoscience and Remote Sensing Symposium IGARSS 2011, Vancouver, Canada (2011)

Persico, R.: On the role of measurement configuration in contactless GPR data processing by means of linear inverse scattering. IEEE Trans. Antennas Propag. **54**(7), 2062–2071 (2006)
Persico, R.: Introduction to Ground Penetrating Radar: Inverse Scattering and data processing. Wiley, New York (2014). ISBN 9781118305003
Persico, R., Prisco, G.: A reconfigurative approach for SF-GPR prospecting. IEEE Trans. Antennas Propag. **56**(8), 2673–2680 (2008)
Persico, R., Leucci, G., Matera, L., Ciminale, M., Dei, D., Parrini, F., Pieraccini, M.: Applications of a reconfigurable stepped frequency GPR in the Chapel of the Holy Spirit, Lecce (Italy). In: Proceedings of VII International Workshop on Advanced Ground Penetrating Radar, Nantes, France, 3–5 July 2013
Persico, R., Sala, J.: The problem of the investigation domain subdivision in 2D linear inversions for large scale GPR data. IEEE Geosci. Remote Sens. Lett. (2014). doi:10.1109/LGRS.2013.2290008 (in print)
Persico, R., Soldovieri, F.: One-dimensional inverse scattering with a Born model in a three-layered medium. J. Opt. Soc. Am. Part A **21**(1), 35–45 (2004)
Persico, R., Soldovieri, F.: Effects of the background removal in linear inverse scattering. IEEE Trans. Geosci. Remote Sens. **46**(4), 1104–1114 (2008)
Persico, R., Soldovieri, F., Pierri, R.: Convergence properties of a quadratic approach to the inverse scattering problem. J. Opt. Soc. Am. Part A **19**(12), 2424–2428 (2002)
Persico, R., Bernini, R., Soldovieri, F.: On the configuration of the measurements in inverse scattering from buried objects under the distorted Born approximation. IEEE Trans. Antennas Propag. **53**(6), 1875–1886 (2005)
Persico, R., Romano, N., Soldovieri, F.: Design of a balun for a bow tie antenna in reconfigurable ground penetrating radar systems. Prog. Electromagnet. Res C. **18**, 123–135 (2011)
Persico, R., Ciminale, M., Matera, L.: A new reconfigurable stepped frequency GPR system, possibilities and issues; applications to two different Cultural Heritage Resources. Near Surf. Geophys. **12**, 793–801 (2014). doi:10.3997/1873-0604.2014035
Pieraccini, M., Noferini, L., Mecatti, D., Atzeni, C., Persico, R., Soldovieri, F.: Advanced processing techniques for step-frequency continuous-wave penetrating radar: the case study of "Palazzo Vecchio" Walls (Firenze, Italy). Res. Nondestr. Eval. **17**, 71–83 (2006)
Prisco, G., Persico, R.: Reconfigurable stepped frequency GPR systems. In: 12th International Conference on Ground Penetrating Radar, GPR2008, Birmingham, UK (2008)
Qin, H.H., Cakoni, F.: Nonlinear integral equations for shape reconstruction in the inverse interior scattering problem. Inverse Prob. **27**, 1–17 (2011)
Roberts, R.L., Daniels, J.J.: Analysis of GPR polarization phenomena. J. Environ. Eng. Geophys. **1**, 139–157 (1996)
Roqueta, G., Jofre L., Feng, M.: Microwave Nondestructive evaluation of corrosion in reinforced concrete structures. In: Proceedings of the 5th European Conference on Antennas and Propagation (EUCAP), pp. 787–791 (2011)
Sala, J., Linford, N.: Processing stepped frequency continuous wave GPR systems to obtain maximum value from archaeological data sets. Near Surf. Geophys. **10**, 3–10 (2012)
Sandmeier K.J.: Reflexw 3.0 manual Sandmeier Software ZipserStrabe1 D-76227 Karlsruhe Germany (2003)
Schneider, W.A.: Integral formulation for migration in two and three dimensions. Geophysics **43**(1), 49–76 (1978)
Sheriff, R.E.: Nomogram for fresnel-zone calculation. Geophysics **45**(5), 968–972 (1980)
Stolt, R.H.: Migration by fourier transform. Geophysics **43**(1), 23–48 (1978)
Utsi, E.: The shrine of Edward the confessor: a study in multi-frequency GPR investigation. Near Surf. Geophys. **10**, 65–75 (2012)

Antennas for GPR Systems

Lara Pajewski, Fabio Tosti and Wolfgang Kusayanagi

Abstract Antennas are a critical hardware component of a radar system, dictating its performance in terms of capability to detect targets. In this Chapter, a wide review on Ground-Penetrating Radar (GPR) antennas is given. Firstly, the general characteristics of GPR antennas are resumed and the requirements they have to satisfy are listed: these are somehow unique and very different than in conventional radar antennas, since GPR antennas operate in a strongly demanding environment, in close proximity to or at a limited distance from the natural or manmade investigated area. Subsequently, an overview on the most frequently used GPR antennas, for both pulsed and stepped-frequency radar systems, is provided; recent studies concerning innovative solutions are presented and information on antenna arrays for GPR applications is included, as well. The Chapter continues with a census of commercial antennas of a number of GPR manufacturers, where the centre frequencies and general characteristics of the antennas available on the market are schematically organised in tables. Aided by measurements and powerful computer modelling techniques, GPR antenna designers are increasingly able to predict and understand the performance of proposed design in realistic electromagnetic environments: the Chapter includes two sections reviewing techniques for the experimental characterisation and numerical modelling of GPR antennas. Finally, conclusions are drawn and research perspectives in the field of GPR antennas are discussed.

L. Pajewski (✉) · F. Tosti · W. Kusayanagi
Department of Engineering, Roma Tre University, Via Vito Volterra 62,
00146 Rome, Italy
e-mail: lara.pajewski@uniroma3.it

F. Tosti
e-mail: fabio.tosti@uniroma3.it

W. Kusayanagi
e-mail: wolfgang.kusayanagi@yahoo.com

A. Benedetto and L. Pajewski (eds.), *Civil Engineering Applications of Ground Penetrating Radar*, Springer Transactions in Civil and Environmental Engineering,
DOI 10.1007/978-3-319-04813-0_2

1 Introduction

Ground-penetrating radar (GPR) provides high-resolution images of subsurface and structures through wide-band electromagnetic waves. It operates in a wide range of frequencies, from 10 MHz to 5 GHz for impulse systems and from 1 to 8 GHz for stepped-frequency systems. The electromagnetic field emitted by the radar interacts with the investigated scenario: reflection, transmission and scattering phenomena occur at any interface corresponding to a change in dielectric properties. The return is detected by the radar receiver.

Antennas are a critical hardware component of a radar system, dictating its performance in terms of capability to detect targets. Nevertheless, most of the research efforts in the GPR area currently focus on the use of this imaging technique in a plethora of different applications, on its combination with other non-destructive methods, on the improvement of modelling, imaging and inversion techniques for GPR, whereas a limited number of studies deal with technological issues related to the design of novel systems, including the synthesis, optimisation and characterisation of innovative antennas.

In order to couple the electromagnetic energy into the ground or structure under test, GPR antennas operate in close proximity (ground-coupled antennas), else at a limited distance above (air-coupled antennas).

The reflected signals can be detected by the same transmitting antenna (monostatic system) or by a second, separate receiving antenna (bistatic system). In multistatic radar systems, there are at least three antennas—for example, one receiving and two transmitting, or two receiving and one transmitting, or else multiple receiving and transmitting antennas. This kind of radar is a generalisation of the bistatic system, with one or more receivers processing returns from one or more transmitters.

Traditionally, most GPR systems are bistatic. However, the antennas are often housed in a single module and their orientation and spacing cannot be changed. Separate antenna modules represent a significant advantage in the applications where transmitter and receiver can be placed on the two opposite sides of the investigated structure, thus halving propagation and attenuation losses (trans-illumination surveys). Examples of such applications are GPR measurements in boreholes for civil-engineering and environmental studies, quality-control of precast structures, or imaging of pillars and walls.

Still concerning bistatic radars with separate antenna modules, the ability to vary the antenna spacing can be a powerful aid in optimising the system for different applications and for the detection of specific types of targets: in fact, to maximise target coupling, the antennas should be spaced such that the refraction focusing peak in both the transmitting and receiving radiation patterns points to the common depth to be investigated. In this respect, it is important to keep in mind that antenna patterns vary as the relative permittivity of the illuminated medium changes (King and Smith 1981; Daniels 2010), as well as that the minimum separation between receiving and transmitting antennas has to be equal to the maximum size of the larger antenna, in order to limit the unwanted electromagnetic interaction between them.

Larger antenna spacing increases the reflectivity of planar targets, which can be advantageous in some applications. The depth resolution of targets decreases as the antenna spacing increases, however this factor is small until the spacing approaches half the target depth. When the angle subtended between the transmitting antenna, the sought target and the receiving antenna is close to zero, a bistatic GPR turns out to perform as a monostatic system (pseudo-monostatic radar).

A further advantage of transmitting and receiving antennas hosted in separate modules, is the possibility to change their orientation and polarisation. Usually, GPR antennas are linearly polarised. In this case, if the sought target has a size prevailing over the others, the electric field of both antennas should be polarised parallel to the long axis of the target, in order to maximise reflections and increase detectability. On the other hand, an orientation of the electric field perpendicular to a long object will allow revealing targets buried deeper. For an equidimensional target there is not an optimal orientation. As far as the mutual orientation of the transmitting and receiving antennas is regarded, they can be arranged as parallel or orthogonal to one another. Parallel orientation of antennas allows maximising the polarisation match between them; in this case, the antennas can be arranged in broadside or end fire configuration with respect to the survey line direction, i.e., with an orientation parallel or perpendicular to the line. On the contrary, when the transmitting and receiving antennas are arranged with orthogonal orientations, they are cross-polarised and target information can be extracted based on the coupling angle.

GPR antennas operate in a strongly demanding environment and should satisfy a number of requirements, somehow unique and very different than in conventional radar antennas. First of them is an ultra-wide frequency band: the radar has to transmit and receive short-duration time-domain waveforms, in the order of a few nanoseconds, the time-duration of the emitted pulses being a trade-off between the desired radar resolution and penetration depth. The fractional bandwidth of a GPR antenna can be as high as 160 %.[1]

Secondly, GPR antennas should have a linear phase characteristic over the whole operational frequency range, constant (or predictable) polarisation and constant gain.

Thirdly, due to the fact that a subsurface imaging system is essentially a short-range radar, the coupling between transmitting and receiving antennas should be low and short in time.

GPR antennas should have quick ring-down characteristics, in order to prevent masking of targets and guarantee a good resolution. Ringing occurs due to the reflections within the antenna as well as at the antenna/ground (or investigated structure) interface.

The radiation patterns should ensure minimal interference with unwanted objects, usually present in the complex operational environment. To this aim, the antennas should provide high directivity and concentrate the electromagnetic energy into a narrow solid angle.

---

[1] The fractional bandwidth is defined as $2(f_{max} - f_{min})/(f_{max} + f_{min})$, where f_{max} and f_{min} are the minimum and maximum frequencies of operation, respectively.

As already mentioned, a specific feature of GPR antennas is that they work very close to the matter or even in contact with it: as a result, the current distribution on the antenna is influenced by the ground or by the investigated structure. In order to achieve a wide applicability of the radar system, changes in electrical properties of the matter should not affect strongly the antenna performance. Moreover, the antenna should provide stable performance at different elevation levels.

For an efficient coupling of electromagnetic waves into the ground/investigated structure, good impedance matching is necessary at the antenna/matter interface.

Another important requirement concerns the weight and size of the antennas: for ease of utilisation and to allow a wide applicability, the antennas shall be light and compact.

To be sure that the emitted signal is stable, a warm-up time should be foreseen before using GPR transmitting antennas. This allows avoiding errors in depth calculations, due to time deviations of the traces in the initial working period. Obviously, such errors have more relevance when materials with higher electromagnetic-wave propagation velocity are investigated (Pereira et al. 2006).

In Sect. 2, an overview on the most frequently used GPR antennas, for both pulsed and stepped-frequency radar systems, is provided; recent studies concerning innovative solutions are presented. Information on antenna arrays for GPR applications is given as well.

Section 3 is devoted to carry out a census of commercial antennas of a number of GPR manufacturers, specifying the different centre frequencies and general characteristics of the antennas available on the market.

Aided by measurements and powerful computer modelling techniques, GPR antenna designers are increasingly able to predict and understand the performance of proposed design in realistic electromagnetic environments. Section 4 deals with experimental characterisation of GPR antennas and Sect. 5 focuses on the electromagnetic modelling of GPR antennas.

Finally, conclusions are drawn in Sect. 6 and research perspectives in the field of GPR antennas are discussed.

2 Most Frequently Used GPR Antennas

In this Section, the most frequently used GPR antennas are presented and their main characteristics are briefly described. The Reader is encouraged to consult (Daniels 2004, 2009, 2010; Turk 2011), including interesting overviews on GPR antennas. Moreover, the Proceedings of recent editions of the biannual events "International Conference on Ground Penetrating Radar" and "International Workshop on Advanced Ground Penetrating Radar" are excellent references for an update on the latest research developments in the field of GPR antennas.

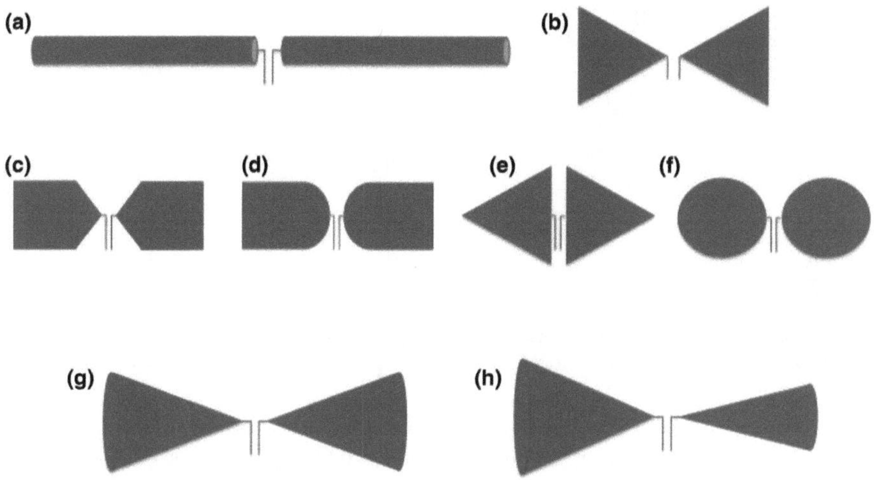

Fig. 1 Examples of dipole antennas: **a** fat dipole; **b** bow-tie; **c** and **d** planar rectangular dipoles; **e** diamond dipole; **f** elliptical dipole; **g** and **h** biconical antennas with equal and unequal cone angles

2.1 Dipole Antennas

Examples of dipole antennas are sketched in Fig. 1, they have been widely used in ground-penetrating radar systems. These antennas generally show linear polarisation, low directivity and relatively limited bandwidth, unless loading or distributed loading techniques are employed to increase the bandwidth at the expense of radiation efficiency.

Conventional dipoles, being constituted by straight and thin metallic elements, are narrowband. Their bandwidth can be enlarged by thickening the wires (fat dipoles, Fig. 1a) or else by adopting planar dipole-like solutions as the bow-tie, rectangular, diamond, and elliptical antennas (Fig. 1b–f). The biconical antenna consists of two cones placed symmetrically along their principal axis and fed at the same point (Fig. 1g); cones with unequal angles can be also used (Fig. 1h).

Among dipole antennas, the bow-tie is the most popular GPR structure, due to its simplicity and relatively ultra-wideband characteristics; it was conceived decades ago (Brown and Woodward 1952). More complicated flare shapes, as in the rectangular and elliptical planar dipoles, aim at achieving a larger bandwidth than with a conventional bow-tie, without increasing the antenna complexity. Both the radiation pattern and input impedance of a bow-tie are strongly affected by the flare angle (see Fig. 2a): by adjusting this geometrical parameter, it is possible to control the radiation pattern and to obtain maximum radiation towards the ground.

In biconical antennas, similarly to bow-ties, the cone angle has a critical effect on the radiation pattern and input impedance. In the case of unequal cones, the cone angles can be adjusted so that most power is radiated in the desired direction, usually occurring along the cone with smaller angle.

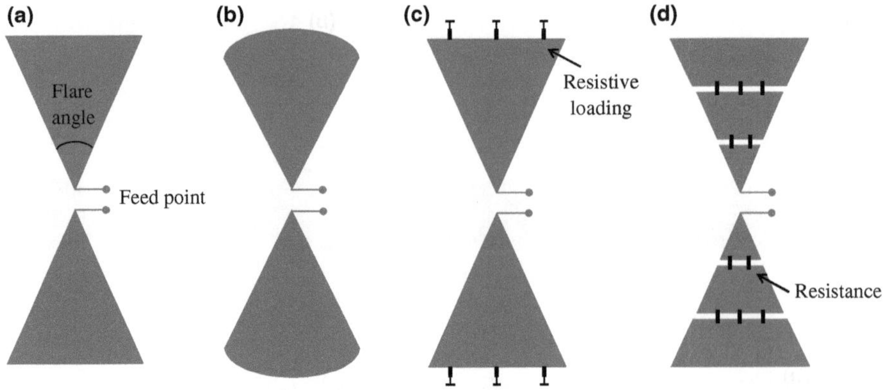

Fig. 2 Examples of bow-tie antennas: **a** basic bow-tie; **b** rounded bow-tie; **c** bow-tie with resistive loading at the far end; **d** bow-tie with sections coupled through resistances

A fairly simple variant of the traditional bow-tie has rounded flares (as in Fig. 2b). The rounding results in a flatter frequency behavior of the impedance. Moreover, the ring-down characteristics of the antenna turn out to be improved, as the rounded shape guarantees that electromagnetic reflections from the flare ends occur at the same time instant.

Reflections from the flare ends of a bow-tie can be eliminated by end loading them (Fig. 2c), or else by dividing the antenna in sections, coupled through resistances (Fig. 2d). Alternatively, the amplitude of the charge and current reaching the far end can be reduced by resistively coating the antenna. Another possibility is that of realising the antenna from a material such as Nichrome, which has a defined loss per unit area (Daniels 2009). Resistive loading is very effective in widening the band, but drastically decreases the antenna efficiency and consequently its gain.

Dipole antennas are omnidirectional, hence they are very sensitive to external electromagnetic interference in receive mode, even more due to their ultra-wideband characteristics. In order to reduce possible interference, a shield is usually placed around the antenna. The shield has to be carefully designed in order to avoid affecting the GPR performance; it turns obviously out to be useful also in protecting the antenna from damage.

In many applications, as cultural-heritage prospecting, estimation of soil properties, and investigation of scenarios presenting several layers of interest at different depths, it is necessary to have at disposal different antennas in order to gather acquisitions on different frequency bands. Moreover, in particular situations, as in the proximity of a tower antenna for broadcast radio communications, narrow-band interferences can be present, affecting some frequency bands more than others. These circumstances make a multiband system potentially more versatile and applicable; consequently, radars equipped with dual antennas working on two different centre frequencies have been introduced in the market by several GPR manufacturers. As an alternative, a reconfigurable antenna has been recently proposed and realised (Persico and Prisco 2008; Matera et al. 2014): it consists of a bow-tie equipped with two series of switches along its arms, so that their size can be varied and the antenna can work at three different frequencies thus covering the 50–1000 MHz band. A bistatic stepped-frequency radar has been equipped with two prototypes of the innovative reconfigurable bow-tie and successfully tested in several situations as well as under different conditions. Of course, a traditional system equipped with three different couples of fixed-length bow-ties would require more space and its size and weight could represent a considerable problem.

Another interesting study involving bow-tie antennas for GPR applications was recently presented in (Ajith and Bhattacharya 2014). A resistively-loaded bow-tie antenna, operating in the 300 MHz–3 GHz range and employing a metamaterial lens to improve the forward gain and front-to-back ratio, was proposed, realised and experimentally characterised.

2.2 End-Fire Tapered Slot Antennas

End-fire tapered slot antennas include linearly and exponentially tapered solutions (see Fig. 3), the latter being also known as Vivaldi antenna. Performance characteristics of these travelling-wave slot radiators include a wide frequency band, a medium directivity, and low sidelobes. They are commonly manufactured on a dielectric substrate and their electromagnetic properties depend on the antenna size, rate of divergence of conducting plates, substrate thickness and permittivity.

The Vivaldi antenna is quite often used in GPR systems and was first introduced in 1979 (Gibson 1979): its slot has an exponentially tapered shape, as in Fig. 3b. Due to the exponential profile, this antenna has a long electrical length compared to its actual physical size, hence compact designs are possible without compromising too much the directivity. Moreover, constant gain characteristics can be achieved over frequency.

In (Zhou and Cui 2011), a metamaterial with meander-line inclusions was applied to improve the directivity of a Vivaldi antenna.

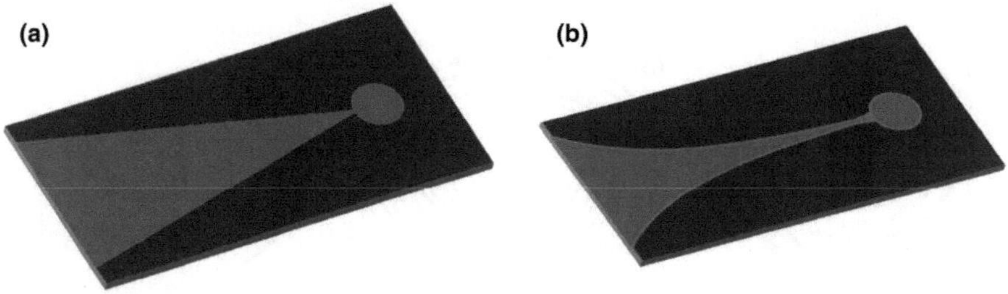

Fig. 3 Examples of end-fire tapered slot antennas: **a** linearly tapered slot antenna; **b** exponentially tapered slot antenna, also called Vivaldi

2.3 Spiral Antennas

The spiral antenna is considered as a frequency-independent device; in practice, the bandwidth is finite, due to the finite length of antenna arms and finite inner gap width. The arms making up the spirals can follow equiangular (logarithmic) (Pastol et al. 1990) or Archimedean (Lacko et al. 2002) shape. The minimum operating frequency is determined by the outer diameter of the spiral, the maximum frequency is determined by the inner diameter and by the precision at the feed region (which usually is located in the center of the spiral). Archimedean spiral antennas are inherently circularly polarised, with relatively constant input impedance and radiation patterns over the operating frequency range. Left- or right-hand circular polarisation can be achieved by applying the appropriate phase shift between the arms at the feed point.

The distinctive property of radiating circular polarisation makes the spiral antenna attractive for the detection of mines or other cylindrically-shaped objects. However, the phase center of the antenna changes with the frequency of operation, thus the spiral is a dispersive device. For this reason, it is usually used within stepped-frequency radar systems, where the dispersion can be corrected by suitable post processing of measured data. The ends of the arms are sometimes terminated with loads, to compensate for the end reflections at low frequency operation.

Figure 4 shows examples of spiral antennas: a two-arm Archimedean spiral, a four-arm Archimedean spiral, and an equiangular spiral. At first glance, the four-arm spiral is only a two-arm with four arms—but it is a lot more than that. The addition of a couple of additional arms improves the axial ratio, but it also introduces some extra capabilities. In particular, the four-arm spiral can operate very much like the two-arm spiral (mode 1), but a different mode (mode 2) is also possible. By using the correct feed network and signal processing, each arm of the spiral can be fed 90° out of phase, allowing mode-1 operation, which produces a radiation pattern with a maximum in boresight. When feeding each arm 180 degrees out of phase, mode-2 operation produces a boresight null. A beamforming network can be used to combine these two modes of operation vectorially and form four-squinted beams. Such an antenna can be steered in the right direction to detect targets in different spatial regions.

Fig. 4 Examples of spiral antennas: **a** self-complimentary 2-arms Archimedean spiral; **b** self-complimentary 4-arm Archimedean spiral; **c** equiangular or logarithmic spiral

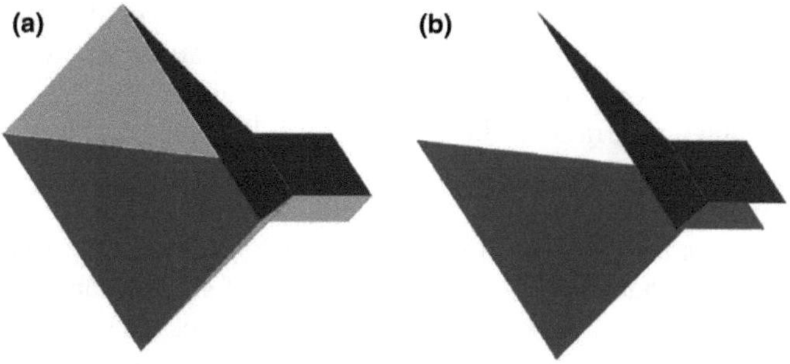

Fig. 5 **a** Conventional horn; **b** TEM horn

Non-planar spiral structures can be realized, as conical spiral antennas (Miller and Landt 1977). Planar structures have bidirectional properties, whereas the conical spiral antenna usually has unidirectional properties. To achieve a unidirectional beam, cavity backing of planar spirals can be employed.

2.4 Horn Antennas

In Fig. 5, the geometry of both a conventional and transverse electromagnetic (TEM) horn is sketched. The TEM horn is very often used in GPR systems, because of its time-domain excellent characteristics. It consists of two metallic plates diverging from a feeding point. The structure is derived from the horn, by removing the side walls.

This antenna does support a TEM mode, hence its name; this implies that the infinite TEM horn does not have any cut-off frequency; in practice the lower cut-off frequency is limited by the finite size of the structure.

The TEM horn is non dispersive, directive and can have ultra-wideband properties. These can be achieved in different ways: by using resistive loading; else, by tapering the width of the plates along the length of the horn, thus tapering the characteristic impedance from the value of the feeding line (usually 50 Ω) to the free-space value (377 Ω); else, by filling the horn with dielectric material (Scheers 2000; Yarovoy et al. 2000b). The dielectric filling gives the possibility to reduce the antenna size in its longitudinal length, without degrading the directivity.

2.5 Antenna Arrays

Array of antennas can be used in GPR systems, in order to enable a faster data collection by increasing the extension of investigated area per time unit. This can be a significant advantage in archaeological prospection (Trinks et al. 2010), road and bridge inspection (Simi et al. 2012), mine detection (Nuzzo et al. 2014), as well as in several other civil-engineering and geoscience applications where the collection of data requires the execution of a large number of profiles.

Moreover, antenna arrays allow collecting multi-offset measurements simultaneously, thereby providing additional information for a more effective imaging and characterisation of the natural or manmade scenario under test.

Two approaches are possible to GPR array design. The simplest is to conceive the array as a multi-channel radar system composed of single-channel radars. Much more can be achieved, if advanced array-design techniques are employed to synthesize the whole system. This second approach is definitely promising, as it gives the possibility to fully exploit the potentiality of antenna arrays: they can provide a high directivity by using simple elements, and the capability of a steerable beam, as in smart antennas. In fact, a fundamental characteristic of an array is the change of its radiation pattern in response to different excitations of its elements. Unlike a single antenna, wherein the radiation pattern is fixed, the radiation properties of an array can be changed upon feeding its elements with different currents: this gives a freedom to synthesize a desired pattern from an array, without changing its physical size.

The first introduction of multi-channel GPR systems dates back to more than twenty years ago (Warhus et al. 1993). Initially, the complete acceptance of multichannel recording was limited by the quality of the data and complex data processing required (Francese et al. 2009): different frequency responses of the antenna element prevented an easy amalgamation of individual profiles into useful images. More recently, the multi-channel manufacturers have provided GPR systems where the antenna responses of the individual elements are much closer (Linford et al. 2010; Simi et al. 2010). Signal processing techniques can be applied to measured data, to compensate for small differences in the frequency responses of the antennas and balance the signal between different elements, before constructing the radar image.

The advanced array presented in (Simi et al. 2010) is of particular interest for the inspection of transport infrastructures: it is equipped with two vertically-polarised arrays of dipoles (with a centre frequency of 200 MHz), devoted to collect data useful to locate longitudinal objects, as pipes and other utility lines; furthermore, it includes a dual-frequency (200 and 600 MHz) horizontally-polarised array, able to detect transversal targets as junctions and connection lines. Overall, 38 radar profiles are simultaneously acquired by this system at a speed of up to 15 km/h, thus guaranteeing a high productivity.

Air- and ground-coupled innovative arrays for stepped-frequency GPR have been recently developed and optimised (Eide et al. 2012, 2014). They are mainly conceived for road and bridge-deck inspection and use boomerang-shaped monopole elements mounted on a layer of microwave absorbing material and backed by a ground plane. The antenna elements provide a flat frequency response from 200 MHz to 3 GHz, thus allowing to take full advantage of the stepped-frequency radar properties of covering a wide frequency range without switching between different antennas.

A linear Vivaldi antenna array for water leaks detection, with a bandwidth between 250 and 2000 MHz, was lately designed (Alkhalifeh et al. 2014).

As far as arrays with steerable beams are concerned, the project of a novel radar system, equipped with an antenna array that can automatically reconfigure itself, was presented in (Bianchieri-Astier et al. 2014): the radar is conceived for planetary exploration and its transmitter is able to suitably feed the array elements, in order to focus the radiated beam on a high-energy zone detected in a previous recorded acquisition, thus optimising the signal-to-noise ratio and the localisation of buried objects.

Another important issue when using antenna arrays is data positioning. The position of array elements has to be recorded during the surveys, by using suitable high-precision positioning systems. As a collateral consequence, with today's multi-channel radar systems it is no longer necessary to follow a recording procedure using parallel lines: the array of antennas can theoretically have any arbitrary orientation while surveying an area. Of course, this makes data processing more challenging.

3 Overview on Commercial GPR Antennas

This Section provides a schematic overview of commercial antennas of a number of GPR manufacturers, specifying the different centre frequencies and general characteristics of the available antennas.

In Tables 1 (radio frequencies) and 2 (microwave frequencies), commercial GPR antennas have been subdivided according to various frequency ranges. As it is well known, the size of a target detectable with a GPR depends on the frequency of the used antenna. On one hand, high-frequency antennas can detect smaller features thus generating a high-resolution image, however they present a shallower penetration depth. On the other hand, low-frequency antennas can penetrate deeper into the ground and sense deeper buried targets, but their resolution is coarse. Hence, in relation to the planned survey and desired target, one should choose the right frequency of investigation accordingly. Radio-frequency antennas are better suited for applications requiring a deep penetration depth, e.g.: geological, geotechnical and geohydrological investigations, stratigraphic surveys, deep pipe and bedrock detection, as well as for river/landfill analyses and lake-bottom mapping. Microwave-frequency antennas are better suited for applications requiring a high resolution and precision measurements, e.g.: concrete inspection, locating rebar and shallow pipes, road mapping, quality assessment of precast structures, forensic applications, bridge condition assurance, utility and voids detections. As a general rule, it is unlikely that one antenna can cover the full range of applications a perspective user is interested to; for this reason, some manufacturers of ground-penetrating radars offer a range of antennas which can be used with a single control unit.

The information provided here is not exhaustive: the tables were completed in January 2015, by using information available on the webpages of GPR manufacturers. For several antennas, we specified whether they are shielded or unshielded, whether transmitting and receiving antennas are hosted in separate modules, as

Table 1 Antennas available from the above-selected GPR manufacturers, with center frequency in the radio-frequency portion of the electromagnetic spectrum

	0–50 (MHz)	51–100 (MHz)	101–200 (MHz)	201–300 (MHz)
Ditch Witch				250[s,d]
Geological Prospecting	20–30		120–150, 150–200, 150–250	150–250
Geoscanners	20[u], 25[u], 33.5[u], 50[u]	64.75, 80, 100[ut]	103[s,u,bt], 119.5[s,bt], 120	210[s,bt]
Geotech		90[sep]	150[sep]	250[df,sep]
GSSI[a]	15–80	15–80, 100[s]	200[s]	270[s], 300[df]
IDS[b]	25[d,u], 40[d,u]	80[d,s], 100[d,s]	150[d,u], 200[df,dp,a]	250[df], 300[u,d]
K–S Analysis	0.00125–50[sf]			
Malå	25[d,u], 30[u], 50[d,u]	80, 100[d,s,u]	160, 200[s,d,sep,u]	250[s,d]
NGS[c]		80		300
NIITEK			200–7000[a,uwb]	200–7000[a,uwb]
Penetradar				300[s,d]
PipeHawk			135–835[uwb], 128–764[uwb]	135–835[uwb], 128–764[uwb]
Radar System	25, 38[d,u], 50	75[d,u], 100[s]	150[d,u]	300[s]
Radarteam	40[d]	52, 70[d], 80, 100[d,s,,df]	124, 150[d]	250[d,s,df]
Radiodetection				250[uwb]
Radiolocation		100	200	300
SenSoft[d]	12.5[u], 25[u], 50[u]	100[u]	110[s], 200[u], 225[s], 250	
Terraplus	25[u], 50[u]	80, 100[s,u]	160, 200[u,a,sep]	250[a,s]
Toikka				
Transient			125	300
US Radar		100		250
Utsi	50[a]	100[a]		250[a]
Vermeer				
3D-Radar	7–30, 15–60, 30			200–3000[sf]

The meaning of the letters in the superscript is the following: *s* shielded antenna; *u* unshielded antenna; *sep* separable antenna; *a* array; *d* dipole antenna; *bt* bow-tie antenna; *h* horn antenna; *dp* dual polarized; *df* dual frequency; *sf* step-frequency; *uwb* ultra wideband
[a]Geophysical Survey Systems, Inc.
[b]IDS Ingegneria Dei Sistemi SpA
[c]Northeast Geophysical Services
[d]Sensors and Software

well as the antenna type (dipole, bow-tie, horn, array, dual-polarised antenna, dual-frequency antenna). According to the available information, antennas commonly employed in commercial GPR devices are essentially dipoles, bow-ties and horns.

Further information is organized in Tables 3, 4, 5, 6, 7 and 8. Multi-channel GPR systems are not considered in this Section.

Table 2 Same as in Table 1, in the microwave-frequency portion of the electromagnetic spectrum

	301–400 (MHz)	401–500 (MHz)	501–600 (MHz)	601–700 (MHz)	701–800 (MHz)
Ditch Witch				700[s,d]	
Geological Prospecting					
Geoscanners	307[s,bt], 380[s,bt], 390[h]	480[s,bt], 485		675[s,bt], 692	
Geotech	400[h,sep]			700[df]	
GSSI	400[s]				800[df]
IDS	400[df]		600[d,df,dp,a]	700[df]	
K-S Analysis					
Malå	350[s], 400[s,sep]	450, 500[s]			800[s]
NGS		500			
NIITEK	200–7000[a,uwb]	200–7000[a,uwb]	200–7000[a,uwb]	200–7000[a,uwb]	200–7000[a,uwb]
Penetradar	400[s,d]	500[s,d h]			
PipeHawk	135–835[uwb], 128–764[uwb]	135–835[uwb], 128–764[uwb]	135–835[uwb], 128–764[uwb]	135–835[uwb], 128–764[uwb]	135–835[uwb], 128–764[uwb]
Radar System	300[s]	500[s]			750[s]
Radarteam	350[s,d]	500[s,d,df]	600[s,d]		
Radiodetection	400	500	600		
SenSoft		450[s], 500			
Terraplus	400[a,sep]	450, 500[s]			750, 800[s]
Toikka					800[h]
Transient		500		700	
US Radar		500			
Utsi	400[a]				
Vermeer	400				
3D-Radar	200–3000[sf]	200–3000[sf]	200–3000[sf]	200–3000[sf]	200–3000[sf]

	801–900 (MHz)	901–1000 (MHz)	1001–2000 (MHz)	>2.0 (GHz)
Ditch Witch				
Geological Prospecting		1.0	1.5	
Geoscanners	890	1.0[s,h]	1.5[s,bt], 1.9[h], 2.0[s,bt,]	
Geotech		1.0[h]	1.2, 1.7[h], 2.0[h]	
GSSI	900[s]	1.0[h]	1.6, 2.0[h]	2.6
IDS Corp.	900[s,df]	1.0[h], 1.2[s], 1.6[s]	2.0[s,d,dp,h]	
K-S Analysis				
Malå		1.0[s]	1.2[s], 1.3[s,sep], 1.6[s]	2.3[s]
NGS	900			
NIITEK	200–7000[a,uwb]	200–7000[a,uwb]	200–7000[a,uwb]	200–7000[a,uwb]
Penetradar		1.0[h]	2.0[h]	2.5[h]

(continued)

Table 2 (continued)

	801–900 (MHz)	901–1000 (MHz)	1001–2000 (MHz)	>2.0 (GHz)
PipeHawk	135–835[uwb]			
Radar System	900[s]	1.0[s,h]	1.5[s], 2.0[u]	
Radarteam	900[d,s,df]	1.0[s,bt]	1.3[a,sep]	
Radiodetection				
SenSoft	900[s]	1.0	1.2[s]	
Terraplus		1.0[s]	1.2[s], 1.3[a], 1.6[s]	2.3[s]
Toikka		1.0[h]	1.5[h]	
US Radar	900	1.0	1.5, 1.6, 2.0	2.5, 2.6
Utsi		1.0[a]	1.5[a]	4.0[a], 6.0[a]
Vermeer				
3D-Radar	200–3000[sf]	200–3000[sf]	200–3000[sf]	200–3000[sf]

4 Experimental Characterisation of GPR Antennas

Any GPR antenna should be measured, in order to verify its characteristics. It may be tested in open space or in a conventional anechoic chamber, and this allows the determination of its radiation pattern, gain and directivity in the air. However, evaluating its actual performance in a realistic working environment is much more important and still rarely done. To the best of our knowledge, few suitable facilities exist for this type of measurement. The antenna operation in the presence of a soil or manmade structure is often checked by simply measuring its internal impedance. However, this test is not enough, since a good antenna matching to the transmission line does not guarantee the proper radiation properties.

A GPR antenna test facility is available at the Technical University of Denmark. It consists of a soil-filled wooden box, with size $3.8 \times 3.3 \times 1\ m^3$, and a mechanical scanning system installed over it, able to move the Antenna Under Test (AUT) along two orthogonal directions in a plane parallel to the air-soil interface. The proximity to the box edges limits the angular range of the scan, depending on the beamwidth of the AUT and its height above the soil. An ultra-wideband sensor for electromagnetic field measurements in time domain is buried at the center of the box, at a depth of about 20 cm; this probe is a loop antenna made of a semi-rigid cable, showing a -10 dB bandwidth ranging from 180 to 3880 MHz (Yarovoy et al. 2000a, b). The semi-rigid cable is connected to one port of a vector network analyser, and the other port is attached to the AUT through a phase-stable coaxial cable; in this way, the measurement is performed just as in a standard anechoic chamber. The choice of soil type depends on the application the AUT is meant for. The moisture content can be varied and a moisture meter is used to keep it reasonably uniform, despite the continual loss of water through evaporation from the upper surface.

This facility was used within the EU FP6 Network of Excellence Antenna Centre of Excellence (ACE), to measure the footprint of eight different antennas (Lenler-Eriksen et al. 2005): three exponentially-tapered slots from the French Centre

Table 3 Further information on Geoscanners antennas

Centre frequency (MHz)	Antenna's model ID	Antenna's type	Antenna's configuration	Shielded or unshielded	Weight (kg)
64.75	Gekko-60/Gekko-60 Bistatic		Monostatic		5.0
			Bistatic		10.0
80	Gekko-80/Gekko-80 Bistatic		Monostatic		3.3
			Bistatic		6.6
103	GCB GCB-100	Quarter wavelength bowtie	Bistatic	Shielded	2.1
119.5	BA-100	Borehole			2.1
120	Gekko-120/Gekko-120 Bistatic		Monostatic		2.5
			Bistatic		5.0
210	GCB-200	Quarter wavelength bowtie	Bistatic	Shielded	3.5
250	FLB-250	Airborne			4.5
307	GCB-300	Quarter wavelength bowtie		Shielded	2.6
380	GCB-400	Quarter wavelength bowtie		Shielded	1.65
390	FLB-390	Airborne			3.0
480	GCB-500	Quarter wavelength bowtie		Shielded	1.7
485	BA-500	Borehole			1.45
675	GCB-700	Quarter wavelength bowtie		Shielded	1.35
692	BA-1000	Borehole			1.4
890	HA-1000	Horn			6.5
1000	GCB-1000	Quarter wavelength bowtie		Shielded	1.05
1500	GCB-1500	Quarter wavelength bowtie		Shielded	1.05
1900	HA-2000	Horn			6.5
2000	GCB-2000	Quarter wavelength bowtie		Shielded	1.05
20, 25, 33.3, 50, 100	MFC-1250/MFC-1250 Bistatic		Monostatic	Unshielded	2.25
			Bistatic		4.5

Table 4 Further information on GSSI antennas

Centre frequency (MHz)	Antenna's model ID	Antenna's type	Antenna's configuration	Shielded or unshielded	Weight (kg)
15–80	3200 MLF				15.0
					23.0
100	3207AP 3207F		Monostatic	Shielded	13.0
			Bistatic		26.0
200	5106/5106A			Shielded	20.5
270	50270S			Shielded	8.6
400	50400S			Shielded	5.0
900	3101A/3101D			Shielded	2.3
1000	41000S	Horn			7.3
1000	5101				1.8
1600	5100B				1.8
1600	51600S				1.8
2000	62000-003/62000-007				1.3
					1.8
2000	4105	Horn			7.3
2000	42000S	Horn			7.3
2600	52600S				1.8
300, 800	D50300/800				5.0

National de la Recherche Scientifique (CNRS) and Laboratoire d'Electronique, Antennes et Télécommunications (LEAT), two loaded bow-ties from the Italian manufacturer IDS Ingegneria dei Sistemi SpA, two spirals from the University of Liverpool, and a patch from the Institute of Communications and Computer Systems in Greece. The near-field antenna footprint is a three-dimensional set of data (two space coordinates and time), characterising the antenna radiation into the ground: from the time slices of the footprint, the spatial distribution of the radiated energy can be determined. By Fourier transforming the data over the time coordinate, a spectral analysis of the radiation can be done; moreover, radiation patterns in the ground can be calculated via near-to-far field transforms of the measured data.

A similar test facility is available at the Delft University of Technology (Yarovoy et al. 2007). It consists of a $2 \times 2 \times 1.5$ m^3 wooden box filled with dry sand, a mechanical scanner, a couple of loop sensors (Yarovoy et al. 2000a, b) buried at a depth of 17 cm in the sand, a set of pulse generators, a stroboscopic receiver, and a vector network analyser. The lowest frequency, at which accurate antenna measurements can be performed, is about 170 MHz, whereas the highest is approximately 9 GHz.

This facility was used during the above-mentioned ACE Network of Excellence to measure both the antenna footprint and elevation profile of two exponentially-tapered slots from CNRS/LEAT and a balanced dielectric-wedge antenna. The elevation profile provides important information about antenna coupling to the ground: to measure it, the AUT has to be placed right above the probe and the

Table 5 Further information on IDS antennas

Centre frequency (MHz)	Antenna's model ID	Antenna's type	Antenna's configuration	Shielded or unshielded	Weight (kg)
25	TR U 25	Dipole	Bistatic	Unshielded	18.0
40	TR U 40	Dipole	Bistatic	Unshielded	18.0
80	TR 80	Dipole	Bistatic	Shielded	23.0
100	TR 100 B	Dipole	Bistatic	Shielded	22.0
	TR 100 M		Monostatic		
150	BA 150	Dipole, borehole		Unshielded	1.5
200	TR 200	Dipole	Monostatic	Shielded	6.0
300	BA 300	Dipole, borehole		Unshielded	1.0
400	TR 400		Monostatic	Shielded	5.5
600	TR 600H	Dipole	Monostatic	Shielded	5.5
600	TR 600V	Dipole	Monostatic	Shielded	3.0
900	TR 900		Monostatic	Shielded	3.0
1000	HR1000	Horn			6.5
1200	TR 1200		Monostatic	Shielded	1.6
1600	TR 1600		Monostatic	Shielded	1.5
2000	HR2000	Horn			6.5
2000	TR-HF	Dipole		Shielded	1.3
2000	TR B IP	Dipole		Shielded	1.5
2 × 600	VA600-DK	Dipole		Shielded	8.0
1 × 200 + 2 × 600	1 × TR 200 + 2 × TR 600 (RIS S)	Array (dipole)	Monostatic	Shielded	17.0
			Bistatic		
1 × 200 + 2 × 400	1 × TR 200 + 2 × TR 400 (RIS S)	Array (dipole)	Monostatic	Shielded	16.5
			Bistatic		
3 × 400	3 × TR 400 (RIS S)	Array (dipole)	Monostatic	Shielded	16.5
			Bistatic		
3 × 600 + 1 × 200	TRMF SMA	Array (dipole)	Monostatic	Shielded	10.0
200, 600	TRMFTX	Array (dipole)		Shielded	6.0
200, 600	TRMFRX	Array (dipole)		Shielded	6.0
200, 600	DUAL F				6.0
4 × 200, 600	DCL 200, DCL 600 (RIS MF Hi-Mod)	Array (dipole)			58.0 (full configuration)
4 × 400, 900	DCL 400, DCL 900 (RIS MF Hi-Mod)	Array (dipole)		Shielded	58.0 (full configuration)

(continued)

Table 5 (continued)

Centre frequency (MHz)	Antenna's model ID	Antenna's type	Antenna's configuration	Shielded or unshielded	Weight (kg)
2 × 200 + 4 × 200, 600 + 2 × 200	DML 200 + DCL 200, DCL 600 + DML 200 (STREAM-EM)	Array (dipole)			90.0 (full configuration)
1000	STREAM MT	Array (dipole)			
1 × 200	DML 200 (STREAM-X)	Array (dipole)			40.0
1 × 600	DML 600 (STREAM X)	Array (dipole)			36.0
2 × 600	DML 600 (STREAM X)	Array (dipole)			40.0
4 × 1600	TR HIRESS	Array (dipole)		Shielded	3.7
600, 2000	TR 600 + TR-HF	Array (dipole)		Shielded	7.0
200, 600	DUAL F	Dipole		Shielded	6.0
400, 900	TR Dual-F 400/900				3.0
600, 1600	TR-AL	Dipole		Shielded	5.0
250, 700	ST Duo			Shielded	15.0

field radiated into the ground is recorded at different antenna elevations. Then, the measured signals are shifted in time, to compensate the time delays due to the different antenna elevations, and compared. If the waveform of the signals remains similar and only its amplitude varies with antenna elevation, this means that the ground is not influencing the antenna performance, which will reasonably remain similar for different types of soils.

Further studies are desirable, in this area. Challenging ideas can definitely come by looking at test facilities specifically designed for medical-imaging antennas. The wideband biomedical imaging is a subject that has much in common with GPR: both fields frequently employ antennas designed to operate very close to the matter or in contact with it, over the widest possible bandwidth, and to radiate pulse waveforms. Medical imaging antennas work at higher frequencies than GPR and radiate into a medium with a much higher permittivity. Despite these dissimilarities, sharing and integrating expertise in GPR and biomedical antenna testing will definitely be fruitful.

Table 6 Further information on Malå antennas

Centre frequency (MHz)	Antenna's model ID	Antenna's type	Antenna's configuration	Shielded or unshielded	Weight (kg)
25		Dipole		Unshielded	3.85
30	RTA 30 MHz			Unshielded	7.8
50		Dipole		Unshielded	2.65
50	RTA 50 MHz			Unshielded	7.0
80	GX80 HDR				24.6
100		Dipole		Unshielded	1.1
100		Dipole			5.5
100	RTA 100 MHz			Unshielded	6.0
100		Borehole (dipole)			6.9 (full configuration)
100				Shielded	25.5
160	GX160 HDR				10.7
200		Dipole		Unshielded	0.55
200				Shielded	4.2
250		Dipole			4.5
250				Shielded	7.85
250		Borehole (dipole)			9.5 (full configuration)
250				Shielded	12.0
350				Shielded	19.0
400				Shielded	2.1
450	GX450 HDR				5.5
500				Shielded	5.0
500				Shielded	18.0
750	GX750 HDR				3.6
800				Shielded	2.6
1000				Shielded	2.4
1200				Shielded	1.0
1300				Shielded	1.5
1600				Shielded	0.6
2300				Shielded	0.5
200, 400, 1300		Array		Shielded	

Table 7 Further information on SenSoft antennas

Centre frequency (MHz)	Antenna's model ID	Antenna's type	Antenna's configuration	Shielded or unshielded	Weight (kg)
12.5		Dipole	Bistatic	Unshielded	7.2
25		Dipole	Bistatic	Unshielded	3.6
50		Dipole	Bistatic	Unshielded	1.8
50		Borehole			3.4
100		Dipole	Bistatic	Unshielded	1.2
100		Borehole			2.9
110			Monostatic	Shielded	9.0
200		Dipole	Bistatic	Unshielded	0.8
200		Borehole			2.3
225			Monostatic	Shielded	1.0
450			Monostatic	Shielded	0.7
900			Monostatic	Shielded	0.7
1200			Monostatic	Shielded	0.5
50, 100, 200			Monostatic		25.0

Table 8 Further information on 3D-Radar antennas

Centre frequency (MHz)	Antenna's model ID	Antenna's type	Antenna's configuration	Shielded or unshielded	Weight (kg)
200–3000	VX0909	Array	Stepped frequency		20.0
200–3000	VX1213	Array	Stepped frequency		24.0
200–3000	VX1821	Array	Stepped frequency		32.0
200–3000	VX2125	Array	Stepped frequency		36.0
200–3000	VX2429	Array	Stepped frequency		42.0
200–3000	VX3341	Array	Stepped frequency		56.0
200–3000	DX1821	Array	Stepped frequency		28.0
200–3000	DX2125	Array	Stepped frequency		33.0
200–3000	DX2429	Array	Stepped frequency		38.0
200–3000	DX3341	Array	Stepped frequency		50.0
200–3000	DXG0908	Array	Stepped frequency		25.0
200–3000	DXG1212	Array	Stepped frequency		32.0
200–3000	DXG1820	Array	Stepped frequency		42.0

5 Electromagnetic Modelling of GPR Antennas

Aided by powerful computer modelling techniques, GPR antenna designers are increasingly able to predict and understand the performance of proposed design in realistic electromagnetic environments.

In order to predict correctly the GPR response from a particular scenario, Maxwell's equations have to be solved, subject to the physical and geometrical properties of the considered problem and to its initial conditions. Several techniques have been developed in computational electromagnetics, for the solution of Maxwell's equations. These methods can be classified into two main categories: differential and integral equation solvers, which can be implemented in the time or frequency domain. The differential solvers include the Finite Difference Time Domain (FDTD) (Yee 1966; Taflove 1980; Taflove and Hagness 2005; Taflove et al. 2013) and Finite-Volume Time-Domain techniques (Shankar et al. 1989), the Finite Element Method (Jin 2014) and Transmission Line Method acronym (TLM) (Christopoulos 1995); the spectral-domain and differential Cylindrical- and Spherical-Wave Approaches may also be mentioned (Di Vico et al. 2005; Frezza et al. 2013). The Method of Moments (Harrington 1993) is a well-known integral-equation approach. All of the different techniques present compromises between computational efficiency, stability, and the ability to model complex geometries.

The FDTD technique has several advantages over alternative approaches: it has inherent simplicity, efficiency and conditional stability; it is suitable to treat impulsive behaviour of the electromagnetic field and can provide either ultra-wideband temporal waveforms or the sinusoidal steady-state response at any frequency within the excitation spectrum; it is accurate and highly versatile; and it has become a mature and well-researched technique. Moreover, the FDTD technique is suitable to be executed on parallel-processing CPU-based computers and to exploit the modern computer visualisation capabilities.

Hybrid methods were developed through the years, to combine advantages and mitigate disadvantages of different techniques (Huang et al. 1999; Monorchio et al. 2004; van Coevorden et al. 2006).

A GPR numerical simulation includes, at some degree of detail, a model of the source or antenna, a model of subsurface or structure, and models of the targets. Implementing an accurate model of the antenna is important for the investigation of new antenna designs, as well as when the goal of the simulation is to calculate the response from a complex structure in order to achieve a better understanding and interpretation of GPR data. In the latter case, if a description of the real antenna is omitted from the model and an ideal source is used (as a Hertzian dipole), the simulations cannot accurately replicate the amplitudes and time-domain shape of real responses, since the interactions between the targets and the antenna are not being taken into account.

Most GPR simulations prior to 1991 were simple two-dimensional models. In Moghaddam et al. (1991) the first 2.5D model was developed. Subsequently, Giannopoulos developed and compared 3D models of GPR by using TLM and FDTD technique (Giannopoulos 1997); later, he developed GprMax

(Giannopoulos 2005), which has been improved through the years and is nowadays a very well-known and largely validated FDTD simulator, available for free public download (both for academic and commercial use) on www.gprmax.com, together with a detailed user guide.

In Shlager et al. (1994), an FDTD model of a resistively loaded bow-tie, fed through an image plane by a coaxial line, was implemented, and the antenna design was optimised for pulse radiation. This work is of particular interest as it faced two important FDTD antenna-modelling problems, namely the issue of how to model the feed region of the antenna and the potential errors associated with staircasing.

Few researchers have combined a realistic antenna model with a realistic model of the investigated subsurface or structure.

In Bourgeois and Smith (1996), for the first time a realistic description of a GPR antenna was included in a wider FDTD model, replicating an experimental scale-model of a GPR system. In particular, bow-ties were simulated, together with their loading resistors and rectangular cavity shielding; the feed was represented through a transmission line. The constitutive parameters of the scale-model emulsions were simulated by fitting a Debye model. E-plane patterns were calculated for different antenna elevations above the surface of the emulsion; scattering responses from buried metallic and plastic pipes were obtained, too, and showed acceptable agreement with measured data.

A second-order accurate in time and fourth-order accurate in space FDTD model was developed in Cassidy (2001), where a pulseEKKO 1 GHz antenna was simulated. Numerical results and measured data showed good agreement.

In Lampe and Holliger (2001), Holliger et al. (2003), Lampe et al. (2003), Lampe and Holliger (2005), FDTD models of low-frequency bow-ties were implemented; a sub-gridding technique was used. Antenna radiation patterns over lossless and lossy half-spaces, with shielded and unshielded antennas, were calculated; the effect of including cavity absorbing material and loading resistors was investigated; how topographic roughness and subsurface heterogeneities affected the GPR responses was studied, too; the consequences of Wu-King loading of bowties were finally explored.

In Klysz et al. (2006), the FDTD technique was employed for the first time to simulate a commercial antenna (GSSI 1.5 GHz). Simulated and measured free-space radiation patterns were compared; moreover, data measured on two concrete cells of different moisture contents were compared with numerical results; the agreement was acceptable in both cases. The simulation did not include the receiving antenna and the concrete was assumed to be a homogeneous material.

Detailed FDTD models of two commercial high-frequency ground-coupled GPR bow-ties, namely a GSSI 1.5 GHz antenna and a MALA 1.2 GHz antenna, were developed in Warren (2009), Warren and Giannopoulos (2011). Taguchi's optimisation method was successfully used to determine the values of some unknown parameters. The result of the optimisation was a 98 % cross-correlation match of the modelled and real crosstalk responses in free-space for both antennas. Further validation was carried out using a series of oil-in-water emulsions, with different steel and composite rebar, as well as a rectangular metallic box, embedded in the emulsions as targets. The achieved accuracy of both the antenna models was very good, therefore they were subsequently used in more advanced studies. In Warren et al. (2014), the radiation characteristics of the GSSI 1.5 GHz antenna were investigated over a range of different lossy dielectrics; FDTD results were compared with measured data. In Giannakis et al. (2014a), the FDTD model of the GSSI 1.5 GHz antenna was incorporated in a highly-realistic model including soil inhomogeneities, vegetation, rough surface and water puddles, for humanitarian demining applications. Finally, an air-coupled horn antenna was implemented in Giannakis et al. (2014b), by using GprMax: it was tested in a realistically modelled pavement scenario and compared with the previously-implemented GSSI 1.5 GHz antenna.

6 Conclusions

Antennas are a fundamental hardware component of a radar system, dictating its performance in terms of capability to detect targets. In this Chapter a review on Ground-Penetrating Radar (GPR) antennas is provided. In Sect. 1, the general characteristics of GPR antennas are resumed and the requirements they have to satisfy are listed. In Sect. 2, an overview on the most frequently used GPR antennas, for both pulsed and stepped-frequency radar systems, is presented, including references to recent studies and information on GPR antenna arrays. In Sect. 3, commercial antennas of a number of GPR manufacturers are surveyed, the centre frequencies and general features of the antennas available on the market are schematically organised in tables. Section 4 of the Chapter is devoted to presenting two antenna test facilities, allowing an experimental characterisation of GPR antennas in different media. Finally, Sect. 5 of the Chapter focuses on the numerical characterisation of GPR antennas.

Activities on GPR antenna design should progress in various directions. These include:

- Design of antenna arrays, with high directivity and the capability of a steerable beam.
- Design of adaptive GPR antennas, with electronic control of characteristics, to adapt to different soils and materials.
- Application-specific antenna designs.
- Miniaturisation of GPR antennas.
- Design of reconfigurable antennas.
- Development of improved measurement facilities, allowing the experimental characterisation of GPR antennas in realistic working environment.
- Improvement of electromagnetic-modelling techniques for an efficient and accurate numerical characterisation of GPR antennas in realistic scenarios.
- Use of metamaterials to improve the characteristics of GPR antennas.
- In Appendix, a list of companies associated with the manufacturing and supply of GPR equipment is provided.

Acknowledgments The Authors are grateful to COST (European COoperation in Science and Technology) for funding the Action TU1208 "Civil Engineering Applications of Ground Penetrating Radar," supporting this work.

Appendix—List of Companies Manufacturing and Offering Ground Penetrating Radars

The following is a list of companies associated with the manufacturing and supply of GPR equipment. No endorsement is implied by the inclusion of these companies in the list.

Ditch Witch, USA—http://www.ditchwitch.com/
Geological Prospecting CO., Ltd., Russia—http://www.trgeo.ru/main_eng.htm
Geophysical Survey Systems Inc., GSSI, USA—http://www.geophysical.com/
Geoscanners AB, Sweden—http://www.geoscanners.com/
Geotech, Russia—http://www.geotechru.com/en/
Ingegneria Dei Sistemi SpA, IDS, Italy—https://www.idscorporation.com/
K-S Analysis, Germany—http://www.ks-groundradar.com/
Malå Geoscience AB, Sweden—http://www.malags.com/home
Non-Intrusive Inspection Technology Inc., NIITEK, USA—http://www.niitek.com/
Penetradar Inc., USA—http://www.penetradar.com/
PipeHawk Plc, U.K.—http://www.pipehawk.com/
Radar Systems Inc., Latvia—http://www.radsys.lv/
Radarteam Sweden AB, Sweden—http://www.radarteam.se
Radiodetection Ltd., U.K.—http://www.spx.com/en/radiodetection/
Radiolocation Ltd., Hong Kong—http://www.radiolocation.com.hk/eng/index/index.php?
Sensors and Software Inc., Canada—http://www.sensoft.ca/
Terraplus USA Inc., USA—http://www.terraplus.ca/
Toikka Engineering OY, Finland—http://personal.inet.fi/business/toikka/ToikkaOy/
Transient Technologies Company, Ukraine—http://viy.com.ua/e
Utsi Electronics Ltd., U.K.—http://www.utsielectronics.co.uk/
Vermeer, USA—http://www2.vermeer.com/vermeer/AP/en/N/

References

Ajith, K.K., Bhattacharya, A.: Improved ultra-wide bandwidth bow-tie antenna with metamaterial lens for GPR applications. In: Proceedings of the 15th International Conference on Ground Penetrating Radar (GPR 2014), pp. 739–744, Oct 2014

Alkhalifeh, K., Craeye, C., Lambot, S.: Design of a 3D UWB linear array of Vivaldi antennas devoted to water leaks detection. In: Proceedings of the 15th International Conference on Ground Penetrating Radar (GPR 2014), pp. 783–786, Oct 2014

Biancheri-Astier, M., Saintenoy, A., Ciarletti, V.: Development of an Agile beam georadar prototype for the investigation of pLanetary environment (AGILE). In: Proceedings of the 15th International Conference on Ground Penetrating Radar (GPR 2014), pp. 837–840, Oct 2014

Bourgeois, J., Smith, G.: A fully three-dimensional simulation of a ground-penetrating radar: FDTD theory compared with experiment. IEEE Trans. Geosci. Remote Sens. **34**(1), 36–44 (1996)

Brown, G.H., Woodward, O.M.: Experimentally determined radiation characteristics of conical and triangular antennas. RCA Rev. **13**, 425–452 (1952)

Cassidy, N.: The application of mathematical modelling in the interpretation of near-surface archaeological ground-penetrating radar. Ph.D. thesis, Keele University, Keele (2001)

Christopoulos, C.: The Transmission Line Modeling Method: TLM. IEEE Press, Piscataway (1995)

Daniels, D.J.: Antennas. Chap. 5 In: Ground Penetrating Radar. IEE Radar, Sonar, Navigation and Avionics Series, 2nd edn. pp. 131–183. The Institution of Engineering and Technology (2004)

Daniels, D.J.: Antennas. Chap. 4 In: Ground Penetrating Radar Theory and Applications, pp. 99–133. Elsevier Science, Amsterdam (2009)

Daniels, D.J.: Antennas. Chap. 3 In: EM Detection of Concealed Targets, Microwave and Optical Engineering, pp. 83–127. Wiley-IEEE Press, USA (2010)

Di Vico, M., Frezza, F., Pajewski, L., Schettini, G.: Scattering by a finite set of perfectly conducting cylinders buried in a dielectric half-space: a spectral-domain solution. IEEE Trans. Antennas Propag. **53**(2), 719–727 (2005)

Eide, E., Valand, P.A., Sala, J.: Ground-coupled antenna array for step-frequency GPR. In: Proceedings of the 15th International Conference on Ground Penetrating Radar (GPR 2014), pp. 756–761, Oct 2014

Eide, E., Kiessling, T., Typpo, J.: Wideband antenna array for step-frequency ground penetrating radar. In: Proceedings of the 14th International Conference on Ground Penetrating Radar (GPR 2012), pp. 152–155 (2012)

Francese, R.G., Finzi, E., Morelli, G.: 3-D high-resolution multi-channel radar investigation of a Roman village in Northern Italy. J. Appl. Geophys. **67**(1), 44–51 (2009)

Frezza, F., Mangini, F., Pajewski, L., Schettini, G., Tedeschi, N.: Spectral domain method for the electromagnetic scattering by a buried sphere. J. Opt. Soc. Am. A **30**(4), 783–790 (2013)

Giannakis, I., Davidson, N., Giannopoulos, A.: Realistic modelling of ground penetrating radar for landmine detection using FDTD. In: Proceedings of the 15th International Conference on Ground Penetrating Radar (GPR 2014), pp. 954–959, Oct. 2014 (Giannakis et al, 2014a)

Giannakis, I., Giannopoulos, A., Pajewski, L.: Numerical modelling of ground penetrating radar antennas. In: Geophysical Research Abstracts, European Geosciences Union (EGU) General Assembly 2014, 27 April–2 May 2014, Vienna, Austria, article ID EGU2014–1553 (Giannakis et al, 2014b)

Giannopoulos, A.: The investigation of transmission-line matrix and finite-difference time-domain methods for the forward problem of ground probing radar. Ph.D. thesis, Department of Electronics, University of York, York, UK (1997)

Giannopoulos, A.: Modelling ground penetrating radar by GprMax. Constr. Build. Mater. **19**, 755–762 (2005)

Gibson, P.J.: The Vivaldi aerial. In: Proceedings of the 9th European Microwave Conference, p. 736 (1979)

Harrington, F.R.: Field Computation by Moment Methods, reprinted edn. Wiley-IEEE Press, Piscataway (1993)

Huang, Z., Demarest, K., Plumb, R.: An FDTD/MoM hybrid technique for modeling complex antennas in the presence of heterogeneous grounds. IEEE Trans. Geosci. Remote Sens. **37**(6), 2692–2698 (1999)

King, R.W., Smith, G.S.: Antennas in Matter. MIT Press, Cambridge (1981)

Klysz, G., Ferrieres, X., Balayssac, J., Laurens, S.: Simulation of direct wave propagation by numerical FDTD for a GPR coupled antenna. Nondestr. Test. Eval. Int. **39**(4), 338–347 (2006)

Holliger, K., Lampe, B., Meier, U., Lambert, M., Green, A.: Realistic modeling of surface ground-penetrating radar antenna systems: where do we stand? In: Proceedings of the 2nd International Workshop on Advanced Ground Penetrating Radar, pp. 45–50 (2003)

Jin, J.: The Finite Element Method in Electromagnetics, 3rd edn. Wiley-IEEE Press, Hoboken (2014)

Lacko, P.R., Charmain, C.F., Johnson, M., Ralston, J.M., Bradley, M.R., McCummins, R.: Archimedean-spiral and log-spiral antenna comparison. In: Proceedings of the SPIE, Detection and Remediation Technologies for Mines and Minelike Targets, 4742(VII), pp. 230–236, Aug 2002

Lampe, B., Holliger, K.: Numerical modeling of a complete ground-penetrating radar system. Proc. SPIE **4491**, 99 (2001)

Lampe, B., Holliger, K., Green, A.: A finite-difference time-domain simulation tool for ground-penetrating radar antennas. Geophysics **68**(3), 971–987 (2003)

Lampe, B., Holliger, K.: Resistively loaded antennas for ground-penetrating radar: a modeling approach. Geophysics **70**, K23–K32 (2005)

Lenler-Eriksen, H.-R., Meincke, P., Sarri, A., Chatelee, V., Nair, B., Craddock, I.J., Alli, G., Dauvignac, J.-Y., Huang, Y., Lymperopoulos, D., Nilavalan, R.: Joint ACE ground penetrating radar antenna test facility at the Technical University of Denmark. In: Proceedings of the IEEE Antennas and Propagation Society International Symposium, 4A, pp. 109–112, 3–8 July 2005

Linford, N., Linford, P., Martin, L., Payne, A.: Stepped frequency ground-penetrating radar survey with a multi-element array antenna: results from field application on archaeological sites. Archaeol. Prospection **17**, 187–198 (2010)

Matera, L., Ciminale, M., Desantis, V., Persico, R., Giannotta, M.T., Alessio, A.: Application of a reconfigurable stepped frequency system to cultural heritage prospecting. In: Proceedings of the 15th International Conference on Ground Penetrating Radar (GPR 2014), pp. 18–24, Oct 2014

Miller, E.K., Landt, J.A.: Short-pulse characteristics of the conical spiral antenna. IEEE Trans. Antennas Propag. **25**(5), 621–625 (1977)

Moghaddam, M., Yannakakis, E., Chew, W., Randall, C.: Modeling of the subsurface interface radar. J. Electromagn. Waves Appl. **5**(1), 17–39 (1991)

Monorchio, A., Bretones, A., Mittra, R., Manara, G., Martin, R.: A hybrid time-domain technique that combines the finite element, finite difference and method of moment techniques to solve complex electromagnetic problems. IEEE Trans. Antennas Propag. **52**(10), 2666–2674 (2004)

Nuzzo, L., Alli, G., Guidi, R., Cortesi, N., Sarri, A., Manacorda, G.: A new densely-sampled ground penetrating radar array for landmine detection. In: Proceedings of the 15th International Conference on Ground Penetrating Radar (GPR 2014), pp. 969–974, Oct 2014

Pastol, Y., Arjavalingam, G., Halbout, J.M.: Transient radiation properties of an integrated equiangular spiral antenna. In: Proceedings of the IEEE Symposium on Antennas and Propagation, pp. 1934–1937. Dallas, Texas (1990)

Pereira, M., Fernando Rial, I., Lorenzo, H., Arias, P., Novo, A.: Setting up a GPR system for road evaluation. In: Proceedings of the 11th International Conference on Ground Penetrating Radar, pp. 1–6. Columbus Ohio, USA, 19–22 June 2006

Persico, R., Prisco, G.: A reconfigurative approach for SF-GPR prospecting. IEEE Trans. Antennas Propag. **56**(8), 2673–2680 (2008)

Scheers, B., Piette, M., Vander Vorst, A.: Development of dielectric-filled TEM horn antennas for UWB GPR. In: Proceedings of Millennium Conference on Antennas and Propagation, 2000

Shlager, K., Smith, G., Maloney, J.: Optimization of bow-tie antennas for pulse radiation. IEEE Trans. Antennas Propag. **42**(7), 975–982 (1994)

Shankar, V., Hall, W.F., Mohammadian, A.H.: A time-domain differential solver for electromagnetic scattering problems. Proc. IEEE **77**(5), 709–721 (1989)

Simi, A., Manacorda, G., Miniati, M., Bracciali, S., Buonaccorsi, A.: Underground asset mapping with dual-frequency dual-polarized GPR massive array. In: Proceedings of 13th International Conference on Ground Penetrating Radar, June 2010

Simi, A., Manacorda, G., Benedetto, A.: Bridge deck survey with high resolution ground penetrating radar. In: Proceedings of the 14th International Conference on Ground Penetrating Radar, June 2012

Taflove, A.: Application of the finite-difference time-domain method to sinusoidal steady state electromagnetic penetration problems. IEEE Trans. Electromagn. Compat. **22**(3), 191–202 (1980)

Taflove, A., Hagness, S.C.: Computational Electrodynamics: The Finite-Difference Time-Domain Method, 3rd edn. Artech House Publishers, Norwood (2005)

Taflove, A., Oskooi, A., Johnson, S.G.: Advances in FDTD Computational Electrodynamics: Photonics and Nanotechnology. Artech House, Norwood (2013)

Trinks, I., Johansson, B., Gustafsson, J., Emilsson, J., Friborg, J., Gustaffsson, C., Nissen, J., Hinterleitner, A.: Efficient, large-scale archaeological prospection using true three-dimensional GPR array system. Archaeol. Prospection **17**, 175–186 (2010)

Turk, A.S.: GPR Antennas. Chap. 3.4 In: Subsurface Sensing, Wiley Series in Microwave and Optical Engineering, pp. 83–96. Wiley, Hoboken (2011)

Yarovoy, A.G., Jongh, R.V.D., Ligthart, L.P.: Ultra-wideband sensor for electromagnetic field measurements in time domain. IEE Electron. Lett. **36**(20), 1679–1680 (2000a)

Yarovoy, A.G., Schukin, A.D., Lightart, L.P.: Development of dielectric filled TEM horn. In: Millenium Conference on Antennas and Propagation (2000b)

Yarovoy, A., Meincke, P., Dauvignac, J., Craddock, I., Sarri, A., Huang, Y.: Development of antennas for subsurface radars within ACE. In: Proceedings of the IEEE International Conference on Ultra-Wideband, pp. 299–304, 24–26 Sept 2007

Yee, K.: Numerical solution of initial boundary value problems involving Maxwell's equations in isotropic media. IEEE Trans. Antennas Propag. **14**(3), 302–307 (1966)

van Coevorden, C., Bretones, A., Pantoja, M., Ruiz, F., Garcia, S., Martin, R.: GA design of a thin-wire bow-tie antenna for GPR applications. IEEE Trans. Geosci. Remote Sens. **44**(4), 1004–1010 (2006)

Warhus, J.P., Mast, J.E., Nelson, S.D., Johansson, E.M.: Ground-penetrating imaging radar development for bridge deck and road bed inspection. U.S. Department of Energy by the Lawrence Livermore National Laboratory, California (Contract W-7405-Eng-48), pp. 1–16 (1993)

Warren, C.: Numerical modelling of high-frequency ground-penetrating radar antennas. Ph.D. thesis, School of Engineering, The University of Edinburgh, Edinburgh (2009)

Warren, C., Giannopoulos, A.: Creating FDTD models of commercial GPR antennas using Taguchi's optimisation method. Geophysics. **76**, Article ID G37 (2011)

Warren, C., Chivaridzo, N., Giannopoulos, A.: Radiation characteristics of a high-frequency antenna in different dielectric environments. In: Proceedings of the 15th International Conference on Ground Penetrating Radar (GPR 2014), pp. 767–772, Oct 2014

Zhou, B., Cui, T.J.: Directivity enhancement to Vivaldi antennas using compactly anisotropic zero-index metamaterials. IEEE Antennas Wirel. Propag. Lett. **10**, 326–329 (2011)

Part II
GPR Surveying of Pavements, Bridges, Tunnels and Buildings; Underground Utility and Void Sensing

Innovative Inspection Procedures for Effective GPR Surveying of Critical Transport Infrastructures (Pavements, Bridges and Tunnels)

Josef Stryk, Amir M. Alani, Radek Matula and Karel Pospisil

Abstract This project is in line with the interests of one of the four working groups' activities within the COST Action project—TU1208. The project concerns diagnostics of transport infrastructure structures—pavements, bridges and tunnels—by ground penetrating radar (GPR). It includes individual applications which are currently in use and those which are still in the phase of research and verification. Furthermore, it introduces issues which need to be dealt with, so that this NDT method can be applied to a greater extent in this area.

1 Introduction

GPR has had a certain tradition in the field of diagnostics of transport infrastructure structures. GPR is usually not used as an acceptance test of new structures, but rather for the identification of weak and damaged parts of a structure, which occur within its use. GPR is often combined with other methods (see Chap. 13). Some applications are standard practice; others are still being verified within research projects.

Not only can it be used for one-off structural condition diagnostics, but it can also be used for a comparison of the development of certain defects over a certain

J. Stryk (✉) · R. Matula · K. Pospisil
CDV—Centrum Dopravního Výzkumu, v.v.i. Líšeňská 33a,
63600 Brno, Czech Republic
e-mail: Josef.stryk@cdv.cz

R. Matula
e-mail: radek.matula@cdv.cz

K. Pospisil
e-mail: karel.pospisil@cdv.cz

A.M. Alani
School of Computing and Technology, University of West London,
8th floor, Villiers House, Ealing Broadway, London, W5 2PA, UK
e-mail: Amir.Alani@uwl.ac.uk

A. Benedetto and L. Pajewski (eds.), *Civil Engineering Applications of Ground Penetrating Radar*, Springer Transactions in Civil and Environmental Engineering,
DOI 10.1007/978-3-319-04813-0_3

time period. Results of GPR surveys can be used to supplement information gained from visual inspections in order to optimize maintenance, repair and reconstruction works.

The first three chapters provide a brief description relating to the diagnostics of pavements, bridges and tunnels by ground penetrating radar, while focusing on individual applications. The remaining two chapters outline the need for more intensive application of this NDT method and open issues.

Retaining walls (Hugenschmidt and Kalogeropoulos 2009) and railway tracks (Hugenschmidt et al. 2013) are important transport infrastructure structures which are also diagnosed by GPR; however they are not included in this chapter.

There exist a large number of papers and research projects concerning these issues but only selected and the most state-of-the-art papers and projects are described here. The results of a European project—Mara Nord (2011a, b, c)—and American reports from the Strategic Highway Research Program 2 (SHRP 2 2013a, b, c) are examples of the more complex studies.

2 Pavement Diagnostics

2.1 Introduction

GPR measurements are performed on flexible pavements (asphalt bound layers laid onto unbound sub-base layers), semi-rigid pavements (asphalt layers laid onto hydraulically bound layers) and rigid pavements (with thick concrete layers on top), as well as on pavements overlaid by new asphalt or concrete layers, which all have specific properties.

Road pavements are line structures; therefore, the accuracy of measurement location plays an important role.

Measurements can be carried out at road network scale or project scale. Network scale means the searching measurements for finding problematic road sections, which is often not very detailed. Project scale means the detailed measurement of a road section which has been selected for intervention (maintenance or repair) or detailed diagnostics. Nowadays, GPR is mainly used for pavement diagnostics at project scale.

One of the first GPR applications in road engineering was the continuous determination of pavement layer thicknesses, with a focus on bound layers—mainly asphalt pavement layers.

Another extended application is the localization of the in-built reinforcement (dowels and tie bars in jointed unreinforced concrete pavement or rebar in continuously reinforced concrete pavement or in-built steel grids).

Figure 1 shows examples of recordings taken from measurements of road asphalt layer thicknesses (radargram with visible surface course, binder course and roadbase) and the localization of the positions of dowels and tie bars in concrete slabs of road pavement.

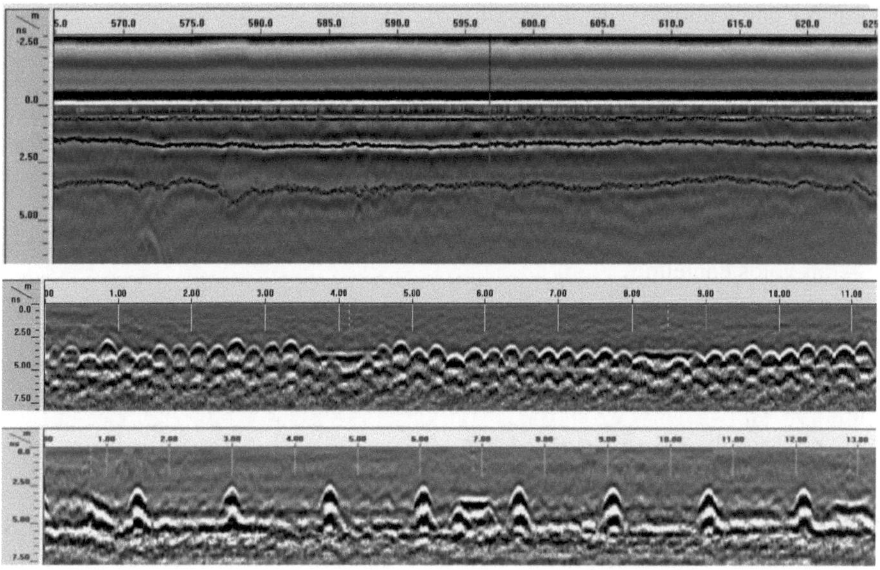

Fig. 1 Examples of records from GPR measurements of road pavements—radargram of 60 m long section of flexible pavement with three selected asphalt pavement layers: surface course, binder course, roadbase (on *top*), radargram of 11.5 m long section near a transversal joint of concrete pavement with visible positions of 41 dowels in three slabs (in the *middle*), radargram of 13 m long section near a longitudinal joint of pavement with visible positions of eight tie-bars in two concrete slabs (at the *bottom*). *Source* CDV

The aim of other applications is the locating of hidden (under surface) distresses, spots with delamination of layers etc. and evaluation of their severity.

Individual applications of GPR pavement diagnostics are shown below, classified into common and research applications:

Common applications:

- thickness of asphalt pavement layers (Saarenketo and Scullion 2000; Muller 2012; Saarenketo 2006; Fauchard et al. 2008; Loizos and Plati 2007; Edwards and Mason 2011);
- thickness of concrete pavement layer (with or without dowels, tie-bars or rebar) (Fauchard et al. 2008; Edwards and Mason 2011);
- thickness of bound or unbound sub-base layers, including depth to the bottom of the embankment (Saarenketo 2006; Edwards and Mason 2011);
- position of reinforcement in concrete pavement (Stryk and Matula 2013);
- de-bonding and delamination of pavement layers (Krysiński and Sudyka 2012);
- heterogeneity of pavement:
  - changing of pavement layer structure;
  - identification of caverns (e.g. under concrete slabs, as a result of floods) (Cassidy et al. 2011);
  - identification of frost heaves (Saarenketo 2006; Berthelot et al. 2010), etc.

Research applications:

- condition of reinforcement in-built in concrete pavement (e.g. corrosion);
- depth of surface cracks (Krysiński and Sudyka 2013);
- localization of bottom-surface initiated cracks (Krysiński and Sudyka 2013);
- pavement pumping prediction (Tosti and Benedetto 2012);
- heterogeneity of pavement (see Chaps. 5–7):
  - moisture content;
  - air voids content;
  - compaction;
  - segregation of aggregate.

2.2 *Equipment and Methods Used*

GPR systems with antennas providing a range of several hundreds of MHz up to 4 GHz are used for the diagnostics. Most of the GPR equipment used in road surveys utilise pulse radar principles. For detailed studies, stepped frequency radar is also used, which measures the phase and amplitude of the reflected signal on each frequency in addition to using an inverse Fourier Transform of these data to build a time domain profile.

Air coupled antennas are mainly of the horn antenna configuration, with central frequency 1 or 2 GHz. These antennas have proven to be effective in cases concerning identification of layer thicknesses of bituminous bound pavements. If road construction details are needed for depths greater than 0.6 m, the ground coupled antennas could potentially be used too.

Measurements at the network scale are performed in the longitudinal direction under high speed, so that traffic is not affected. In these cases, the measurement is mainly carried out with the use of one or several air coupled antennas. On the contrary, measurements at the project scale, including special applications, require a measurement at slow/walking speed, which is normally performed with one or more ground coupled antennas, e.g. measurements of cross sections or crack depth identification. 3D data recording is normally performed using an array of antennas, allowing the measurement of the whole width of the traffic lane in its entirety leading to 100 % survey of the road pavement.

Distance measuring and referencing instruments are used to trigger the control unit as the system covers the required survey length. GPS has also been used widely in recent years as an integral part of the GPR data collection process.

Digital video or still images are often collected to help with the data interpretation and to avoid confusion with positioning.

Examples of typical GPR systems used for the diagnostics of road pavements are shown in Fig. 2.

The biggest manufacturers and operators of GPR systems are associate members of the European Association EuroGPR (http://www.eurogpr.org). A number of these manufacturers and operators have in-house developed software for

Fig. 2 GPR systems for road diagnostics—a two-channel pulse radar system with a combination of ground coupled and air coupled antennas (*upper left*), a cart for measuring the position of dowels and tie bars with two ground coupled antennas (*upper right*), stepped frequency systems for 3D measurements—ground-coupled (*lower left*) and air-coupled (*lower right*). *Source* Roadscanners, cart—CDV

processing and evaluating collected field data within the context of road pavement surveys (RoadScan Module, Interactive 3D module, SliceView, etc.).

In addition, there is further software which is not related to any specific manufacturer and which is able to work with the data collected with different GPR systems, or they combine diagnostic results with other data, e.g. software Road Doctor, which combines GPR data with results of pavement deflection measurements by a falling weight deflectometer, parameters of road surface properties, performed bore holes and image recording. (Fig. 3)

The velocity of the radar signal within a layer is related to relative dielectric permittivity ε_r of layer material (in short, the dielectric constant). Examples of the most common road materials, their ε_r values and influencing factors are presented in Table 1 (ASTM D6432-11 2011).

Wet concrete and concrete with a high chloride ion content (or other conductive materials) can significantly affect a GPR signal's penetration and/or attenuation.

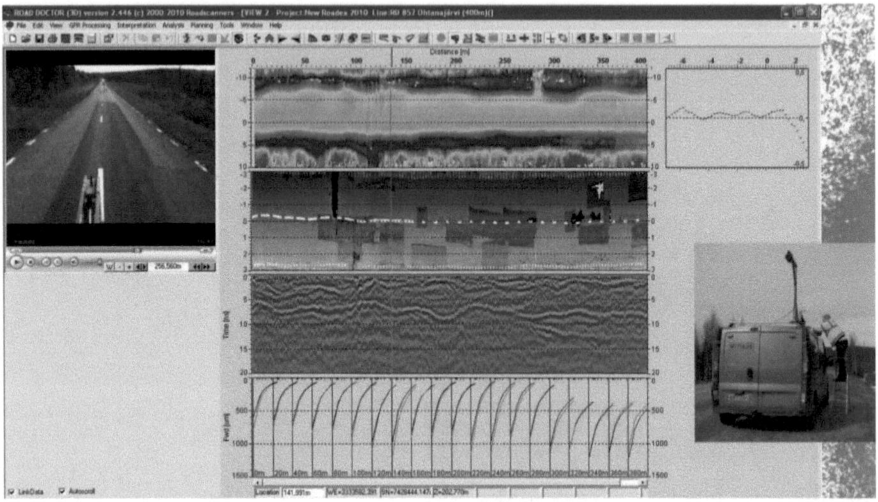

Fig. 3 Software road doctor—form of presentation of measurement results for laser scanning, georadar and falling weight deflectometer FWD. *Source* Roadscanners

Table 1 Approximate electromagnetic properties of selected materials used in road pavements (ASTM D6432-11 2011)

Material	Relative permittivity ε_r	Signal velocity (m/ns)
Air	1	0.3
Asphalt	3–5	0.173–0.134
Quartz	4	0.15
Sand—dry (d)	4–6	0.15–0.12
Granite—dry	5	0.134
Concrete (w, f, age)	5–10	0.134–0.095
Dolomite	6–8	0.122–0.106
Limestone—dry	7–9	0.113–0.1
Sand—saturated (d, w, f)	25	0.055
Water	81	33

d = function of density, w = function of porosity and water content, f = function of frequency

Survey during the winter period is not recommended due to the freeze and thaw effect of concrete as well as compromise in the contrast due to moisture ingress in concrete. This situation is exacerbated by the presence of sprayed salt/grit which in turn can result in significant signal attenuation.

The velocity of signal propagation through a layer is required for calculation of the layer's depth. The most commonly used methods for determining signal velocities are as follows:

- utilization of available published data of mean velocities for pavement materials—the least accurate method for preliminary interpretations only (the presence of moisture in the material will greatly affect the velocity);

Table 2 Typical depth of penetration for different GSSI antennas

Center frequency (MHz)	Depth of penetration (m)	Travel time (ns)	Typical applications
2600	0.4	8–15	Thickness and condition assessment for shallow depths
1600	0.5	10–15	
900	1.0	10–25	Subbase evaluation and layer condition assessment
400	4.0	20–100	Void detection, utilities, thick concrete structures, subgrade evaluations
270	6.0	50–200	Geotechnics
2000 HORN	0.75	8–15	Pavement thickness and road condition assessment
1000 HORN	0.9	10–20	Pavement and bridge deck evaluations

HORN = type of air coupled antenna

- drilling core methods—provide point specific information at selected areas of pavement (localized ideally at sections with homogenous features, often used for calibration purposes);
- Common Mid Point (CMP) technique—use ground coupled antennae which measure in line with different positions of transmitter and receiver to a central point (point specific information or continual information in case of set of antennas);
- Wide-Angle Reflection and Refraction (WARR) technique—use set of ground coupled antennas which measure in line with different positions of receiver from transmitter/s (point specific information or continual information in case of set of antennas);
- Reflection Coefficient Technique—calculates velocity from horn antenna measurement using a metal plate on the pavement surface.

The selection of antenna is of paramount importance for specific applications. For example, typical penetration depths for GSSI antennas with different central frequencies are presented in Table 2 together with their recommended application areas.

2.3 Selected Results and Recommendations

The American standard ASTM D4748-10 entitled *Standard Test Method for Determining the Thickness of Bound Pavement Layers Using Short-Pulse Radar* (ASTM D4748-10 2010) concerns exclusively the application of GPR when determining the thickness of bound pavement layers. It mentions the fact that antennas with the central frequency of 1 GHz are used, which are able to identify surface layer thickness up to 40 mm with the accuracy of ±5 mm, or in some cases antennas with the central frequency of 2 GHz which are able to identify a layer of thickness lower than 25 mm with the accuracy of ±2.5 mm. Three methods of finding dielectric value/signal velocity (metal plate calibration, taking cores, or CMP method) are mentioned. The standard contains a table with dielectric constants for different road materials (asphalt $\varepsilon_r = 2–4$, concrete $\varepsilon_r = 4–10$). According to this standard, the error of this method for determination of thickness should be lower than 15 %.

The American standard ASTM D6432-11 entitled *Standard Guide for Using the Surface Ground Penetrating Radar Method for Subsurface Investigation* (ASTM D6432-11 2011) is probably the most detailed standard concerning the use of GPR for diagnosing subsurface properties of soils, rocks and other materials, including those used for the constructions of roads. It also contains a method for the localization of inbuilt objects, particularly engineering networks, reinforcements, etc. The standard describes individual measurement steps and evaluation of measured data, potential sources of disturbances, 3 methods of determination of EM signal velocity (CMP or WARR mode, hyperbolic geometry, depth to the known reflector methods), and effects which are crucial for correct evaluation (temperature, frequency, density, water content, salinity, and other conditions). The standard shows tables of average electromagnetic properties of selected materials (asphalt $\varepsilon\ r = 2–5$, concrete $\varepsilon\ r = 5–10$) and resolution properties of antennas of different frequencies for different environments.

The technical specification of Great Britain DMRB 7.3.2: *Data for pavement assessment*—annex 6 HD 29/2008: *Ground-Penetrating Radar* (DMRB 7.3.2 2008) was issued by Highway Agency in 2008. This specification amends the use of GPR for the evaluation of road conditions. It defines four classes of accuracies of the use of GPR A to D, assigning them with individual applications:

A—routine measurement; sufficient accuracy and reliability;
B—it is necessary to verify measured characteristics with another method;
C—only in exceptional cases;
D—unverified.

Furthermore, measurement speed is important, where we distinguish low speed (<30 km/h) and traffic flow speed (>80 km/h). Class A corresponds with the identification of changes in construction and measurement of bound and unbound road layer thickness. Class B for measuring under low speeds corresponds with the identification of voids under non-reinforced concrete pavement and identification of changes in subbase layer moisture (under repeated measurements). Class B is not applied for the traffic flow speed. Determination of depths of cracks propagating from the asphalt pavement surface is classified in Class C (for slow measurements). The accuracy of measurements of bound layer thickness under slow speed of measurement is ±5 % (regarding measurements under high speeds, it is ±9 %). Regarding subbase layers, it is ±15 % (±30 % under high measurement speeds).

Mara Nord is an international cooperative project, which has been initiated among Finland, Sweden and Norway. The level of knowledge, awareness and experience regarding the use of GPR in the Road Administrations vary in Finland, Sweden and Norway and there was a need to develop procedures to ensure a better quality of GPR services.

The Mara Nord project report entitled *Recommendations for guidelines for the use of GPR in road construction quality control* Mara Nord (2011a) was issued in 2011.

The current practice is described precisely here so that new contractor will know how the GPR surveys should be done and what the basic quality level for the results should be. It describes the technology, equipment, survey planning and performance, data processing, data interpretation and reporting. The following important information is mentioned there:

- air coupled antennas will be used especially when the interest is bituminous pavement and unbound base thickness and their quality;
- normally, only pavement, base course and total thickness of pavement structure is surveyed;
- normally, accuracy in GPR surveys is ± 10 % of the depth without reference samples and ± 5 % if accurate reference samples are available;
- preliminary interpretation of construction thickness is made using dielectric values;
- normally, the minimum number of longitudinal lines is one line per lane;
- for ground coupled systems with 300–600 MHz antennas, distance from the road surface may not exceed 8 cm and with 1.0 GHz or higher frequency antenna the distance may not exceed 5 mm;
- reference data can typically be drilling or sampling information or excavator test pits, the dielectric value of different structures can also be verified using WARR or CMP soundings;
- GPR interpretations must be recorded at least at one meter intervals;
- the results and collected data should always be delivered also as data that has been used in interpretations;
- the thickness of the measured structures should be presented in longitudinal profiles, longitudinal profiles presentencing the difference of the measured data with the designed data or in GIS format;

The Mara Nord project report entitled *The Use of GPR in Road Rehabilitation Projects* Mara Nord Project (2011b) was issued in 2011. It specifies separately the procedure that should be followed in surveys projects made in 2 dimensional (2D) and 3 dimensional (3D) GPR projects. The following important information is mentioned there:

- digital video or digital images collected during the GPR data collection have proven to be essential during GPR data analysis;
- surveys with ground coupled antenna systems can typically be made any time of the year;
- when using air coupled antennas in the winter, measurements should be when the frost front is deeper than 0.8 m from the road surface;
- 2D GPR surveys can be made with a multi-channel system, for example with a 1.0–2.5 GHz air coupled antenna and a 300–600 MHz ground coupled antenna;
- the basic principle is that the measurement is done in outer wheel path in the main direction of the road;

- maximum longitudinal difference is 2.5 m presented in the data with individual objects;
- the sampling interval in GPR surveys should be at least 10 scan /m in longitudinal survey profiles and at least 40 scans /m in cross section profiles;
- normally reference coring should be made at 2–3 km intervals, with a minimum 1/10 km road section.

The American report from Transportation research board's second Strategic highway research program (SHRP 2) entitled *Nondestructive Testing to Identify Delaminations Between HMA Layers* (SHRP 2 2013a) was issued in 2013.

It compares 7 different NDT techniques in identifying anomalous reflection activity within asphalt layers or at asphalt layer boundaries, or both, that may be associated with asphalt mixture stripping or moisture infiltration into debonded areas.

Stripping develops when the aggregates and asphalt binder are incompatible, adhesion is lost, and water separates the asphalt binder from the aggregate.

Three different GPR systems were used (GSSI, MALA, 3D-Radar).

GPR is particularly sensitive to moisture, so a GPR system can pick up areas of moisture infiltration and damage. The detection of the debonded interface was enhanced by introducing moisture.

The conclusion based on laboratory slab testing and field testing is that GPR can identify variations in the pavement, isolate the depth of a discontinuity in the pavement, and provide a relative degree of severity. Severe conditions, such as stripping, can be observed with conventional analysis software. Detecting debonding between asphalt layers is only possible when there is moisture trapped in the debonded area between the layers using current analysis methodology.

None of the NDT technologies can conclusively distinguish between types of pavement discontinuities.

GPR, impact echo, and spectral analysis of surface waves can be valuable project-level tools used independently or in series. GPR equipment should have an array of antennae and frequency sweep pattern ranging up to 3 GHz.

Saarenketo and Scullion (2000) summarize in their paper the principles of operation of both ground coupled and air coupled GPR systems together with a discussion of both signal processing and data interpretation techniques. In the area of subgrade soil evaluation GPR techniques have been used to non-destructively identify soil type, to estimate the thickness of overburden and to evaluate the compressibility and frost susceptibility of subgrade soil. In road structure surveys, GPR has been used to measure layer thickness, to detect subsurface defects and to evaluate base course quality. In quality control surveys, GPR techniques have been used for thickness measurements, to estimate air void content of asphalt surfaces and to detect mix segregation.

The Australian paper from 2010 (Muller 2012) describes a network-level road investigation trial with a new type of traffic-speed noise-modulated ground penetrating radar (NM-GPR) technology. The system is capable to measure using its array of 24 ground-coupled antennas at up to 100 km/h.

The aim of the trial was to assess if this new equipment could operate effectively at network-level scales for typical Queensland road pavements, and reliably produce information of interest to road asset managers. It was targeted on total pavement depth, depth of structural asphalt, pavement homogeneity and location of changes of construction.

All the examples of the NM-GPR data shown in this paper are raw, gain-adjusted data.

The technology has the capability of determining pavement layering information where none currently exists, while reducing the amount of time, cost, public disruption and effort compared with current physical pavement investigations.

3 Bridge Diagnostics

3.1 Introduction

Diagnostics of bridges by the GPR method is usually carried out on concrete bridges, or stone-masonry arch bridges, respectively (Solla et al. 2012). Measurement is performed when problems are identified within a visual inspection (crack occurrence, evidence of rebar corrosion, leaking etc.). This can be done directly from the pavement surface or separately on individual bridge elements (bridge girders, piers or columns).

Measurements can be carried out for the whole length of the bridge or only for selected areas of interest.

GPR application on bridges usually concerns a condition evaluation of a bridge deck. Another significant application is location of reinforcement in the form of construction bars and pre-stressed or post-tensioned tendons (or their ducts).

Individual applications are shown below, classified into common and research applications.

Common applications (Tarussov et al. 2013; Hugenschmidt and Mastrangelo 2006; Sbartaï et al. 2007; Benedetto et al. 2012; Benedetto 2013):

- position of reinforcement in bridge deck (spacing);
- concrete cover of reinforcement in bridge deck;
- thickness of bridge deck;
- position of pre-stressed or post-tensioned tendons or tendon ducts (Dérobert and Berenger 2010);
- de-bonding and delamination of pavement layer;
- bridge deck deterioration (cracks, caverns, etc.);
- bridge girder diagnostics (Beben et al. 2012).

Research applications:

- diameter of reinforcement in-built in concrete (Chang et al. 2009);
- condition of reinforcement in bridge deck (e.g. corrosion) (Hubbard et al. 2003);

- evaluation of sealing course (waterproofing layer) on bridge deck (Dérobert et al. 2011);
- moisture content (see Chap. 7).

3.2 Equipment and Methods Used

Most commonly dipole antennas with frequencies higher than 1 GHz are used. Measurement of the bridge deck is usually performed at low speed in longitudinal as well as transversal directions. Devices with sets of antennas are preferred because they allow wider width measurements and therefore shorten the time required for a survey, see Fig. 4. Some systems allow the measurement of nearly the whole width of a traffic lane with higher speed, see Fig. 2. Some systems use antennas arranged with different polarisation.

Presentation of GPR data is usually carried out in the form of maps at different depths, see example at Fig. 4, or in the form of longitudinal or transversal cuts, see Fig. 5, where at least the following parameters are identified:

- pavement surface;
- pavement bottom;
- top of the concrete deck (if it is not the same as the pavement bottom; the waterproofing layer is normally too thin to be interpreted as a separate layer);
- top reinforcement of concrete deck;
- bottom reinforcement of concrete deck;
- bottom of concrete deck.

Measurement of vertical areas such girders, piers, etc., is commonly performed with a single antenna, often manually held.

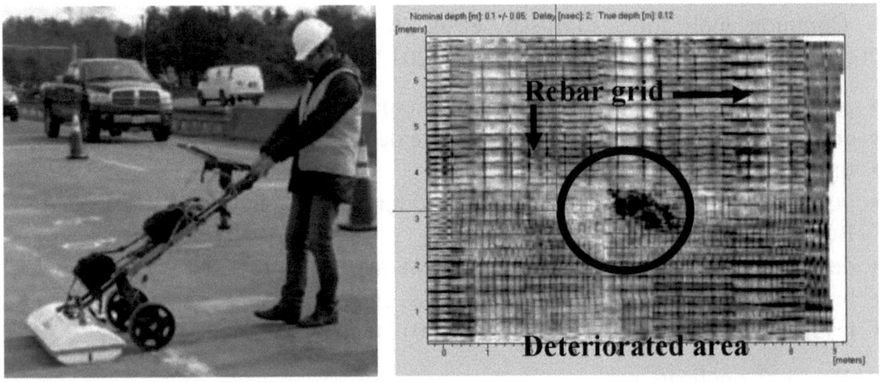

Fig. 4 GPR system RIS Hi-BrigHT specially designed for bridge deck surveying (*left*), visualisations of measured data at a depth of 10 cm with visible rebar grid and deteriorated area in the centre (*right*). *Source* IDS and (Alani et al. 2014)

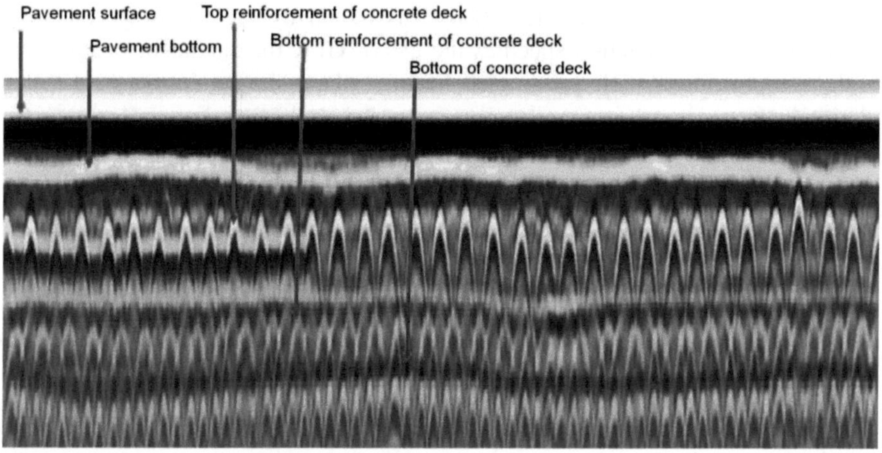

Fig. 5 An example of a typical GPR data profile from a bridge. *Source* Mara Nord project (2011a)

A number of GPR system produces develop their own software for the evaluation of data measured on bridges (BridgeScan Module, GRED HD Bridge, SliceView 3D Visualization, BridgePro etc.).

3.3 Selected Results and Recommendations

American standard ASTM D6087-08 entitled *Standard Test Method for Evaluating Asphalt-Covered Concrete Bridge Decks Using Ground Penetrating Radar* (ASTM D6087-08 2008) shows the measurement procedure and subsequent evaluation. The measurement should be performed in a grid with the step in the transversal direction <1 m (for air coupled antenna) <0.6 m (for ground coupled antenna) and with the step in the longitudinal direction $\leq$0.15 m. Two methods of evaluating the measured data are described here. In both cases the signal attenuation is taken into account in the propagation through the structure. The first one evaluates the ratio of amplitudes corresponding with the lower and upper bridge deck surface, which is used for determining spots where reinforcement is separated from concrete. The latter evaluates relative changes in amplitudes corresponding with individual bars of the upper layer of the steel reinforcement, from which damage of the concrete cover layer is derived. The error rate of this method, determined on the basis of the evaluation of 10 bridges, should be lower than 11.2 % according to the standard. This procedure can only be applied for some tapes of monolithic reinforced concrete bridge decks.

The technical specification of Great Britain DMRB 3.1.7 entitled *Advice notes on the non-destructive testing of highway structures*—3.5 BA 86/2006: *Ground*

Penetrating Radar (DMRB 3.1.7 2006) was issued by the Highway Agency in 2006. This recommendation specifies the use of GPR for diagnostics of masonry bridges and bridges with prestressed beams, where the tendons are placed in plastic ducts.

It contains alternative measurement methods in the bistatic mode, when the receiver and transmitter are separated and moved along the tested structure surface. The annex contains specific examples of evaluation for:

- testing of masonry bridges;
- laboratory experiments of GPR on concrete beams with plastic ducts.

Mara Nord's project report entitled *Recommendations for guidelines for the use of GPR in bridge deck surveys* (Mara Nord Project 2011c) was issued in 2011. It describes GPR equipment, survey planning and performance, data processing and reporting. It states that in most cases, GPR alone cannot provide reliable enough information of damage and its nature, but it is an excellent tool for the initial mapping and specifying of locations, whereupon other non-destructive evaluation methods and limited ground truth testing can be used to verify the problems.

Detailed survey using 2D or 3D ground coupled antenna systems is recommended with a longitudinal sampling rate of 100 scans/m and a distance between parallel profiles of 50 cm with the 2D system and a maximum of 12 cm with the 3D system. With horn antenna systems the sampling rate should be at least 10 scans/m. The qualitative interpretation and analysis includes detection of the areas of abnormal behaviour of the signal amplitude or frequency content of the signal. In addition to layer thicknesses, deterioration in the deck slab can be analysed through the signal amplitude strength, attenuation and dispersion calculated from the GPR data.

Results are presented as structural surface maps of the layer and deterioration maps of the bridge concrete deck slab.

Structural maps should be presented as a basic standard of a report (thickness of the asphalt pavement, concrete thickness to the top reinforcement bars).

Deterioration maps or time slices should be calculated and presented from different depths in the concrete deck slab, in order determine the extent of deterioration.

The following deterioration maps should at least be presented:

- amplitude/attenuation at the level of concrete deck surface (0–2.5 cm);
- amplitude/attenuation at the level of reinforcement;
- amplitude/attenuation below the reinforcement.

The dispersion/frequency content decay maps should also be presented from the same levels if possible. The maps can be very useful in detecting areas with horizontal cracks in the concrete deck (delamination).

Using the same dispersion/frequency data, maps of moisture problems can also be produced.

The American SHRP 2 report entitled *Nondestructive Testing to Identify Concrete Bridge Deck Deterioration* (SHRP 2 2013a) was issued in 2013. It compares 10 different non-destructive methods, including GPR. Evaluation of NDT technologies was carried out for the following four deterioration types: delamination, corrosion, cracking and concrete deterioration. The most serious deterioration types were identified as delamination and corrosion.

For categorizing and ranking the technologies, the following five performance measures were selected:

- accuracy;
- repeatability;
- ease of data collection, analysis and interpretation;
- speed of data collection and analysis, and
- cost of data collection and analysis.

Field validation was conducted on a two-span concrete deck (203 mm thick) of a steel girder structure. The laboratory validation testing was carried out on two test decks (a newly fabricated concrete deck with simulated defects, and a test deck removed from a bridge).

GPR maps are based on signal attenuation of the top rebar level. The delaminated areas are indirectly detected based on the areas with high-energy attenuation. The most significant results were obtained on the newly fabricated concrete deck with simulated defects. The artificial delamination was created through the placement of synthetic inserts and not through the penetration of moisture and chlorides that would create a corrosive environment leading to rebar corrosion. Results of this testing also confirm that corrosion-induced delamination in real bridge decks is not easily reproducible in the laboratory using inserts.

GPR was identified as the technology that provides the highest value in bridge deck deterioration detection in this study. It obtained the highest score for delamination detection. In corrosion detection it holds third position, after the electrical resistivity and half-cell potential method.

GPR is the recommended tool because of its speed and ability to identify delamination and describe the corrosive environment.

The German document B 10 entitled *Recommendation for non-destructive testing of civil engineering structures by GPR* (Bauwesen et al. 2008) was issued in 2008 by the German Society for Non-Destructive Testing (DGZfP). It is a very detailed document covering the whole issue of GPR diagnostics. Apart from information on the measurement principle, used equipment, and processing of measured data, applications in the following areas (materials) are described: concrete, masonry, foundations (geotechnics), asphalt and rail superstructures. Accuracy for the determination of layer thickness is determined to be 5 % and higher, depending upon the situation and on whether homogeneous material is present. Accuracy of the determination of reinforcement location (steel meshes) is shown at intervals of 5–10 %, regarding the fact that in case meshes in several layers is concerned, the distance between rebar in individual layers is crucial. In case this distance is >10 cm, it is possible to obtain information to the depth of 30–40 cm under these meshes. If this distance is <5 cm, no information on material condition under this reinforcement can be obtained. The mentioned localization accuracy of dowels and tie bars in concrete pavements is 5–10 % (taking into account the effects of concrete moisture on the determination of the position).

The German BASt-report B 55 entitled *Examination of GPR in combination with magnetic techniques for the determination of moisture and salinity of concrete bridge decks with asphalt cover* (BASt-report B 55 2007) was issued in 2007 by the Federal Highway Research Institute (BASt). A combination method involving GPR and magnetic field measurement was developed. If the dielectric constant of a construction material in the microwave frequency range is known, it is possible to derive its moisture and salinity using calibration curves. With a 1 GHz GPR, the traveling time of the electromagnetic waves from the road surface to the upper reinforcement layer is measured. The depth of these reinforcement bars is determined independently using a static magnetic field measurement. From the comparison of the radar reflection time with the magnetically determined depth, the real component of the permittivity is derived. The analysis of the reflected radar amplitude allows an estimate of the salinity.

A novel GPR system that is designed and developed for bridge applications is presented by A. Benedetto in his paper (Benedetto et al. 2012). The system called RIS Hi Bright is a pulse GPR system which integrates two arrays (8 plus 8) of ground coupled antennas, having a central frequency of 2 GHz. This configuration makes it possible to optimise the survey increasing the reliability and the productivity of the system because the number of scan is reduced and, using a different polarisation in the array, it is possible to avoid any transversal scan.

The GPR inspection of bridges to detect corrosion and cracking problems is based on the concept that as the moisture and chloride concentration increase, the relative dielectric constant and conductivity of the concrete increase. This leads to certain characteristics of the radar waveform that can be reasonably associated with the presence of deterioration.

The automatic tracking of the cracks is done by processing the signal amplitudes all over the 3D matrix. Three-dimensional visualisations of cracks and two-dimensional tomographies and sections are shown to demonstrate the potentialities of the GPR system associated with the signal-processing algorithm. The effectiveness and accuracy of the procedure have been calibrated and validated on four different bridges in Italy. A three-dimensional approach for tracking cracks in bridges using this system is described in more detail in (Benedetto 2013).

Alani et al. (2014) describes in his paper two case studies on two major bridges in the UK also carried out using the GPR system RIS Hi-BrigHT. The first case study was focused on the identification of possible structural defects including damaged rebar and moisture ingress at specific locations of the bridge deck. The second case study was focused on a full assessment of the bridge with particular emphasis on the identification of possible defects including structural cracks within the deck structure and establishing the layout of the upper and lower rebar positions throughout the bridge. It reports a remarkable similarity in the processed data concerning areas affected by ingress of moisture within the deck structures of the two bridges under investigation. The observed similarities in the processed data between the two reported case studies present an interesting concept within the general context of the interpretation of GPR data, with the potential for use in many other forthcoming cases.

The application of GPR to determine the parameters of beams is presented by D. Beben in his paper (Beben et al. 2012). The subject of this study was a three-span road viaduct made of simply supported reinforced concrete beams with unknown geometrical and strength parameters. The main objective of experimental tests was to determine the geometric parameters of viaduct beams and establish the steel reinforcements appearing in them. A ground coupled antenna with a frequency of 2000 MHz was used for the measurement and data were processed using the GRED 3D Utilities software. The locations of the main upper and bottom reinforcing bars, the stirrups and their spacing and an internal hole in the main beams were identified. GPR proved to be an effective method of determining the parameters of the reinforced concrete beams.

4 Tunnel Diagnostics

4.1 Introduction

In relation to the applications of GPR diagnostics, the least documented field is the evaluation of the condition of tunnel structures (mainly focused on tunnel linings). The reason for this is mainly due to difficulty of access to tunnels (not easily affordable in terms of traffic interruption) as well as the adaptability of the antenna systems for tunnel survey purposes.

Different types of tunnel linings can be surveyed, including unreinforced concrete, reinforced concrete, shotcrete lining (sprayed concrete) and brick.

GPR cannot be used in a tunnel where steel fibres have been used in the shotcrete.

The examined areas are tunnel walls and tunnel roofs. The presence of cables and conduits on the sides of the tunnel roof made it impossible to collect GPR data in those areas.

Measurements can be carried out for the whole length of the tunnel with the use of air coupled GPR systems or only for selected areas of interest with ground coupled GPR systems.

The speed of measurement is walking or slow, up to 30 km/h, depending upon the nature of the tunnel structure and accessibility. Generally speaking, referencing during data collection in GPR applications during tunnel surveys is rather complex. No doubt, utilization of GPS and similar referencing devices is limited in such cases. Referencing is mostly realized by using survey wheels connected to the antenna system.

The main focus of tunnel surveys is predominantly related to the determination of the thickness and condition assessment of the tunnel lining, including the location of reinforcement positions. In addition to this, the case studies reported within the literature demonstrate that the gap between concrete and the rock face, as well as the verification of structural detailing of the tunnel, is of great interest to engineering practitioners and researchers.

Individual applications are shown below, classified into common and research applications.

Common applications (Silvast and Wiljanen 2008; Lalagüe and Hoff 2010; Zhang et al. 2010; Xiang et al. 2013; Alani and Banks 2014):

- location of reinforcement in tunnel lining;
- thickness of tunnel lining;
- homogeneity of tunnel lining;
- structural detailing;
- moisture ingress detection.

Research applications:

- condition of reinforcement in tunnel lining;
- evaluation of space between lining and rock;
- moisture content (see Chap. 7), etc.

4.2 Equipment and Methods Used

Both air coupled and ground coupled systems can be used in tunnel diagnostics.

The measurement is performed in lines at different heights with the use of special mounts built in the measuring vehicle, or with a manually held antenna in case of smaller areas of interest.

Air coupled GPR systems provide good quality structural information from the concrete tunnels. The most critical speed-reducing factor is the antenna-wall distance, which had to stay roughly the same between measurements. The standard GPR data analysis consists of reflection amplitude and dielectric value calculations and their analysis.

One example of GPR system used for the diagnostics of tunnel lining is shown in Fig. 6.

A typical GPR data profile from a tunnel survey is presented in Fig. 7. Two layers of bars are visible in reinforced concrete lining together with an interface between two different materials (rockwall).

The same software is used for evaluation of GPR data as in case of bridges with the help of 3D analyses and slide views.

4.3 Selected Results and Recommendations

The American SHRP 2 report entitled *Mapping Voids, Debonding, Delaminations, Moisture, and Other Defects Behind or Within Tunnel Linings* (SHRP 2 2013c)

Fig. 6 The survey van for tunnel diagnostics equipped with GSSI horn 1.0 GHz antennas. *Source* Roadscanners

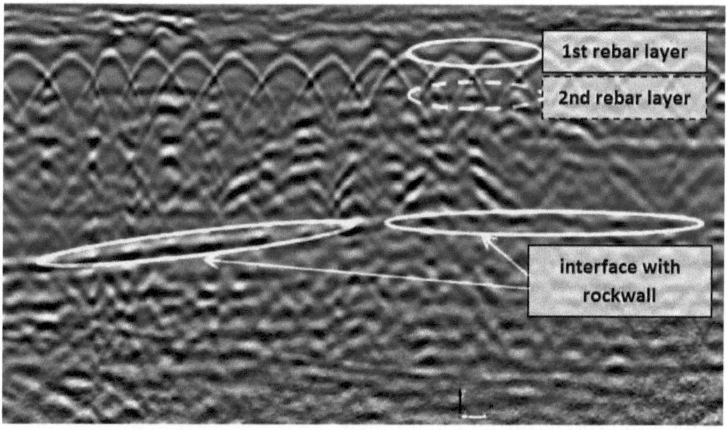

Fig. 7 An example of a typical GPR data profile from a tunnel, 11 cm deep: transversal layer of reinforcing bars with a spacing between 16 and 20 cm, 17 cm deep: second layer which indicates a second layer of rebar, 40 cm deep: interface between two different materials, likely to be the rockwall. *Source* (Alani and Banks 2014)

was issued in 2013. It compares 6 different NDT methods, including GPR (separately for ground coupled and air coupled systems). Deterioration detectability, detection depth and accuracy were evaluated for each method.

A ground coupled system with 1 GHz central frequency antenna and an air coupled system with 1500 and 900 MHz central frequency antenna were used. They were tested on laboratory specimens and real tunnels.

The report states that ground-coupled GPR results can possibly detect defects at different depths within or immediately behind tunnel linings. The air coupled system results indicate areas of high moisture or low density (high air voids). Such areas may represent problems within or behind the tunnel lining. Data should be collected every 304 mm along the tunnel lining. It is recommended that the surface dielectric data from the air coupled GPR be used for determining where to conduct more in depth tests.

Both systems are suitable for the detection of delamination, air filled voids, water filled voids and moisture intrusion. However, for ground coupled GPR, the defects can only be detected if they contain significant air pockets or significant moisture.

Finally the report suggests the following sequence of testing:

- start with collection of thermal images and air coupled GPR data on the tunnel lining;
- analyse the data;
- select areas for in depth testing;
- conduct in depth testing with the ground coupled GPR and either the ultrasonic tomography, ultrasonic echo, or portable seismic property analyser device;
- evaluate the data collected from these devices.

It was stated that the GPR equipment requires considerably more training and experience than the other devices.

The pilot project results from various GPR tests performed on bedrock (tunnel without lining) are presented in the Roadscanners report from 2008 (Silvast and Wiljanen 2008).

The purpose of the tests was to determine the capacity of GPR in defining the depth and continuation of the excavation damaged or disturbed zones (EDZ).

Ground coupled systems with antenna frequency—400–1500 MHz, air coupled systems with antenna frequency 1 and 2.2 GHz and GeoScope 3D step frequency radar were used.

The scans were collected at 1–5 cm horizontal intervals depending on the measurement system. The bedrock surface was interpreted first and after that the reflections from single fractures were picked. The reflection zone of the possible EDZ was then interpreted from the profiles.

All systems provided good quality GPR data useful to the investigations of bedrock properties near the excavated surface. 3D and air-coupled GPR systems need to be measured by vehicle and are therefore more compatible for large-scale measurements. Ground coupled GPR systems are effective for smaller scale projects and for research purposes.

The determination of space behind pre-cast concrete elements in tunnels using GPR is described in Lalagüe paper (Lalagüe and Hoff 2010). It is focused on tunnels in Nordic countries which are lined with prefabricated concrete elements to protect against frost and leakages. This protective lining is fixed to the rock face in a few points. Safety inspections have hitherto consisted of random drilling into the concrete lining. However, such random inspection is both unreliable and expensive. Therefore GPR has been introduced in the vault walls to map the contours of the gap more systematically and to pinpoint potential rockfalls in the vault roof.

GeoScope 3D step frequency radar with the frequency range from 100 MHz to 2 GHz was used in this study with scan width equals to 52.5 cm. To fit vertical surfaces, an arm has specially been designed. This was assembled to a lightweight frame which is directly mounted to the vehicle's bumper.

Each tunnel side was measured at two different heights: approximately 1.5 and 2 m. Thirteen tunnels were investigated, this way. For all these tunnels (except one where the vaults were built of concrete with lightweight expanded clay aggregate), it was relatively easy to find bedrock surface and the back of the vault.

The pilot study gave promising results and it was shown that the equipment is well suited to find material that is in contact with the vault.

The application of GPR in grouting evaluation for shield tunnel construction presents Zhang in his paper (Zhang et al. 2010). In order to evaluate the effectiveness of the grouting treatment before the tunnel operation, a nondestructive testing method using GPR was proposed to detect the grout thickness behind the lining segments. GPR has shown to be a viable approach due to the facts that:

- the detecting objects (concrete segments, grout and soil) were in the depth of one meter or less;
- dielectric parameters of all the materials can be obtained from the laboratory; the contrasts in the dielectric properties among these three materials were large enough;
- only the boundary between the grout and the soil needed to be found since the concrete segments had a known even thickness of 0.35 m.

Chinese paper of Xiang et al. (2013) demonstrates a case study of the Damaoshan Tunnel located in Fujian province to perform a condition assessment combining GPR and finite-difference time-domain (FDTD) techniques based on prior information regarding the designed tunnel structure. This combination was used to assist and improve the interpretation of field data. The aim of this survey was to locate the rebar, estimate the thickness of the second lining, and determine the presence and distribution of any damage for an annual inspection. Additionally, a symmetry-based algorithm and a hyperbola match method were combined to achieve these goals and determine the wave velocity inversion.

5 Common Needs

The issues which need to be dealt with in order to improve the use of ground penetrating radar in practice are shown below. These issues need to be explained to the common users of this NDT method as well as to administrators of civil engineering structures, who order diagnostics by this method.

- optimization of device settings for specific applications—number of channels, antenna frequency, measurement speed, way of localization of the measurement spot, etc. (it is connected to optimization of costs and required time of individual GPR applications);
- determination of measurement accuracy (in depth)—acceptance of this accuracy for basic applications, particularly for the determination of individual road pavement layer thicknesses and reinforcement location;

- raster optimization of the measurement—3D measurement versus line measurement (in two perpendicular directions);
- minimization of the number of drill holes for calibration purposes—the use of CMP (Common Mid Point), WARR (Wide-Angle Reflection and Refraction) and others techniques;
- the method how the measurement results should be fed into databases and road administrators' management systems (e.g. layer thicknesses).

6 Open Issues

The issues mentioned in the previous chapter are being dealt with at a certain level. Apart from them, it is necessary to tackle other topics which need to be dealt with in the future:

- increase level of automation when evaluating the measured data for specific applications;
- widen training of personnel for different GPR applications in terms of measurement and data evaluation (increased requirements for skilled personnel with sufficient practice);
- integrate this non-destructive method and its different applications into European standards and national technical specifications (in Europe, there is no equivalent to standards ASTM D6432-11 (2011), ASTM D4748-10 (2010) and ASTM D6087-08 (2008); only in some European countries, there are technical specifications covering NDT diagnostics of civil engineering structures by GPR, e.g. in Nordic countries—Finland, Sweden and Norway (Mara Nord Project 2011a, b, c), in UK (DMRB 7.3.2 2008; DMRB 3.1.7 2006), in Germany (Bauwesen et al. 2008; BASt-report B 55 2007) and their range and level of details is quite different);
- organize comparative tests of individual devices at national and international level for different GPR applications (at least for pavement diagnostics, where similar comparative tests are carried out for other devices measuring variable parameters e.g. surface unevenness, surface skid resistance or bearing capacity of pavement);
- harmonize prices of individual GPR applications and include them into officially issued price lists.

7 Conclusions

Ground penetrating radar is already in use for all three types of civil engineering structures (pavements, bridges and tunnels). A number of applications are standard procedures, but even for them there is still a room for improvement and there is still need to set more strict requirements. The application of this method either separately or in combination with other NDT methods is being tested. The results of COST action TU1208 should support this effort.

Acknowledgments The authors acknowledge the COST Action TU1208: *Civil Engineering Applications of Ground Penetrating Radar* and the project of the Technology Agency of the Czech Republic No. TA02030759: *New diagnostics methods as a supporting decision tools for maintenance and repair of road pavements—their contribution and ways of their usage*, supporting this work.

References

Alani, A.M., Banks, K.: Applications of Ground Penetrating Radar in Medway Tunnel—Inspection of Structural Joints. In: 15th International Conference on Ground Penetrating Radar (GPR), Brussels, Belgium (2014)

Alani, A.M., Aboutalebi, M., et al.: Integrated health assessment strategy using NDT for reinforced concrete bridges. NDT&E Int. **61**, 80–94 (2014)

ASTM D4748-10.: Standard Test Method for Determining the Thickness of Bound Pavement Layers Using Short-Pulse Radar (2010)

ASTM D6087-08.: Standard Test Method for Evaluating Asphalt-Covered Concrete Bridge Decks Using Ground Penetrating Radar (2008)

ASTM D6432-11.: Standard Guide for Using the Surface Ground Penetrating Radar Method for Subsurface Investigation (2011)

BASt-report B 55.: Überprüfung des Georadarverfahrens in Kombination mit magnetischen Verfahren zur Zustandsbewertung von Brückenfahrbahnplatten aus Beton mit Belagsaufbau, Bundesanstalt für Straßenwesen (2007)

B 10: Merkblatt über das Radarverfahren zur Zerstörungsfreien Prüfung im Bauwesen, Deutsche Gesellschaft für Zerstörungsfreie Prüfung e.V., DGZfP (2008)

Beben, D., Mordak, A., et al.: Identification of viaduct beam parameters using the ground penetrating radar (GPR) technique. NDT&E Int. **49**, 18–26 (2012)

Benedetto, A.: A three dimensional approach for tracking cracks in bridges using GPR. J. Appl. Geophys. **97**, 37–44 (2013)

Benedetto, A., Manacorda, G., et al.: Novel perspectives in bridges inspection using GPR. Nondestr. Test. Eval. **27**(3), 239–251 (2012)

Berthelot, C., Podborochynski, D., et al.: Ground-penetrating radar evaluation of moisture and frost across typical Saskatchewan road soils. Adv. Civ. Eng., 9 (2010)

Cassidy, N.J., Eddies, R., et al.: Void detection beneath reinforced concrete sections: The practical application of ground-penetrating radar and ultrasonic techniques. J. Appl. Geophys. **74**(4), 263–276 (2011)

Chang, C.W., Lin, C.H., et al.: Measurement radius of reinforcing steel bar in concrete using digital image GPR. Constr. Build. Mater. **23**(2), 1057–1063 (2009)

Dérobert, X., et al.: Pathologies, diagnostic et réparation des chapes d'étanchéité d'ouvrages d'art, Techn. et Méthodes LPC, chapter 6, annex 2, 199 p (2011)

Dérobert, X., Berenger, B.: Case study: Expertise and reinforcement of a particular ribbed slab post-tensioned structure. Non-destr. Eval. Reinf Concr. Struct. **2**, 574–584 (2010)

DMRB 3.1.7.: Design Manual for Roads and Bridges, Advice notes on the non-destructive testing of highway structures—advice note 3.5 BA 86/2006: Ground Penetrating Radar (GPR), UK, Department for Transport (2006)

DMRB 7.3.2.: Design Manual for Roads and Bridges, Data for pavement assessment—annex 6 HD 29/2008: Ground-Penetrating Radar (GPR), UK, Highway Agency (2008)

Edwards, L., Mason, Q.: Evaluation of nondestructive methods for determining pavement thickness, final report, (prepared for Headquarters Air Force Civil Engineer Support Agency) (2011)

Fauchard, C., Rejiba, F., et al.: Step frequency radar applied for asphalt thickness measurements with various interface conditions. In: 12th International Conference on Ground Penetrating Radar (GPR), Birmingham, UK, 16–19 June 2008

Hubbard, S.S., Zhang, J., et al.: Experimental detection of reinforcing bar corrosion using nondestructive geophysical techniques. ACI Mater. J. **100**(6), 501–510 (2003)

Hugenschmidt, J., Kalogeropoulos, A.: The inspection of retaining walls using GPR. J. Appl. Geophys. **67**(4), 335–344 (2009)

Hugenschmidt, J., Mastrangelo, R.: GPR inspection of concrete bridges. Cem. Concr. Compos. **28**(4), 384–392 (2006)

Hugenschmidt, J., Kasa, C., et al.: GPR for the inspection of industrial railway tracks. Near Surf. Geophys. **11**(5), 485–491 (2013)

Krysiński, L., Sudyka, J.: Typology of reflections in the assessment of the interlayer bonding condition of the bituminous pavement by the use of an impulse high-frequency ground-penetrating radar. Nondestr. Test. Eval. **27**(3), 219–227 (2012)

Krysiński, L., Sudyka, J.: GPR abilities in investigation of the pavement transversal cracks. J. Appl. Geophys. **97**, 27–36 (2013)

Lalagüe, A., Hoff, I.: Determination of space behind pre-cast concrete elements in tunnels using GPR. In: 13th International Conference on Ground Penetrating Radar (GPR), pp. 1–5. Lecce, Italy (2010)

Loizos, A., Plati, C.: Accuracy of pavement thicknesses estimation using different ground penetrating radar analysis approaches. NDT&E Int. **40**(2), 147–157 (2007)

Mara Nord Project.: Recommendations for guidelines for the use of GPR in bridge deck surveys, June 2011a

Mara Nord Project.: Recommendations for guidelines for the use of GPR in road construction quality control, June 2011b

Mara Nord Project.: The Use of GPR in Road Rehabilitation Projects, June 2011c

Muller, W.: A network-level road investigation trial using Australian-made traffic-speed 3D ground penetrating radar (GPR) technology. In: 25th ARRB Conference, Perth, Australia, 23–26 Sept 2012

Saarenketo, T., Scullion, T.: Road evaluation with ground penetrating radar. J. Appl. Geophys. **43**(2–4), 119–138 (2000)

Saarenketo, T.: Electrical properties of road materials and subgrade soil and the use of ground penetrating radar in traffic infrastructure surveys, Faculty of Science, Department of Geosciences, University of Oulu, Ph.D. dissertation work, p. 121 (2006)

Sbartaï, Z.M., Laurens, S., et al.: Using radar direct wave for concrete condition assessment: Correlation with electrical resistivity. J. Appl. Geophys. **62**(4), 361–374 (2007)

SHRP 2, Strategic Highway Research Program 2: Nondestructive Testing to Identify Delaminations Between HMA Layers, vol. 1 and 2. Transportation Research Board, Washington (2013a)

SHRP 2, Strategic Highway Research Program 2: Nondestructive Testing to Identify Concrete Bridge Deck Deterioration. Transportation Research Board, Washington (2013b)

SHRP 2, Strategic Highway Research Program 2: Mapping Voids, Debonding, Delaminations, Moisture, and Other Defects Behind or Within Tunnel Linings. Transportation Research Board, Washington (2013c)

Silvast, M., Wiljanen, B.: ONKALO EDZ—Measurements using ground penetrating radar (GPR) method, working report, Posiva Oy, p 66 (2008)

Solla, M., Lorenzo, H., et al.: Ground-penetrating radar for the structural evaluation of masonry bridges: Results and Interpretational tools. Constr. Build. Mater. **29**, 458–465 (2012)

Stryk, J., Matula, R.: Possibilities of ground penetrating radar usage within acceptance tests of rigid pavements. J. Appl. Geophys. **97**, 11–26 (2013)

Tarussov, A., Vandry, M., et al.: Condition assessment of concrete structures using a new analysis method: Ground-penetrating radar computer-assisted visual interpretation. Constr. Build. Mater. **38**, 1246–1254 (2013)

Tosti, F., Benedetto, A.: Pavement pumping prediction using ground penetrating radar. Soc. Behav. Sci. **53**(3), 1045–1054 (2012)

Xiang, L., Zhou, H., et al.: GPR evaluation of the Damaoshan highway tunnel: A case study. NDT and E Int. **59**, 68–76 (2013)

Zhang, F., Xie, X., et al.: Application of ground penetrating radar in grouting evaluation for shield tunnel construction. Tunn. Undergr. Space Technol. **25**(2), 99–107 (2010)

Inspection Procedures for Effective GPR Surveying of Buildings

Vega Pérez-Gracia and Mercedes Solla

Abstract A considerable number of studies about GPR applications in building inspection can be found in the literature. New advances in software development, laboratory tests under controlled conditions and numerous cases studies are representative works in this field of knowledge. Some applications are focused on rebar detection, on concrete building assessment, and in modern masonry structures. However, the majority of the works are focused on cultural heritage buildings evaluations, presenting interesting and diverse cases studies. Remarkable results can be found about cracks detection and inspection of masonry walls and columns. Software development has been focused, in many cases, to the enhancement of radar images to facilitate data interpretation. In other cases, synthetic models have been developed to compare results with GPR images from complex scenarios. Evaluations of quantitative properties of constructive materials have been developed based on laboratory tests. Other special works have been also based on laboratory tests: damp measures, concrete degradation due to corrosion, and damages due to tree roots are tested in laboratory specimens under controlled conditions. Although it is a promising subject, few studies have been applied in buildings, revealing the difficult inherent to these complex scenarios. Open issues have been defined as a final conclusion based on the revision of different works. Developments of radar imaging, models and new applications seem to be the most relevant future lines in the GPR building inspection, probably based in a proper and complete definition of casuistic and requirements in structures evaluations.

V. Pérez-Gracia (✉)
Department of Strengthen of Materials and Structures, EUETIB/CEIB,
Technical University of Catalonia, Catalonia, Spain
e-mail: vega.perez@upc.edu

M. Solla
Defense Center University (University of Vigo), Spanish Naval Academy,
Pontevedra, Spain
e-mail: merchisolla@uvigo.es

A. Benedetto and L. Pajewski (eds.), *Civil Engineering Applications of Ground Penetrating Radar*, Springer Transactions in Civil and Environmental Engineering,
DOI 10.1007/978-3-319-04813-0_4

1 Introduction

Buildings assessment is a decisive task before maintenance of the structures. All building materials deteriorate with age and exposure to the weather. Also, natural hazards can cause damage, although sometimes they cannot be detected by visual inspection. There are buildings that look in good conditions, but they are affected by hidden problems as damp in basements, deterioration of wood structures, corrosion, etc. Examples of damages in different buildings are shown in Fig. 1. The money and interests investing in a building is considerable, and care use to be required to preserve its value. However, inconveniences and cost of inspections and maintenance cause that, in many cases, owners do nothing until great visual damages or failure. Particular remark is done in the case of ancient buildings or cultural heritage. In these cases, preventive maintenance is mandatory to safeguard our legacy. But maintenance requires extensive information about the real condition of the different members of the building. This information must be collected in broad inspections that usually should provide a realistic description about the building state, particularly referred to its fabric and its structure, and should define the extension of damages and defective parts. Based on this information, further plans for preventive maintenance could be designed, and decisions about repairing parts of the building can be made. Building inspection is performed for diverse reasons. One of them is that, in many cities, a periodic inspection is required in some old buildings. Other possible reason for inspection is to conduct careful evaluations after an earthquake or other natural hazard. Also, inspection could

Fig. 1 Different damage in assorted parts of buildings

be planned as a previous task in restoration of cultural heritage or special buildings, or to define risk maps in cities. In refurbished buildings, inspection could be made in order to evaluate restorations or changes in the structure of the building. However, one of the most usual causes for inspection is to evaluate structures when damages are already detected. A careful and minimally invasive inspection use to be required in all these cases, because buildings often should be kept in service.

Inspection information is crucial in the evaluation of strength and resistance of structural members. In order to obtain this information, visual inspection use to be the first evaluation, consisting of a visual scrutiny that provides information about external elements and damages. Conclusion about possible internal conditions and deterioration is defined with the external data. Bearing in mind the decisions of the expert evaluator, further intensive studies could be considered, selecting the most significant areas that must be fully evaluated. Slightly invasive testing is usually applied as more meticulous evaluation. Therefore, random destructive tests are executed in selected areas, providing valuable and punctual information about the building. However, in many cases, a more intensive scrutiny is required. Then, it is usual to perform complementary studies involving non-destructive testing (NDT) that supplies indirect information about the internal elements of the structure. Several of the NDT technologies applied in buildings are seismic or impact-echo, acoustic emission, ultrasounds, natural and modal frequency analysis, resistivity, infrared thermography, and GPR. A complete and interesting description of masonry building damage and defects, together to the most common and useful testing methodology can be found in Binda et al. (2000). A noteworthy review of the state of the art in non-destructive building inspection techniques is presented in Hola and Schabowicz (2010) and in McCann and Forde (2001).

Ground-penetrating radar (GPR) is one of the most usual NDT methods in buildings inspection because of the low cost and on the expeditious data acquisition, although in many cases this technique is applied combined with other inspection methods. Two of the most usual applications are in the detection of rebar and reinforcement (e.g., Barrile and Pucinotti 2005; Pérez-Gracia et al. 2008a), and in cultural heritage studies (e.g., Ranalli et al. 2004; González-Drigo et al. 2008; Hemeda 2012a). In these cases, GPR surveys use to be applied as a support or preliminary analysis after visual scrutiny, in order to determine information needed in more extensive evaluations. Sometimes, GPR the result is a part of the data to be used in simulations of building behavior. Other times, GPR information is decisive to select parts of the building that require further analysis with other methods. Consequently, precise data are usually required and, therefore, inspection in buildings needs accurate data acquisition, processing and interpretation.

GPR survey is an especially attractive technique in cultural heritage inspection, where noninvasive evaluations should be required. Examples as Barrile and Pucinotti (2005), Gonzalez-Drigo et al. (2008), Hemeda (2012a), Pérez-Gracia et al. (2009a) and Binda and Saisi (2009) highlight the importance of historical buildings radar explorations. Nevertheless, GPR usually is a complementary method often combined with others, applied to determine hidden elements in the structures or damaged parts. Other proposed GPR applications are in the inspection of actual buildings that reach to a critical age (Ferrieres et al. 2009).

The great differences in structural solutions, as well as the different requirements depending on the projects, yield to a many different strategies and solutions and, in many cases, the potential of the method is still being verified, mainly in experimental laboratory tests or in simulations.

2 Classifying Buildings

Different building typologies require usually, different inspection methodologies. Until today, all applications depend on the construction materials and on the problems that must be evaluated. Perhaps, a broad association could separate between modern and ancient buildings. Modern buildings are concrete or mix structures, being concrete structures with steel reinforcements or concrete slabs with steel beams. Bricks are habitual materials in walls, but also concrete slabs. Ancient buildings include a wide number of typologies. Masonry buildings are usual constructions but materials could diverge greatly. Stones are usual in many cultural heritage constructions, but the nature of stone and its quality depends mainly on the achievable materials close the construction. Therefore, limestone, sandstone, marble, granite, are some of the rocks that can be found in these buildings. However, a large number of historical buildings are built with clay bricks, mudbrick or wood. Supporting structures are, in many cases, wooden beams. In other cases, arches, columns and vaults support the structure. Metallic beams exist also in many buildings and, in many cases, combined typologies and modern materials as a consequence of restorations, can be found. This great diverse typology involves a large number of different evaluation procedures.

Nowadays, a great number of applications related to buildings are conducted in cultural heritage. Several examples of these studies could be found in Ranalli et al. (2004), Gonzalez-Drigo et al. (2008), Hemeda (2012a), Leucci et al. (2012), Masini et al. (2010), Leucci et al. (2007). In some cases, when modern constructive elements coexists with ancient masonry ones, the evaluation of reconstructed buildings involves the detection of the contacts between old and modern parts or between structures dating from different periods (Pérez-Gracia et al. 2008b, 2009a). The evaluation could be also performed to detect metallic connecting elements between the structures of different epochs, or to confirm the existence of old remains under later structures (Imposa 2010; García-García et al. 2007).

Other buildings commonly studied by means of GPR are modern concrete structures. In these cases, frequent applications are detection of rebar, damages or beams under enclosures. Many examples can be found in the literature, especially based on rebar detection (Barrile and Pucinotti 2005; Pérez-Gracia et al. 2008a). The application does not differ much from evaluations of bridges and roads, especially since some of the most common studied parts are the flooring, walls and basement floors. Therefore, most of the analysis of the methodology and their results, performed in order to examine the ability of GPR in concrete evaluations, can be applied in modern buildings surveys. These analysis conceive the ability of GPR to determine the radius of steel rebars embedded in concrete (Chang et al. 2009; Leucci 2012; Zanzi and Arosio 2013). However, other analysis are focused on damp analysis (Laurens et al. 2002, 2005; Klysz and Balayssac 2007; Dérobert et al. 2008; Kalogeropoulos et al. 2011), and on cracks and voids detection (Maierhofer et al. 2003; Zhu and Popovics 2005).

GPR applications in wooden buildings or in wooden members of combined buildings are unusual. However, buildings have structural wood elements in many cases. Numerous masonry buildings have wooden beams or wood supports. Deterioration of these timber elements occurs generally by dampness, woodworms and fungi, and it can affect roofs, beams, floors, etc. (Fig. 1). In the few studies on this subject (Leucci et al. 2013; Muller 2002), the importance of combined techniques is highlighted to define clearly the cause of the anomalies and the resistant characteristics of the wooden elements.

3 Common GPR Studies in Buildings Assessment

Nowadays, the greatest part of GPR applications in buildings is performed in historic constructions. Different examples can be found related to Basilicas and Cathedrals (Ranalli et al. 2004), XIX century factories (Gonzalez-Drigo et al. 2008), palaces (Hemeda 2012a), mediaeval highly modified houses (Pérez-Gracia et al. 2009a), or describing applications in Cultural Heritage and evaluating different NDT techniques (Binda and Saisi 2009).

Furthermore, it is also necessary to emphasize that actual buildings are also evaluated with GPR and other NDT technologies, mainly when damages are detected after visual inspections (Pérez-Gracia et al. 2008a), and concrete and materials currently used in the construction are subject to careful studies (Klysz and Balayssac 2007; Dérobert et al. 2008; Maierhofer et al. 2003).

Main applications in buildings are based on common offset mode. Other radar acquisition methodologies (as common midpoint mode) are not usual in these evaluations, most likely because the studies are performed in small enclosures, and the space available is not large enough. Moreover, the existence of columns, stairs and furniture that is usually found in many buildings complicates the study. Sometimes three-dimensional images are obtained from the interpolation profiles. This is common in studies that aim to detect rebar and reinforcements. Tomography is one of the more unusual methods of study, although interesting results are presented in some studies comparing laboratory tests with mathematical models (Valle et al. 1999), and combined with sonic methods and common offset mode (Binda et al. 2003).

Sometimes the study with GPR is made solely on a structural element in which damage has been observed. This means that are studied in an isolated manner floors, beams, walls or columns, for example. The reduced size of many of these members forces the use of high frequency antennas. Therefore, antennas with frequencies higher than 500/400 MHz are usually are the most usual in buildings applications. Lower frequencies (400/500 MHz) are applied in the case of soils, while higher frequencies (900 MHz, 1, 1.6, 2.3 GHz...) are used in the evaluation of walls, columns and beams.

4 Application in Building Diagnosis: Detection of Damages

Landslides, earthquakes and the age of the buildings are usual causes for diagnosis studies, in order to evaluate the level of damage and deterioration of the structures. Other natural hazards, as subsidence sinkholes, gypsum dissolution, or floods are also cause of damage in man-made structures. Recent earthquakes have emphasized the importance of detailed assessments of the impaired buildings, prior its rehabilitation. Several proposals of diagnosis procedures include GPR as a supporting methodology, being applied in different cases (Imposa 2010; García-García et al. 2007; Chang et al. 2009). However, GPR is not actually a common tool in these evaluations, being applied mainly in the study of natural structures (Martín-Crespo and Gómez-Ortiz 2007), in order to determine the possible causes of visible damage. Examples of these applications are the studies reported by Pueyo-Anchuela et al. (2010). Nevertheless, promising studies demonstrate the viability of monitoring structures using GPR as a combined method. One suggesting evaluation is the study of the progressive damage of a concrete beam applying the integration of several electromagnetic sensing techniques (Catapano et al. 2012). Other interesting research evaluates the progress of damages in bricks during and after floods (Binda et al. 2010), combining thermography, sonic pulse velocity tests and GPR. Other compared evaluation in constructive materials during moist process changes determines the correlation between GPR changes in wave velocity and resistivity results, even conclude that GPR provides only qualitative results because no direct calibration between radar velocities and water content were obtained (Valek et al. 2010).

One of the most popular GPR applications is the diagnosis of damages in Cultural Heritage projects. Many Heritage buildings are additionally damaged after natural hazards by inadequate emergency interventions (Binda et al. 2010). Moreover, the age of these buildings, as well as possible external hazards, causes the appearance of visible damage in the structure or cladding, such as damp patches, cracking and detachment of materials. Often, damage to the structure has an extension greater than those observed in visual inspections. In these cases, GPR is applied as a successful tool, mainly as a combined methodology. This technique has been proved appropriate to detect damages in different members of historic buildings. In many cases, GPR is applied to detect internal cracks and unfilled joints in masonry buildings (Leucci et al. 2012; Massini et al. 2010). Therefore, it is usual to evaluate walls and, in some cases, other supporting structures, as columns (Leucci et al. 2007; Pérez-Gracia et al. 2013; Masini et al. 2010). Results

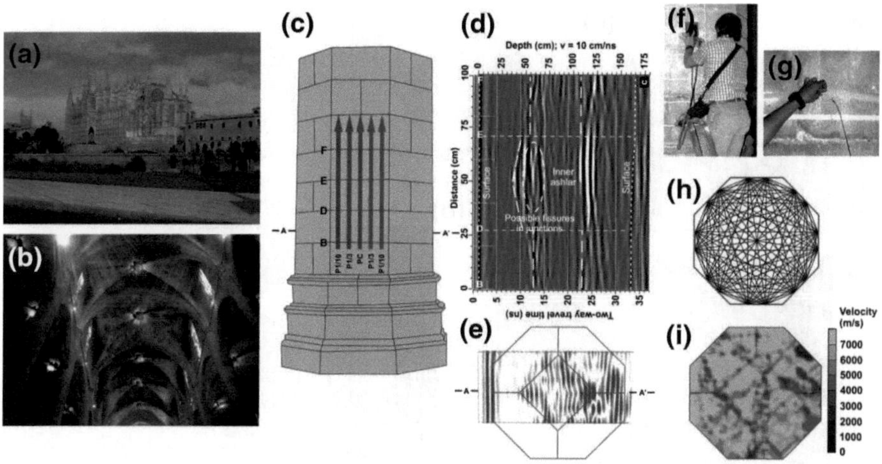

Fig. 2 Radar and seismic inspection of the supporting structures in Cultural Heritage building. **a** Gothic cathedrals are one of the hugest cultural heritage buildings. **b** In many cases slender columns support loads. **c** Radar data profiles to study one column. **d** Radar data obtained along one of the profiles. **e** Synthetic profile obtained interpolating GPR data from vertical profiles. **f** Radar data acquisition with a 1.5 GHz center frequency antenna. **g** Accelerometer on the column surface. **h** Seismic configuration and coverage in a hypothetical homogeneous column. **i** Seismic tomography showing the p-wave velocity distribution inside the column (From Pérez-Gracia et al. 2013)

allow defining the inner part of the structure and the possible existence of cracks or damp. However, accurate interpretation requires usually the application of other methodologies. In several cases, seismic data use to be applied combined with GPR to define internal structures (Fig. 2). In Fig. 2, radar images from stone columns are shown. The homogeneous material (columns are built with one internal and four external stones was expected to produce homogeneous images. However, contacts between stones are visible, as well as other random anomalies. The knowledge of the internal contacts provides valuable information about the shape of the construction, helping in further models to determine the behavior of the member. Images could presumably be representative of the health and thickness of junctions. Random anomalies could be associated to cracks or inner defects of the stones. In general, radar data is clean and regular enough to allow accurate interpretations.

In other cases, GPR data provides obscure images showing a great number of irregular anomalies. In such cases, other studies as thermography, endoscopy or even other geophysical surveys provide useful and complementary results that facilitate the interpretation of the images (Fig. 3). It is remarkable that radar data processing allows suggesting images of the inner structures that facilitates the interpretation of the anomalies. Therefore, B-scans could be processed and interpolate in order to obtain pseudo three-dimensional images that represent the most

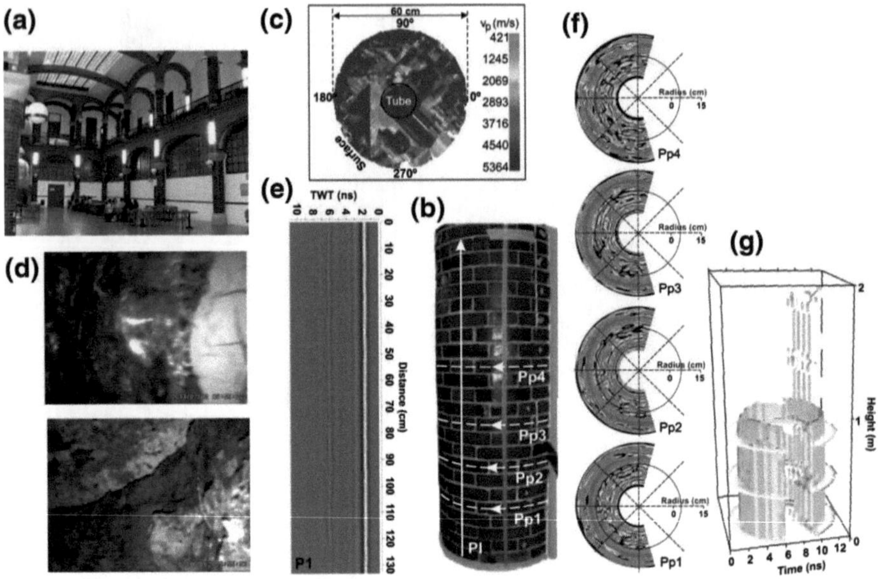

Fig. 3 Study of masonry columns that present high irregular internal shape and materials. **a** These masonry columns are the main load supporting structures of the building. **b** One of the studied columns, showing five of the GPR lines on its surface. **c** Seismic tomography results showing the irregular p-wave velocities distribution. **d** Endoscopy images that confirm the irregular distribution of materials inside the column. **e** Radar data obtained in one of the profiles along the shaft of the column. **f** Radar data from four of the perimetric parallel profiles. **g** Three dimensional images of the most important GPR anomalies showing almost one concentric layer of internal elements in the lower part (data from Santos-Assunçao 2014)

remarkable anomalies (Fig. 3). These results allow to visualize elements and, in many cases, could be a useful tool in restoration projects.

Figure 3 shows data from the study of several masonry columns. These columns have an internal irregular structure and radar data is not capable to define entirely the shape and the state of all the internal elements. Therefore, it is not possible to use only one method of study, as additional information is required. Useful evidences are provided by endoscopy that provides a clear picture of the complex interior of the columns. The complexity of the system is consistent with the results offered by the GPR and seismic tomography. Both methodologies provide irregular images that change highly with the column and the position of the measurement in the same column. Only the most continuous reflector—a rainwater drainage pipe found inside some of the columns studied—is detected with GPR as a clear anomaly. Three-dimensional images also reflect this target and it is even possible to define the state of conservation of the tube.

In several cases, suggestive applications in historical buildings reveal the correlation between GPR depth B-scans, time slices and damages observed visually in floors and walls (Orlando and Slob 2009). The study of floors allow also provide significant information about the construction solutions (Pérez-Gracia et al. 2009a). It is crucial in the case of ancient buildings that must be modified or restored, because usually they are buildings that have been modified several times throughout its life. Figures 4 and 5 show different flooring supporting structures in a second floor room of an old house in Barcelona (Spain). Brick vaults are frequent constructions used to support roofs and floors in XVIII, XIX and XX centuries. Usually, vaults are joined with metallic beams.

In Fig. 4, beams are concrete members as a consequence of a later restoration. Over each one of the vaults, formed with clay massif bricks, several layers of leaky bricks support a mortar layer underlying pavement tiles. Under the vaults, equidistant metallic bars crosses between concrete beams. Moreover, the coffered ceiling of the floor below is supported by small wooden girders embedded in the concrete beams. There is a considerable space between vaults and wood paneling of the lower deck. This space was used in some cases to pass wires and climate control tubes. Radar data shows a well-defined image of the elements that configure the whole structure. It is possible to define the concrete beams and reinforcement, the vaults, the metallic bars the bricks and different elements from the wooden coffered ceiling. The ability of the GPR to define also lateral changes between elements is also a suggesting possibility in the evaluation of masonry members and buildings.

Figure 5 shows images obtained in a second floor room supported by metallic beams and small vault. That is a common structural support in many buildings from the beginning of the XX century, still in use. The study of vaults and arches are also applied and developed in different building studies (Pérez-Gracia et al. 2009a; Mast et al. 1992). Radar data acquisition techniques and results are similar to those obtained in arch bridges, more widely developed (Diamanti et al. 2008; Solla et al. 2012a). Perhaps the main difference is that arches in buildings are, in many cases, structures hidden by other elements.

A great number of historical and actual buildings are masonry structures or contain masonry parts. Two of the most usual problems in masonry are the unfilled joints and internal cracks. Other problem is damp. An interesting evaluation about GPR abilities and limits in the study of different damages in masonry, can be found in Maierhofer and Leipold (2001). Interesting works shows its potential by means of simulation models and laboratory tests (Hamrouche et al. 2012), and several research projects are focus on the develop of new processing methodologies and strategies to obtain satisfactory results in the study of masonry damages, with promising results. These projects lead to a methodological approach for the detection of multiple defects inside dielectric or conductive media (Benedetti et al. 2007a, b). In other case, effectiveness of processing techniques is evaluated, focused in obtaining qualitative imaging of damages due to cracks. Advances to define crack shape and position are presented in different works. One of these evaluations, (Caorsi et al. 2001), present the possibility of using a microwave approach based on a genetic algorithm to detect defects inside a known host object.

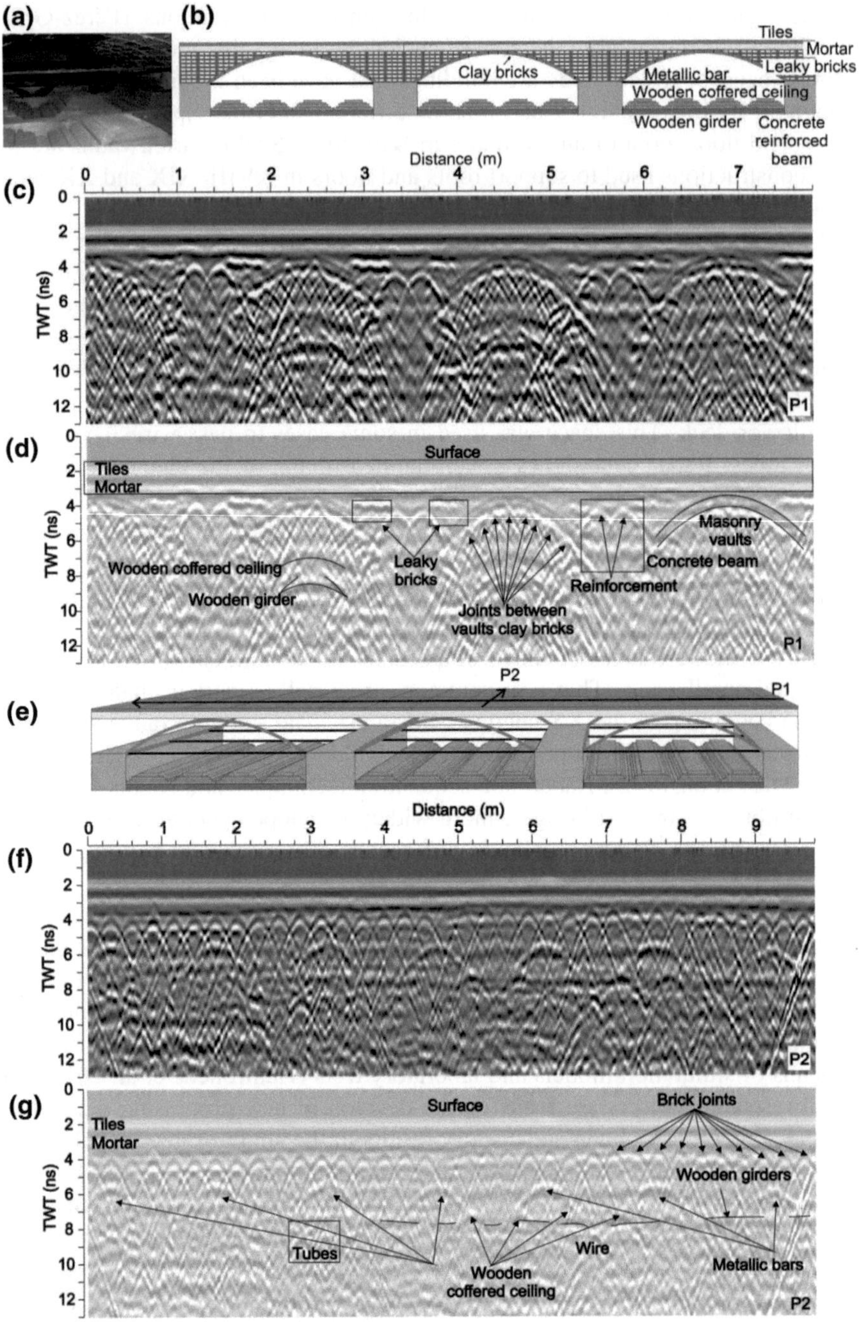

Fig. 4 Radar exploration of building floors (from Pérez-Gracia et al. 2009a). **a** Supporting structure based on vaults. **b** Schematic representation of the supporting structures. **c** Radar data obtained in a profile that crosses the vaults (*P*1). **d** Possible interpretation of *P*1. **e** 3D schematic drawing of the different structural elements and the two radar lines *P*1 and *P*2. **f** Radar imaging from *P*2 line, along the upper part of one vault. **g** Interpretation of the anomalies detected in *P*2

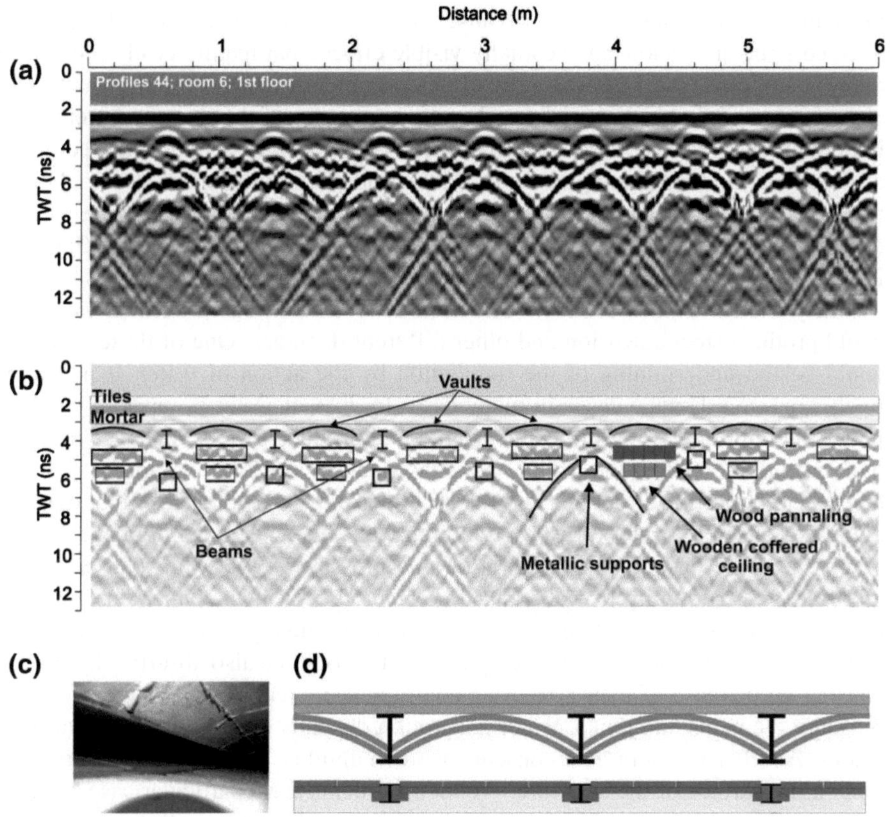

Fig. 5 Radar images of supporting structure in buildings from the beginning of the XX century (from Pérez-Gracia et al. 2009a). **a** Radar image. **b** Possible radar data interpretation. **c** Photograph of the supporting structures. **d** Possible reconstruction of the supporting members attending to radar information and endoscopy

Many existing buildings are constructed of reinforced concrete. The study of concrete buildings is successfully reported in different works. For example, GPR was applied to determine the concrete degradation in a factory floor, demonstrating that changes in concrete conditions could be clearly detected (Lorenzo et al. 2001). Also, GPR was applied to determine damage in the reinforced slab of a residential building (Pérez-Gracia et al. 2008a).

One of the most frequent causes of damage in these buildings is the steel corrosion as a consequence of dampness. Several researches propose the application of GPR to detect this kind of damage. The content of ions in the porous water produces alteration in parts of the structure affected by corrosion. Therefore, the electromagnetic properties are modified, enabling the detection due to differences in travel times, amplitudes and peak frequencies (Ferrieres et al. 2009; Lai et al. 2013). The GPR ability to detect those altered parts could provide information to determine areas affected by the corrosion of internal elements. A preventive evaluation of this damage is an important task because corrosion uses to be developed inside the affected structures, showing eventually visible effects that require costly repairs. In many cases, effects as a consequence of deterioration of the concrete in which steel elements are embedded are visible when the extension of the damage is relatively important. Moreover, reinforcement corrosion causes loss of section and, therefore, decreases of the load bearing capacity. The action of carbonation, presence of chlorides, existence of fissures in concrete members and concrete porosity are usually principal causes of corrosion, combined with water. GPR ability to detect chlorides is also a subject of evaluation (Kalogeropoulos et al. 2011).

Many works are based on the detection of water and damp because humidity could produce steel corrosion and other different damages. One of these damages could be the undermining of the foundation by the action of water. It is a relatively common damage cause in many cases by loss of water in pipes in walls or under the building. Water drags the ground and produce serious damages in foundations and the base of the walls. As a consequence, cracks could appear in walls and other supporting structures and the building may even collapse. Other damage could be caused by water thrust on retaining walls. Defects in the drainage system could cause the ground water saturation and pressure on the walls causes cracks and fractures. Humidity in walls and floors can also produce damage as a consequence of the expansion of ceramic pieces. This problem produces the appearance of fissures and cracks in the affected parts. Water could be also absorbed by bricks, causing an increase of their size. As a consequence, cracks and fissures appear in masonry elements and joints. Water absorption depends on the type and quality of bricks. Another problem as a consequence of humidity is the appearance of fungus on the surface of the structure. Wooden elements are a particular case because water could produce rapid and important damages. Therefore, other usual application is the detection of parts of the structure affected by humidity. Remarkable applications are developed in buildings (Pérez-Gracia et al. 2000, 2008a; Hemeda 2012b; Cataldo et al. 2005), being one of the objectives the study the humidity in walls, in the floor or in other structural elements. Analogous studies are also successfully applied to other different kind of structures such as bridges (Solla et al. 2013) and towers (Binda et al. 2005), obtaining similar results and applying comparable methodologies.

5 Detection of Internal and Buried Elements

Detection of internal elements embedded in a matrix is a common application of GPR in buildings. This is the case of steel reinforcements, pipes, several structural supports and other elements forming part of a building but which are positioned inside the structure. Therefore, rebar, reinforcements, changes in constructive materials or embedded components use to be the most common targets detected with GPR (Barrile and Puccinoti 2005; González-Drigo et al. 2008; Topczewski et al. 2007; Mazurek and Lyskowski 2012; Chang et al. 2009). In a vast number of

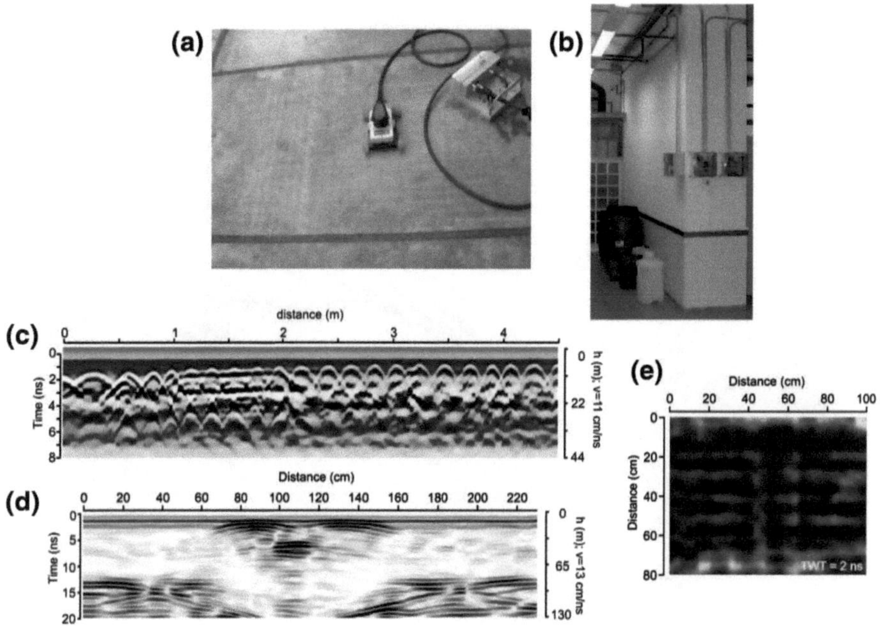

Fig. 6 Study of concrete slabs in a house outdoor patio **a** and study of load walls in a basement **b**, with high frequency antennas. Typical images obtained in these studies are hyperbolas due to reinforcement of walls **c**, reflections in inner members embedded in walls **d**, and time slices obtained from several parallel radar data acquisition lines **e** (Data provided by the GIES research group of the Polytechnical University of Catalonia)

applications, data is obtained from isolate radar acquisition lines, and elongated—and usually metallic—members are detected as hyperbolas in B-scans (Fig. 6). Three-dimensional diagrams and time slices are also common image presentation in rebar detection (Bala et al. 2011).

Mazurek and Lyskowski (2012) compare the results obtained with two antennas commonly used in walls and floors building evaluations, concluding that 800 MHz could penetrate 1.5 times deeper than 1.6 GHz antenna, even detection of rebar is clearer with the second one. In general, center frequencies antennas of 1.5 and 1.6 GHz are wherever used in all works about rebar detection.

GPR detection of buried elements combined with other techniques, like thermography, offers successful results, showing wide and worthwhile information as a consequence of the employment of complementary methodologies. A relevant example is the study of thermal floors pipes. GPR and thermography were use in the analysis and characterization of thermal floors in as-built buildings. Figure 7 shows the 3D GPR results obtained with a central frequency of 2.3 GHz, and the thermography mosaics of the radiant floor. GPR images illustrate the existence of 3 heating pipelines, clearly visible in the three-dimensional representations. Combination of those results with thermography data allows for the identification of a non-working pipe (the one in the middle).

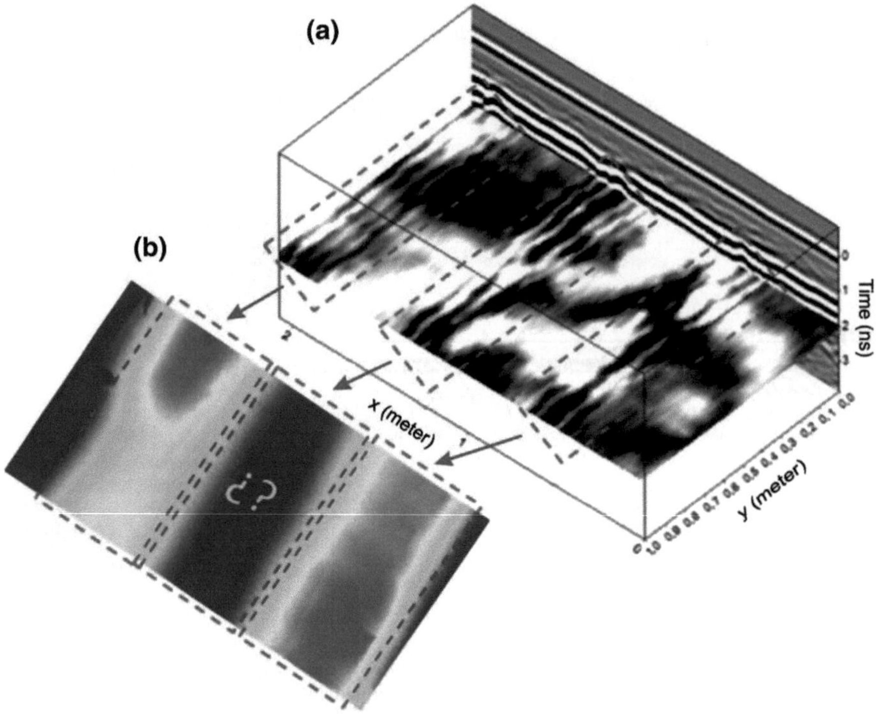

Fig. 7 3D GPR data obtained for the radiant floor with a central frequency of 2.3 GHz (**a**) and thermographic mosaics (**b**). The comparison of both data has revealed the existence of a non-working heating pipeline (Data provided by the applied geotechnologies research group of the University of Vigo)

Geophysical surveys, and predominantly GPR evaluations, are habitually applied in the detection of older constructions or constructive elements embedded inside the walls or buried under the actual building structure (Pérez-Gracia et al. 2000, 2009b; Lorenzo et al. 2002). Figure 8 presents a typical radar imaging obtained in older remains under actual building structures. In many cases, B-scans are used to infer the existence of hidden elements, but currently are more habitual 3D images and time slices, which make available tridimensional patterns of the embedded structures that facilitate the identification of the remains.

6 Foundations and Ground

The study of the ground and the building foundations is also a regular task in several geophysical explorations of buildings. The objective of studying the ground under and around the building is, frequently, the detection of geological structures that could cause damages or failure of the building (Pérez-Gracia et al. 2009b;

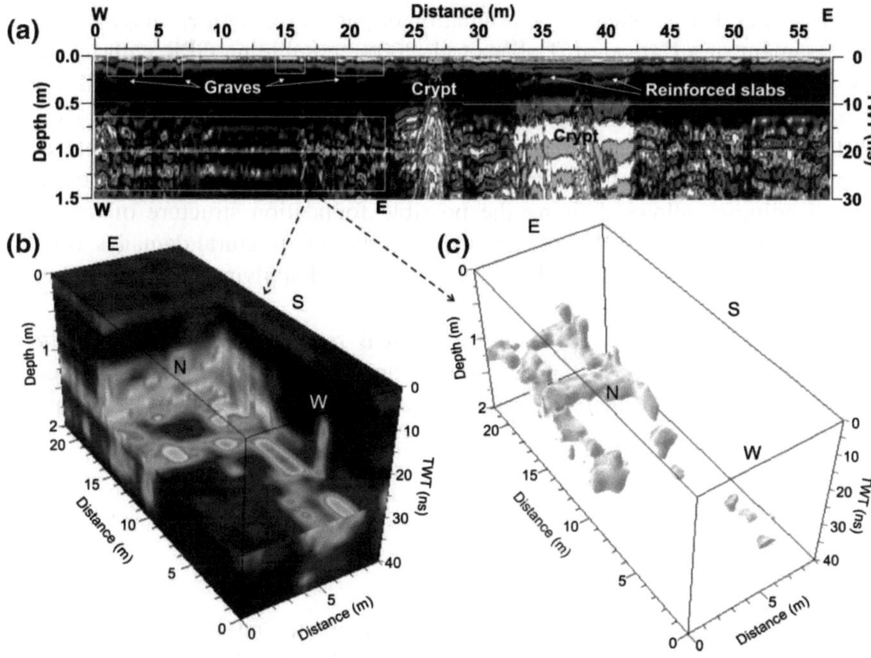

Fig. 8 GPR images corresponding to ancient Arabian remains under a gothic cathedral. B-scans show irregular reflections caused by several man-made structures (**a**). Interpolation of parallel profiles (**b**) and the study of isoamplitudes (**c**) provide a three-dimensional pattern associated to the older elements (From Pérez-Gracia et al. 2009b)

Leucci 2006; Ramirez-Blanco et al. 2008; Elawadi et al. 2006), the finding of man-made structures that could affect the safety of the structure (González-Drigo et al. 2008; Lorenzo et al. 2002), or the location of wet ground areas that could be the cause of damp damages (Pérez-Gracia et al. 2008a; Leucci et al. 2006). Kannan (1999) proposes a ground evaluation prior building new constructions in areas close active sinkholes, being GPR one of the feasible methodologies. Data will assist in the computation of new building foundations. Otherwise, Gutierrez et al. (2009) applies GPR combined with electric survey, to determine the possible causes of a collapse affecting several buildings in a city. GPR was also proposed to study the ground because of root damages in buildings (Satriani et al. 2010). It is usual, in the above mentioned applications, the use of combined methodologies, being resistivity surveys the most common technique used in conjunction with GPR. Seismic surveys are also shared, in some cases, with GPR and resistivity, showing promising results.

However, even the ground is commonly evaluated by means of GPR surveys, few GPR applications deal with the evaluation of foundations, perhaps due to the difficult in the access, an important limit in GPR surveys. Examples of these kind of applications can be found in Dabas et al. (2000) and in Abbas et al. (2005). In both cases, GPR is applied to the evaluation of part of Cultural Heritage buildings: the basement of a Cathedral (Dabas et al. 2000), combining GPR with resistivity surveys, and the ground under a museum (Abbas et al. 2005), being GPR used as a single method. In a complete GPR evaluation of a historical palace, Hemeda (2012a) also include part of the foundations, obtaining conclusions that could help in further restorations. In Perez-Gracia et al. (2009a), the study of the ground around columns allows defining the possible foundation structure of these columns. Kadioglu et al. (2013) present the detection of structural damages affecting foundations, based on enhanced data processing, and applying the methodology to a mosque.

Detection of foundations of archaeological remains in urban environments (e.g., Booth et al. 2010) is a more common application. In these cases, integrated studies and combined methodologies—usually GPR and resistivity techniques—are the most usual surveys.

7 Standard Methodologies

Radar data acquisition methodologies depend greatly on the building element being studied. In many evaluations, ground, walls and floor are scrutinized to detect buried elements or zones affected by water or cracks. In these cases, single radar lines use to be the most habitual data acquisition method, and many examples highlight its successful and valuable results (e.g., Pérez-Gracia et al. 2008a; Ranalli et al. 2004; González-Drigo et al. 2008; Hemeda 2012a).

Depending on the dimensions of the building, parallel radar profiles could be also used to obtain pseudo three-dimensional radar images, three-dimensional radar images and time slices of floors and walls, providing good visualization of cracks (Orlando and Slob 2009), detection of changes in materials (Masini et al. 2010) and location of the internal elements such as rebar (Bala et al. 2011). These applications are mostly in ground, basements and floor studies. In some cases, authors discuss about the density of the grid to obtain better images (Novo et al. 2010) by using a single antenna. Although it is possible to utilize an array of antennas in occasions in particular sites, it is not the most usual case inside building evaluations because of the dimensions of the arrays.

Other authors propose the tomography as a proper methodology to detect embedded objects within walls, showing impressive results and even detecting wooden beams and polystyrene bodies inside masonry (Topczewski et al. 2007). Application of microwave tomography processing is also proposed, developed and applied in interesting works in Kadioglu et al. (2013) and in Catapano et al. (2012).

Combined evaluations by using other geophysical techniques are also a good option proposed and used by different authors in the search of underneath remains under the building or in the ground evaluation to determine the cause of building damages. The method most often combined with GPR is resistivity survey

(Grangeia et al. 2008; Leucci 2006; Dabas et al. 2000). In some cases, other additional geophysical techniques are used together with these two methods. It is possible to find works where GPR and resistivity are applied jointly with very low frequency electromagnetic technique (e.g., Elawadi et al. 2006), and seismic methods (e.g., Pérez-Gracia et al. 2009b) or ultrasonic methods (e.g., Cassidy et al. 2011).

8 New and Experimental Developments

The previous review of the state of the art demonstrates that GPR is often used in non-destructive detection of structural elements and in building appraisal, primarily in the case of Cultural Heritage structures. Detection of hidden targets or evaluation of damage by means this geophysical technique provide valuable information for further restorations, modifications or construction works affecting the building. However, although this information is very useful, sometimes a more quantitative knowledge is necessary to enhance the study of buildings and other man-made structures.

The challenge of getting to know structural characteristics about the materials that are part of buildings has been focused from laboratory tests. Simultaneously, synthetic models and new data processing methodologies have been developed. Usually, new developments have been tested in laboratory and simulation models. In a small amount of cases, the new methodologies have been utilized in more extended building evaluations. These few applications highlight the difficulties in obtaining quantitative results in building evaluations. Although significant quantitative results are obtained in homogeneous media, in the study of buildings, results use to be more ambiguous. One of the problems could be complicate radar images produced by the non-homogeneous structures, due to the scattering in multiple and different targets. Reconstruction of the shapes of the scatterers is, in many cases, a complex task. Development and evaluation of numerical methodologies that facilitates pattern reconstruction (Benedetti et al. 2010; Oliveri et al. 2011; Hajihashemi and El-Shenawee 2010; Soldovieri and Solimene 2007) is an interesting working line that could get an effective improvement of the results and interpretations of GPR data from surveys in buildings. Synthetic and experimental data have been used to evaluate the new processing techniques, outperforming the most common techniques and obtaining in many cases better or equivalent resolution and a non-negligible improvement of results (Benedetti et al. 2010).

Innovative software tools used in pattern reconstruction could be applied to the detection of cracks and structural damages in different parts of the buildings. In many cases, pattern reconstruction is focused to the detection and reconstruction of internal cracks and damages (Benedetti et al. 2007a, b, c) and on the detection of its position and its orientation (Caorsi et al. 2000). Processing strategies are focused in pattern reconstruction with the objective of future evaluations of thick walls and other supporting structures. Although radar systems based on wideband antennas provide high-resolution images, a strong attenuation occurs as a result of its high spectral components. Therefore, the study thick member in structures is difficult. Integrating multi-focusing strategies with stochastic and deterministic reconstruction approaches appear to be a strategy effective in mitigating effects due to the non-linearity and poor conditions of the media, obtaining high resolution images without losing high numerical efficiency (Benedetti et al. 2010; Salucci et al. 2013).

Other interesting and relatively new GPR processing technique that provides good results is microwave tomography (Crocco and Soldovieri 2003; Catapano et al. 2010). Tests in and developments indicate that it is an effective tool to enhance radar images and pattern reconstructions to develop careful diagnostic of damages in structural members (Catapano et al. 2012). This technique seems to be a promising processing method that could provide quantitative information about the studied media and mitigate uncertainties. Different signal processing and models have been also applied in order to determine characteristics concerned to constructive designs, for example, thickness of concrete covers (Bourdi et al. 2013) or voids inside reinforced concrete (Xie et al. 2013).

Calibration and validation of the GPR ability to determine thicknesses under different conditions, and its resolution are other interesting lines of study, partially applied in building studies. Usually, those evaluations are in-situ tests in which GPR data is compared to results from different methodologies. One example of these evaluations is presented in Solla et al. (2012b), where GPR data is compared to laser-scanning results during the study of stone masonry walls. In other cases, experimental controlled tests are in quasi homogeneous media are the tool to determine GPR resolution to detect embedded targets, in many cases rebars or other construction elements (Pérez-Gracia et al. 2008c; Rial et al. 2009). Laboratory tests have been especially effective since they provide significant results that help in the interpretation of radar images obtained from applied cases study in buildings. Some lab experiences were developed in order to define accurate methodologies in the detection of voids, combining GPR with other techniques like thermography (Dérobert et al. 2008) or capacitive techniques (Maierhofer et al. 2003). However, some of the most interesting experimental measures under controlled conditions are focused on to establish methods for measuring constructive materials properties. The knowledge of these properties could be crucial in building assessment. In this line, remarkable laboratory analysis are dedicated to the characterization of the electromagnetic concrete properties (Ferrieres et al. 2009; Topczewski et al. 2007; Ihamouten et al. 2012; Robert 1998), to the detection of steel corrosion (Lai et al. 2011, 2013; Hubbard et al. 2003), and to the estimation of moisture in structures or in constructive materials (Laurens et al. 2002, 2005; Klysz and Balayssac 2007; Sass and Viles 2006).

Together with new processing signal developments and lab or in-situ tests, other important advances are based on signal modeling. GPR response in front of complex media could be modeled with those developed tool (Giannopoulos 2005). Comparing results from synthetic numerical models to tests or field data, it is

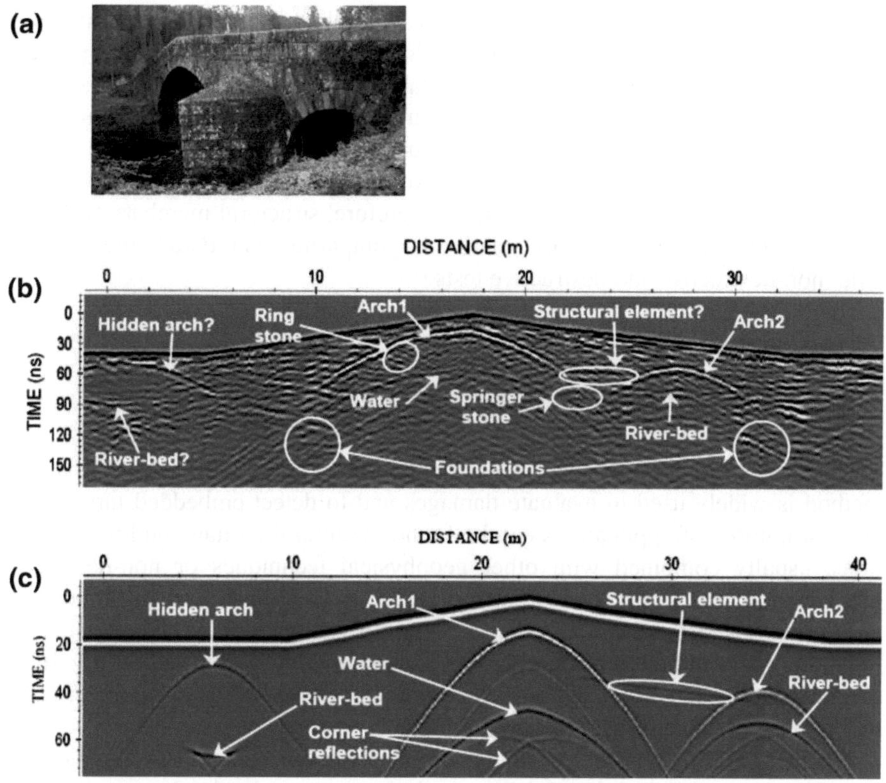

Fig. 9 GPR data provided by a central frequency of 200 MHz showing the presence of a probable previously unknown vault (**b**) now hidden from view by later construction (**a**). Considering this reflection as a hidden vault, FDTD modelling was elaborated to analyze the pattern of reflections to obtain, and it corroborates the existence of a most probable hidden vault (**c**). In addition, the polarity of the signal produced informs that this unexpected vault would be not empty and probably filled (from Solla et al. 2010)

possible to determine patterns to help in radar data interpretation. These numerical tools could help also to understand the GPR detection mechanisms and the signal behavior in front of multifaceted structures. This fact could be highly useful mainly in building evaluations, because data is obtained from ill-posedeness media. As a result, in many studies and tests related to building and structures appraisal, numerical models are usually determined and compared to GPR data (Orlando et al. 2010; Mayer et al. 2003; Czaja 2012). Numerical modeling has demonstrated its capabilities to provide meaningful results about the behavior of the electromagnetic signal from the presence of a hidden vaults, as shown in Fig. 9 (Solla et al. 2010), or even the existence of internal faults such as cracks or cavities (Solla et al. 2014). Comparing GPR data to synthetic images is a useful tool to help and improve radar data interpretation and diminish uncertainties inherent to the methodology.

Summarizing, a great part of the newest developments are mainly based in enhance data interpretation. The tools developed for this purpose are new processing techniques that improve radar images, laboratory tests to reproduce in simplest media targets and constructive elements and, finally, radar synthetic models to compare simulated images to radar data from buildings and other structures.

Other ongoing research lines are focused on obtaining quantitative data to better assess the condition of the structures. Therefore, structural members and materials are studied under laboratory conditions, comparing radar data to the results of other non-destructive and destructive tests.

9 Conclusion and Open Issues

The review of the GPR applications in buildings inspection reveals that this method is widely used to evaluate damages and to detect embedded targets. The greatest number of applications can be found in cultural heritage building inspections, usually combined with other geophysical techniques or non-destructive measures. In many cases, radar data is used as complimentary information to decide wider and more extensive studies. In other cases, radar data assessment is required as a preliminary evaluation in damaged buildings after natural hazards as earthquakes or landslides. In Heritage structures, GPR assessment is an effective tool to limit invasive inspections after disaster events.

In other building typologies, studies can be found in concrete structures inspection. The two main objectives of these evaluations use to be the detection of rebar and the location of zones probably affected by damp. Occasional works undertake corrosion detection in reinforce concrete, with successful result. Modern masonry buildings are also investigated by means of GPR. Detection of voids, humidity or unfilled joints are the most common evaluations. Three dimensional images as well as new processing techniques are utilized to enhance radar images.

Hence, concrete and masonry buildings are mostly the objective of GPR evaluations. However, few applications can be found in wood structures, although GPR is sometimes used to determine the existence of wooden beams or structural elements in masonry buildings. In few cases, damages in wooden members are also analyzed, but these inspections are infrequent.

Evaluation of buildings usually means the study of some specific parts or members. Typical applications are in walls, with the aim of acquire information about constructive techniques, defects or reinforced concrete. Thermography is a method that is frequently combined with GPR to determine shallow defects and targets in walls and floors, because offers valuable complementary information. Assessing the robustness and thickness of walls is other classic GPR application. Other conventional GPR building evaluations concerned to the ground under the construction. Detection of hidden structures and voids, architectural elements, archaeological remains, or shallow stratigraphy are the common reasons for those explorations. Resistivity imaging is the method most often applied in conjunction with the radar in ground and floor evaluations. In some cases, diagnosis of pillars and columns are tasks performed by using GPR tests combined with sonic measurements which permit to define, in some cases, mechanical properties of materials. The study of walls and pavement is, in several cases, accompanied by thermography imaging. Those evaluations provide complimentary details about the shallowest members and targets.

However, these applications offer always qualitative results. In an effort to determine possible quantitative data, laboratory experiences under controlled conditions have been developed, obtaining interesting results. Different structural members (masonry and reinforced concrete walls, beams or arches) have been evaluated under controlled conditions. Also, in some cases, defects and structural elements have been inserted in simple and homogeneous media, to determine physical properties or to observe radar pattern images. Regardless, few applications executed in existing buildings involve quantitative results, perhaps due to the inherent complexity of the studied structures, being one of the open issues in the study of buildings.

Noticeable advances in data processing, in pattern reconstruction and in the acquisition data technologies have facilitated the application of GPR in different problems associated to building inspection. Those innovative technologies have been developed with the purpose of reducing uncertainties and improving radar imaging, and preliminary results demonstrate that are promising solutions. Should be said that these new methods neither have been applied to the study of buildings extensively, although in some cases experimental results have been obtained from real test sites.

Different cases studies and advances in acquisition, processing and imaging, compared with actual requirements in building inspection, focus to the aspects to be improved in future developments. It is possible to define several open issues in GPR building assessment concerning to different aspects:

- Definition and classification of the most habitual casuistic in building inspection could be a useful tool for future evaluations. Recompilation of examples from cases studies could be the first step in this task. The classification could be complemented by the proposal of the most suitable radar data acquisition procedures, associated to each case or problem. Defining clearly the requirements in building inspection could be also an open question.
- Definition of a normative or regulation with suggestions for the different casuistics.
- GPR imaging could be still enhanced. Significant advances show the possibility to obtain accurate imaging diagnosis, by using characteristic radar patterns and models.
- Laboratory developments and tests to obtain electromagnetic parameters and properties of constructive materials must be applied to real buildings. Those new cases studies, could define the scope of the proposed methodologies, and open new test lines.
- Foundation studies could be an interesting application that has been slightly developed until today. Few cases studies are related to foundation assessment,

although this information could be crucial in buildings inspection, mainly in those ancient and historical structures. Perhaps the use of borehole GPR data, from directional drilling could be used in the evaluation of foundations.

- Development in the radar imaging processing, to define clear images of defects affecting structural elements.
- Advances in the analysis and detection of corrosion, applying the methodologies to real buildings, considering also the detection of oxidizing and corrosion agents.
- Development of advanced data processing algorithms based on suitable and complete mathematical models of scattering phenomenon.
- Integration of multi-focusing strategies with stochastic and deterministic reconstruction approaches in order to mitigate the effects of the non-linearity and the indetermination inherent to geophysical studies.
- Development and application of efficient inversion algorithms, testing results under laboratory conditions.
- Application to real buildings the inversion algorithms developed until today, to define the effects of the complex media and complicate scenarios, and to check possible improvements.
- Development of GPR systems capable of provide images and models from the gathered data combined with suitable mathematical models.
- Development of enhance radar equipments adapting the antennas to the different structural elements and spaces inside buildings.
- Development of arrays of antennas to acquire multi-view and multi-static data adapted to the specific requirements in building inspection.

Probably more issues will be opened in next years, based on the research tasks and the applications developed until today. As a previous step, in order to combine efforts and to improve the development of GPR building assessment, it could be useful to establish a catalog of significant publications related to building inspection and to gather the most significant results and methodologies associated to different studies.

Acknowledgments The authors acknowledge the COST Action TU1208 "Civil Engineering Applications of Ground Penetrating Radar", supporting this work. This work was also partially supported by the Spanish Government and the European Comission with FEDER funds, through the research projects CGL-2008-00869/BTE and CGL2011-23621. Authors are highly grateful to the support of Dr. A. Massa and Dr. I. Catapano, participants in the COST Action TU1208 in the Work Group 2, who provide us information about their research activities and interesting suggestions about the actual open issues in the GPR building inspection field.

References

Abbas, A.M., Kamei, H., Helal, A., Atya, M.A., Shaaban, F.A.: Contribution of geophysics to outlining the foundation structure of the Islamic Museum, Cairo. Egypt. Archaeol. Prospection **12**(3), 167–176 (2005)

Bala, D.C., Garg, R.D., Jain, S.S.: Rebar detection using GPR: an emerging non-destructive QC approach. Int. J. Eng. Res. Appl. **1**(4), 2111–2117 (2011)

Barrile, V., Pucinotti, R.: Application of radar technology to reinforced concrete structures: a case study. NDT E Int. **38**, 596–604 (2005)
Benedetti, M., Donelli, M., Massa, A.: Multicrack detection in two-dimensional structures by means of GA-based strategies. IEEE Trans. Antennas Propag. **55**(1), 205–215 (2007a)
Benedetti, M., Franceschini, G., Azaro, R., Massa, A.: A numerical assessment of the reconstruction effectiveness of the integrated GA-based multicrack strategy. IEEE Antennas Wirel. Propag. Lett. **6**, 271–274 (2007b)
Benedetti, M., Lesselier, D., Lambert, M., Massa, A.: Multiple shapes reconstruction by means of multi-region level sets. IEEE Trans. Geosci. Remote Sens. **48**(5), 2330–2342 (2010)
Benedetti, M., Donelli, M., Lesselier, D., Massa, A.: A two-step inverse scattering procedure for the qualitative imaging of homogeneous cracks in known host media—preliminary results. IEEE Antennas Wirel. Propag. Lett. **6**, 592–595 (2007c)
Binda, L., Saisi, A., Tiraboschi, C.: Investigation procedures for the diagnosis of historic masonries. Constr. Build. Mater. **14**, 199–233 (2000)
Binda, L., Saisi, A., Tiraboschi, C., Valle, S., Colla, C., Forde, M.: Application of sonic and radar tests on the piers and walls of the Cathedral of Noto. Constr. Build. Mater. **17**, 613–627 (2003)
Binda, L., Cardani, G., Zanzi, L.: Nondestructive testing evaluation of drying process in flooded full-scale masonry walls. J. Perform. Constructed Facil. **24**, 473–483 (2010)
Binda, L., Zanzi, L., Lualdi, M., Condoleo, P.: The use of georadar to assess damage to a masonry Bell Tower in Cremona, Italy. NDT E Int. **38**, 171–179 (2005)
Binda, L., Saisi, A.: Application of NDTs to the diagnosis of historic structures. In: Proceedings of the NDTCE'09, Non-Destructive Testing in Civil Engineering, Nantes, France, 30th June–3rd July (2009)
Booth, A.D., Clark, R.A., Hamilton, K., Murray, T.: Multi-offset ground penetrating radar methods to image buried foundations of a medieval town wall, Great Yarmouth, UK. Archaeol. Prospection **17**(2), 103–116 (2010)
Bourdi, T., Boone, F., Rhazi, J.E., Ballivy, G.: Use of Jonscher model for estimating the thickness of a concrete slab by technical GPR. Prog. Electromagnet. Res. **28**, 89–99 (2013)
Caorsi, S., Massa, A., Pastorino, M.: A crack identification microwave procedure based on a genetic algorithm for nondestructive testing. IEEE Trans. Antennas Propag. **49**(12), 1812–1820 (2001)
Caorsi, S., Massa, A., Pastorino, M., Righini, F.: Crack detection in lossy two-dimensional structures by means of a microwave imaging approach. Int. J. Appl. Electromagnet. Mech. **11**(4), 233–244 (2000)
Cassidy, N.J., Eddies, R., Styles, P., Brightwell, S., Dods, S.: Combining ground-penetrating radar and ultrasonic survey techniques: new tools for old problems? First Break **29**(8), 85–91 (2011)
Cataldo, R., De Donno, A., De Nunzio, G., Leucci, G., Nuzzo, L., Siviero, S.: Integrated methods for analysis of deterioration of cultural heritage: the crypt of "Cattedrale di Otranto. J. Cult. Heritage **6**, 29–38 (2005)
Catapano, I., Di Napoli, R., Soldovieri, F., Bavusi, M., Loperte, A., Dumoulin, J.: Structural monitoring via microwave tomography-enhanced GPR: the montagnole test site. J. Geophys. Eng. **9**(4), 100–107 (2012)
Catapano, I., Crocco, L., Isernia, T.: A feasibility study of a quantitative microwave tomography technique for structural monitoring. Near Surf. Geophys. **8**(5), 389–395 (2010)
Chang, C.W., Lin, C.H., Lien, H.S.: Measurement radius of reinforcing steel bar in concrete using digital image GPR. Constr. Build. Mater. **23**(2), 1057–1063 (2009)
Crocco, L., Soldovieri, F.: GPR prospecting in a layered medium via microwave tomography. Ann. Geophys. **46**(3), 559–572 (2003)
Czaja, K.: Application of electromagnetic field modeling in GPR investigation on an historic tenement. Geol. Geophys. Environ. **38**(4), 395–410 (2012)
Dabas, M., Camerlynck, C., Freixas, P., Camps, I.: Simultaneous use of electrostatic quadrupole and GPR in urban context: investigation of the basement of the Cathedral of Girona (Catalunya, Spain). Geophysics **65**(2), 526–532 (2000)

Dérobert, X., Iaquinta, J., Klyszc, G., Balayssac, J.P.: Use of capacitive and GPR techniques for the non-destructive evaluation of cover concrete. NDT E Int. **41**(1), 44–52 (2008)
Diamanti, N., Giannopoulos, A., Forde, M.C.: Numerical modelling and experimental verification of GPR to investigate ring separation in brick masonry arch bridges. NDT E Int. **41**, 354–363 (2008)
Elawadi, E., El-Qady, G., Nigm, A., Shaaban, F., Ushijima, K.: Integrated geophysical survey for site investigation at a New Dwelling Area, Egypt. J. Environ. Eng. Geophys. **11**(4), 249–259 (2006)
Ferrieres, X., Klysz, G., Mazet, P., Balayssac, J.P.: Evaluation of the concrete electromagnetics properties by using radar measurements in a context of building sustainability. Comput. Phys. Commun. **180**, 1277–1281 (2009)
García-García, F., Ramírez Blanco, M., Rodríguez Abad, I., Martínez Sala, R., Tort Ausina, I., Benlloch Marco, J., Montalvá Conesa, J.L.: GPR technique as a tool for cultural heritage restoration: San Miguel de los Reyes Hieronymite Monastery, 16th century (Valencia, Spain). J. Cult. Heritage **8**(1), 87–92 (2007)
Giannopoulos, A.: Modelling ground penetrating radar by GPRMax. Constr. Build. Mater. **19**, 755–762 (2005)
González-Drigo, R., Pérez-Gracia, V., Di Capua, D., Pujades, L.G.: GPR survey applied to Modernista buildings in Barcelona: the cultural heritage of the college of industrial engineering. J. Cult. Heritage **9**, 196–202 (2008)
Grangeia, C., Senos Matias, M.J., Figueiredo, F., Hermozilha, H., Carvalho, P.: High resolution geophysics Inside Machado de Castro Museum – Coimbra, Centre Portugal. In: Proceedings of the 14th European Meeting of Environmental and Engineering Geophysics, Kraków, Poland, 15–17 Sept 2008
Gutiérrez, F., Galve, J.P., Lucha, P., Bonachea, J., Jordá, L., Jorda, R.: Investigation of a large collapse sinkhole affecting a multi-storey building by means of geophysics and the trenching technique (Zaragoza city, NE Spain). Environ. Geol. **58**, 1107–1122 (2009)
Hajihashemi, M.R., El-Shenawee, M.: The level set shape reconstruction algo-rithm applied to 2d pec targets hidden be-hind a wall. Prog. Electromagnet. Res. **25**, 131–154 (2010)
Hamrouche, R., Klysz, G., Balayssac, J.P., Rhazi, J., Ballivy, G.: Numerical simulations and laboratory tests to explore the potential of ground-penetrating radar (GPR) in detecting unfilled joints in brick masonry structures. Int. J. Architectural Heritage **6**, 648–664 (2012)
Hemeda, S.: Ground penetrating radar (GPR) investigations for architectural heritage preservation: the case of Habib Sakakini Palace, Cairo. Egypt. Open J. Geol. **2**, 189–197 (2012a)
Hemeda, S.: Ground penetrating radar investigations for architectural heritage preservation of the Habib Sakakini palace, Cairo. Egypt. Int. J. Conserv. Sci. **3**(3), 153–162 (2012b)
Hoła, J., Schabowicz, K.: State-of-the-art non-destructive methods for diagnostic testing of building structures—anticipated development trends. Arch. Civ. Mech. Eng. **10**(3), 5–17 (2010)
Hubbard, S.S., Zhang, J., Monteiro, P.J., Peterson, J.E., Rubin, Y.: Experimental detection of reinforcing bar corrosion using nondestructive geophysical techniques. ACI Mater. J. **100**(6), 501–510 (2003)
Ihamouten, A., Villain, G., Dérobert, X.: Complex permittivity frequency variations from multioffset GPR data: hydraulic concrete characterization. IEEE Trans. Instrum. Meas. **61**(6), 1636–1648 (2012)
Imposa, S.: Infrared thermography and georadar techniques applied to the "Sala delle Nicchie" (Niches Hall) of Palazzo Pitti, Florence (Italy). J. Cult. Heritage **11**(3), 259–264 (2010)
Kadioglu, S., Kadioglu, Y.K., Catapano, I., Soldovieri, F.: Ground penetrating radar and microwave tomography for the safety management of a cultural heritage site: Miletos Ilyas Bey Mosque (Turkey). J. Geophys. Eng. **10**, 11 (2013). doi:10.1088/1742-2132/10/6/064007
Kalogeropoulos, A., van der Kruk, J., Hugenschmidt, J., Busch, S., Merz, K.: Chlorides and moisture assessment in concrete by GPR full waveform inversion. Near Surf. Geophys. **9**(3), 277–285 (2011)
Kannan, R.C.: Designing foundations around sinkholes. Eng. Geol. **52**, 75–82 (1999)

Klysz, G., Balayssac, J.P.: Determination of volumetric water content of concrete using ground-penetrating radar. Cem. Concr. Res. **37**(8), 1164–1171 (2007)

Lai, W.L., Kind, T., Stoppel, M., Wiggenhauser, H.: Measurement of accelerated steel corrosion in concrete using ground-penetrating radar and a modified half-cell potential method. J. Infrastruct. Syst. **19**, 205–220 (2013)

Lai, W.L., Kind, T., Wiggenhauser, H.: Using ground penetrating radar and time–frequency analysis to characterize construction materials. NDT E Int. **44**(1), 111–120 (2011)

Laurens, S., Balayssac, J.P., Rhazi, J., Arliguie, G.: Influence of concrete relative humidity on the amplitude of ground-penetrating radar (GPR) signal. Mater. Struct. **35**(4), 198–203 (2002)

Laurens, S., Balayssac, J.P., Rhazi, J., Klysz, G., Arliguie, G.: Non-destructive evaluation of concrete moisture by GPR: experimental study and direct modeling. Mater. Struct. **38**(9), 827–832 (2005)

Leucci, G., Persico, R., Soldovieri, F.: Detection of fractures from GPR data: the case history of the Cathedral of Otranto. J. Geophys. Eng. **4**(4), 452–461 (2007)

Leucci, G.: Ground penetrating radar: an application to estimate volumetric water content and reinforced bar diameter in concrete structures. J. Adv. Concr. Technol. **10**(12), 411–422 (2012)

Leucci, G., Melica, D., Quarta, G.: The Foggia Cathedral: an in situ integrated geophysical and mechanical study on the wooden structures of the ceiling. In: Proceedings of the Built Heritage 2013 Monitoring Conservation Management, Milan, Italy, 18–20 Nov 2013

Leucci, G.: Contribution of ground penetrating radar and electrical resistivity tomography to identify the cavity and fractures under the main church in Botrugno (Lecce, Italy). J. Archaeol. Sci. **33**(9), 1194–1204 (2006)

Leucci, G., Cataldo, R., De Nunzio, G.: Subsurface water-content identification in a crypt using GPR and comparison with microclimatic conditions. Near Surf. Geophys. **4**(4), 207–213 (2006)

Leucci, G., Masini, N., Persico, R.: Time–frequency analysis of GPR data to investigate the damage of monumental buildings. J. Geophys. Eng. **9**(4), 81–91 (2012)

Lorenzo, H., Cuéllar, V., Hernández, M.C.: Close range radar remote sensing of concrete degradation in a textile factory floor. J. Appl. Geophys. **47**(3–4), 327–336 (2001)

Lorenzo, H., Hernández, M.C., Cuéllar, V.: Selected radar images of man-made underground galleries. Archaeol. Prospection **9**(1), 1–7 (2002)

Maierhofer, C., Brink, A., Röllig, M., Wiggenhauser, M.: Detection of shallow voids in concrete structures with impulse thermography and radar. NDT E Int. **36**(4), 257–263 (2003)

Maierhofer, C., Leipold, S.: Radar investigation of masonry structures. NDT E Int. **34**, 139–147 (2001)

Martín-Crespo, T., Gómez-Ortiz, D.: Collapse hazard assessment in evaporitic materials from ground penetrating radar: a case study. Environ. Geol. **53**, 57–66 (2007)

Masini, N., Persico, R., Rizzo, E.: Some examples of GPR prospecting for monitoring of the monumental heritage. J. Geophys. Eng. **7**(2), 190–199 (2010)

Mast, J.E., Lee, H., Murtha, J.P.: Application of microwave pulse-echo radar imaging to the non-destructive evaluation of buildings. Int. J. Imaging Syst. Technol. **4**, 164–169 (1992)

Mayer, K., Zimmer, A., Langenberg, K.J., Kohl, C., Maierhofer, C.: Nondestructive evaluation of embedded structures in concrete: modeling and imaging. In: Proceedings of the International Symposium in Non-Destructive Testing in Civil Engineering (NDT-CE 2003), Berlin, Germany, 16–19 June 2003

Mazurek, E., Lyskowski, M.: Practical application of high resolution ground penetrating radar method inside buildings. Geol. Geophys. Environ. **38**(4), 439–448 (2012)

McCann, D.M., Forde, M.C.: Review of NDT methods in the assessment of concrete and masonry structures. NDT E Int. **34**(2), 71–84 (2001)

Muller, W.: Trial of ground penetrating radar to locate defects in timber bridge girders. In: Proceedings of the Riding the Wave to Sustainability: IPWEAQ 2002 State Conference, Noosa Lakes, Queensland, Australia, 6–10 Oct 2002

Novo, A., Lorenzo, H., Rial, F., Solla, M.: Three-dimensional ground-penetrating radar strategies over an indoor archaeological site: convent of Santo Domingo (Lugo, Spain). Archaeol. Prospection **17**, 213–222 (2010)

Oliveri, G., Randazzo, A., Pastorino, M., Massa, A.: Imaging of separate scatterers by means of a multiscaling multiregion inexact-Newton approach. Progress Electromagnet. Res. **18**, 247–257 (2011)

Orlando, L., Slob, V.: Using multicomponent GPR to monitor cracks in a historical building. J. Appl. Geophys. **67**, 327–334 (2009)

Orlando, L., Pezone, A., Colucci, A.: Modeling and testing of high frequency GPR data for evaluation of structural deformation. NDT E Int. **43**(3), 216–230 (2010)

Pérez-Gracia, V., Canas, J.A., Pujades, L.G., Clapés, J., Caselles, O., García, F., Osorio, R.: GPR survey to confirm the location of ancient structures under the Valencian Cathedral, Spain. J. Appl. Geophys. **43**, 167–174 (2000)

Pérez-Gracia, V., García García, F., Rodríguez, I.: GPR evaluation of the damage found in the reinforced concrete base of a block of flats: a case study. NDT E Int. **4**, 341–353 (2008a)

Pérez-Gracia, V., García, F., Pujades, L.G., González-Drigo, R., Di Capua, D.: GPR survey to study the restoration of a Roman monument. J. Cult. Heritage **9**(1), 89–96 (2008b)

Pérez-Gracia, V., González-Drigo, R., Di Capua, D.: Horizontal resolution in a non-destructive shallow GPR survey: an experimental evaluation. NDT E Int. **41**(8), 611–620 (2008c)

Pérez-Gracia, V., Caselles, J.O., Clapés, J., Martinez, G., Osorio, R.: Non-destructive analysis in cultural heritage buildings: evaluating the Mallorca cathedral supporting structures. NDT E Int. **59**, 40–47 (2013)

Pérez-Gracia, V., Caselles, O., Clapés, J., Osorio, R., Canas, J.A., Pujades, L.G.: Radar exploration applied to historical buildings: a case study of the Marques de Llió palace, in Barcelona (Spain). Eng. Fail. Anal. **16**, 1039–1050 (2009a)

Pérez-Gracia, V., Caselles, J.O., Clapes, J., Osorio, R., Martínez, G., Canas, J.A.: Integrated near-surface geophysical survey of the Cathedral of Mallorca. J. Archaeol. Sci. **36**, 1289–1299 (2009b)

Pueyo-Anchuela, O., Casas-Sainz, A.M., Soriano, M.A., Pocoví Juan, A., Ipas-Lloréns, J.F., Ansón-López, D.: Integrated geophysical and building damages study of karst eff ects in the urban area of Alcalá de Ebro, Spain. Zeitschrift für Geomorphologie **54**(2), 221–236 (2010)

Ramírez-Blanco, M., García-García, F., Rodríguez-Abad, I., Martínez-Sala, R., Benlloch, J.: Ground-penetrating radar survey for subfloor mapping and analysis of structural damage in the Sagrado Corazón de Jesús Church, Spain. Archaeol. Prospection **15**(4), 285–292 (2008)

Ranalli, D., Scozzafava, M., Tallini, M.: Ground penetrating radar investigations for the restoration of historic buildings: the case study of the Collemaggio Basilica (L'Aquila, Italy). J. Cult. Heritage **5**, 91–99 (2004)

Rial, F.I., Pereira, M., Lorenzo, H., Arias, P., Novo, A.: Resolution of GPR bowtie antennas: An experimental approach. J. Appl. Geophys. **67**(4), 367–373 (2009)

Robert, A.: Dielectric permittivity of concrete between 50 MHz and 1 GHz and GPR measurements for building materials evaluation. J. Appl. Geophys. **40**(1), 89–94 (1998)

Salucci, M., Sartori, D., Anselmi, N., Randazzo, A., Oliveri, G., Massa, A.: Imaging Buried Objects within the Second-Order Born Approximation through a Multiresolution Regularized Inexact-Newton Method". In: International Symposium on Electromagnetic Theory (EMTS), Hiroshima, Japan, 20–24 May 2013

Santos-Assunçao, S.: Assessment of the viability of subsurface radar as support for studies of seismic vulnerability. Ph.D. thesis (in process), Universitat Politecnica de Catalunya, Barcelona, Spain (2014)

Sass, O., Viles, H.A.: How wet are these walls? Testing a novel technique for measuring moisture in ruined walls. J. Cult. Heritage **7**, 257–263 (2006)

Satriani, A., Loperte, A., Proto, M., Bavusi, M.: Building damage caused by tree roots: laboratory experiments of GPR and ERT surveys. Adv. Geosci. **24**, 133–137 (2010)

Soldovieri, F., Solimene, R.: Through-wall imaging via a linear inverse scattering algorithm. IEEE Trans. Geosci. Remote Sens. **4**(4), 513–517 (2007)

Solla, M., Caamaño, J.C., Riveiro, B., Arias, P.: A novel methodology for the structural assessment of stone arches based on geometric data by integration of photogrammetry and ground-penetrating radar. Eng. Struct. **35**, 296–306 (2012a)

Solla, M., González-Jorge, H., Álvarez, M.X., Arias, P.: Application of non-destructive geomatic techniques and FDTD modeling to metrical analysis of stone blocks in a masonry wall. Constr. Build. Mater. **36**, 14–19 (2012b)

Solla, M., Lagüela, S., Riveiro, B., Lorenzo, H.: Non-destructive testing for the analysis of moisture in the masonry arch bridge of Lubians (Spain). Struct. Control Health Monit. **20**, 1366–1376 (2013)

Solla, M., Lorenzo, H., Novo, A., Rial, F.I.: Ground-penetrating radar assessment of the medieval arch bridge of San Antón, Galicia, Spain. Archaeol. Prospection **17**, 223–232 (2010)

Solla, M., Riveiro, B., Lorenzo, H., Armesto, J.: Ancient stone bridge surveying by ground-penetrating radar and numerical modeling methods. J. Bridge Eng. **19**, 110–119 (2014)

Topczewski, L., Fernandes, F.M., Cruz, P.J.S., Lourenço, P.B.: Practical implications of GPR investigation using 3D data reconstruction and transmission tomography. J. Build. Appraisal **3**(1), 59–76 (2007)

Válek, J., Kruschwitz, S., Wöstmann, J., Kind, T., Valach, J., Köpp, C., Lesák, J.: Nondestructive investigation of wet building material: multimethodical approach. J. Perform. Constructed Facil. **24**, 462–472 (2010)

Valle, S., Zanzi, L., Rocca, F.: Radar tomography for NDT: comparison of techniques. J. Appl. Geophys. **41**, 259–269 (1999)

Xie, X., Li, P., Qin, H., Liu, L., Nobes, D.C.: GPR identification of voids inside concrete based on the support vector machine algorithm. J. Geophys. Eng. **10**(3) 034002 (2013). doi: 10.1088/1742-2132/10/3/034002

Zanzi, L., Arosio, D.: Sensitivity and accuracy in rebar diameter measurements from dual-polarized GPR data. Constr. Build. Mater. **48**, 1293–1301 (2013)

Zhu, J., Popovics, J.S.: Non-contact imaging for surface-opening cracks in concrete with air-coupled sensors. Mater. Struct. **39**(9), 801–806 (2005)

Inspection Procedures for Effective GPR Sensing and Mapping of Underground Utilities and Voids, with a Focus to Urban Areas

Christina Plati and Xavier Dérobert

Abstract Ground Penetrating Radar (GPR) has proved its ability to act as a powerful geophysical non-destructive tool for subsurface investigations. The remarkable technological developments have increased, among others, the practice of GPR in sensing and mapping utilities and voids. In particular, GPR is effectively used to locate and map objects such as pipes, drums, tanks, cables and underground features or to detect subsurface voids related to subsidence and erosion of ground materials. Furthermore, deploying GPR methods prior to directional drilling prevents damage to existing utilities, thus resulting in cost effective installations. In that frame, this paper presents some studies showing the GPR performances and limitations, from single-channel systems to the potential of multi-channel 3D imaging and integrating systems.

1 Introduction

Transportation agencies need accurate methods for measuring the near-surface and subsurface conditions of their transportation facilities. Determining pavement thickness, detecting voids beneath pavements and measuring the moisture content in pavement layers are examples of subsurface pavement conditions for which data are necessary. One promising technology for addressing these issues is Ground Penetrating Radar (GPR) (NCHRP 1998).

C. Plati (✉)
National Technical University of Athens, Athens, Greece
e-mail: cplati@central.ntua.gr

X. Dérobert
IFSTTAR, GERS Department,
LUNAM Université, 44344 Bouguenais, France
e-mail: xavier.derobert@ifsttar.fr

A. Benedetto and L. Pajewski (eds.), *Civil Engineering Applications of Ground Penetrating Radar*, Springer Transactions in Civil and Environmental Engineering,
DOI 10.1007/978-3-319-04813-0_5

It is of considerable interest to achieve as much information as possible about the subsurface before starting any underground construction work. In directional drilling it is important to avoid existing utilities as well as obstacles. In general means of utility locating, it can also be of interest to detect voids and pipe leakage. The most common solutions for detecting existing utilities are cable and pipe locators, based on the principle of electromagnetic induction. These devices are quite efficient for locating electric conducting cables and pipes. Most products are based on an external transmitter that is galvanically or inductively coupled to the target. A hand-held device is then used to locate the object. The major drawback of these locators is the fact that they do not detect non-metallic material such as plastics, concrete, ceramics or fiber optic cable. They will usually not save data, providing the possibility of mapping and further post-processing. In addition, electromagnetic (EM) devices require prior knowledge of a utilities approximate location or access to a portion of the utility to excite the utility for detection by the receiver. GPR can detect utilities without any prior knowledge of their respective location and hence is a very powerful reconnaissance tool for areas containing complete unknowns (Nissen et al. 2001).

GPR is a non-invasive and non-destructive tool that has been successful in some transportation applications, such as profiling asphalt pavement thickness and detecting air filled voids. Furthermore it is used for a variety of applications, including: mapping underground utilities, profiling ice thickness, bathymetric (depth measurements) surveys of fresh water lakes, archaeological investigations, shallow bedrock profiling, measuring pavement thickness, measuring pavement base and subbase thickness, locating voids beneath pavements, detecting bridge deck delimitation, mapping soil stratigraphy and characterizing environmental contamination (NCHRP 1998).

In France, some legal accidents in 2007 and 2008 have led to modify the regulation of work in the neighbourhood of utility networks which has been official since July 01, 2012, including one legislation, four decrees and a standard (NF S 70-003-Part 1 to Part 3) (French Standard NF 2012).

In one hand, the responsibility of the owners is increased while three levels define the accuracy of localisation of a utility, from class A when the uncertainty of positioning is below 40 cm, to class C when it is above (for this last case, complementary investigations being required). On the other hand, the SIG mapping of utility network has become the central subject of the territorial collectivities, which have to invest in geomatic science.

The first trials followed by two pilot cities have shown mitigated results. Indeed, three businesses are gathered: the knowledge of the utilities, the detection and geo-localisation techniques, but the professional community is not yet well prepared to such accuracy requirements. Indeed, training courses are under construction, supervised by the Ministry of Ecology, Sustainable Development and Energy, due to the fact that two certifications in 2017, for detection and geo-localisation, will be necessary for services companies.

Similar actions have been also performed in other countries. For example the American Society of Civil Engineers (ASCE) has created guidelines for the collection and depiction of existing subsurface utility data.[1] These guidelines present a credible system for classifying the quality of utility location information that is placed in design plans. This assists engineers, project and utility owners, and constructors in developing strategies to reduce risk by improving the reliability of information on existing subsurface utilities in a defined manner.

Also, in 2006, the Department of Survey and Mapping Malaysia (JUPEM) produced a Standard Guideline for Underground Utility Mapping with the aim to provide standard procedures for collection, compilation and presentation of underground utility data (Jamil et al. 2012). This guideline is the culmination of concerted efforts of the Underground Utility Mapping Technical Committee chaired by JUPEM comprising underground utility stakeholders functioning under the auspices of the National Committee for Mapping and Spatial Data, being the responsible body for the coordination of mapping activities in Malaysia.

This guideline addresses issues such as the roles of stakeholders comprising utility agencies, land surveyors and JUPEM; utility quality levels which are categorized into quality level A, quality level B, quality level C and quality level D with quality level A being the most accurate and quality level D being the least accurate. This guideline also provides specifications on underground utility maps and creation and maintenance of underground utility database which stores related utility data that can be made available to utility agencies and all other relevant parties whenever underground information is required. These data can then be shared and utilized by various agencies involved in the re-development of the area.

In such frame, GPR imaging is one of the promising non-destructive methods that have offered new opportunities for mapping the subsurface structures of shallow earth in highly urbanized regions[2] (Jeng and Chen 2012). The present chapter refers to the procedures used for an effective GPR sensing and mapping of underground utilities and voids with a focus to urban areas.

2 Mapping of Underground Utilities

The often-repeated infrastructural improvements in basic infrastructure facilities are road widening and relocation of utilities. If there are no proper care and systematic work approach observed, the road works may disrupt and cause damage to underground utilities. Hence, mapping of these underground utilities is necessary to avoid great loss and accidents (Ismail et al. 2013).

Underground utility mapping is a process of identifying the position and labelling public utility mains which are located underground. These mains may include lines for telecommunication, electricity distribution, natural gas, water mains and

[1] ASCE, American Society of Civil Engineers, "Standard guideline for the collection and depiction of existing subsurface utility data", (CI/ASCE 38-02).

[2] ASTM Designation D 6432-11, "Standard Guide for Using the Surface Ground Penetrating Radar Method for Subsurface Investigation".

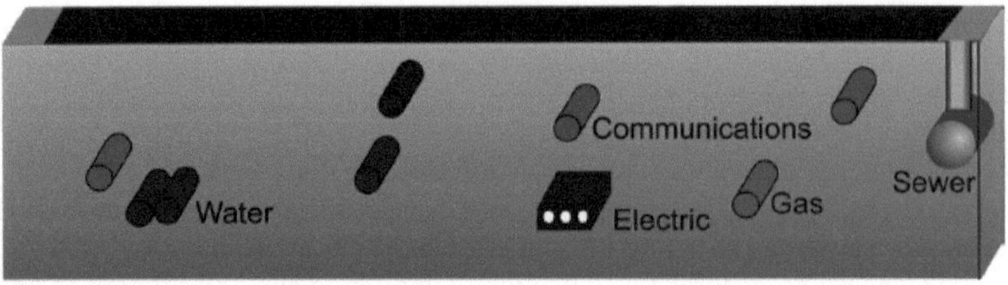

Fig. 1 Underground utilities (Annan 2003)

wastewater pipes (Fig. 1). In some location, major oil and pipe lines, national defence communication lines, mass transit, rail and road tunnels also compete for space underground (Jamil et al. 2012).

Underground utility mapping refers to the detection, positioning and identification of buried pipes and cables beneath the ground, corresponding to three businesses. It deals with features mainly invisible to the naked eyes. While the determination of position can be obtained with conventional or modern survey equipment, the detection and identification of underground utilities require special tools and techniques (Jamil et al. 2012).

GPR is commonly used to locate and map objects such as pipes, drums, tanks, cables and underground features or utilities. In application of utilities detection, radar data are used to detect the existence of underground utilities which mainly have different conductivity and dielectric properties from its surrounding. Several studies have been carried out concerning the mapping of underground utilities from last 90s to now-a-days.

2.1 Case Studies

2.1.1 Detection of a Main Water Supply Pipe in Stockholm, Sweden

In 2001 the GPR technique was used for the detection of a main water supply pipe in Stockholm, Sweden (Nissen et al. 2001). The presence of the pipe was well known, although the exact location was unknown. It was of great importance to locate the exact position in advance of the construction of a big warehouse close to the pipe. The survey was performed over a road in which several pipes and cables were buried. In order to locate the pipe a large number of short profiles (16 parallel profiles) were conducted perpendicular to the presumed strike of the pipe, using a 500 MHz shielded antenna. The data were loaded into a 3D visualization and processing software for analysis.

2.1.2 Locating Underground Utilities in Malaysia

Another study was carried out in main campus of Universiti Sains Malaysia, Penang, specifically located at School of Social Sciences (Ismail et al. 2013). The aim of the study was to locate and map underground utilities or pipes that existed at the study area that could be used as a preliminary study for utilities mapping. Eight parallel ground penetrating radar survey lines were executed with total length of 8 m and line spacing of 2 m. In this case, a shielded antenna with frequency 250 MHz was used to detect and map utilities of the study area with suitable parameter setting. Anomalies were detected at depth <2.5 m at several survey lines that may be due to underground pipe, manhole trench, and underground cable that crosses the study area. Orientations of these utilities were successfully detected and well mapped.

In Kuala Lumpur, Malaysia, a GPR survey was performed using a 450 MHz antenna on a major road. The contractor needed to know where he could safely dig without hitting a utility while installing fiber optic communication cables. The survey was carried out in a few minutes, and locations of the utilities were marked on the pavement as the survey was being conducted. The fiber optic cable contractor quickly and safely trenched between sites of buried utilities (Annan 2003).

2.1.3 Workshop at Georgia, Atlanta

At the Underground Focus Utility Workshop which was held at Dekalb College in Covington, Georgia near Atlanta, the participants had the opportunity to see utility locating in action. The objectives of the workshop were to expose utility locators to new technologies and practices in the industry and to provide hands-on experience with the various techniques. After construction of the Dekalb College site, no as-builts were provided by the contractor. As a result, no records of the underground utility pipes and cables were available. The technologies showcased at the workshop helped locate some of the buried utilities (Annan 2003).

GPR proved to be very effective at the site. One interesting exercise involved location a storm water drainpipe under the circular driveway outside the lecture hall. While the intake could be seen at the curb, the location of the storm drainpipe was unknown. Several GPR traverses around the intake located a large pipe. A series of GPR transects were then carried out to track the alignment of the pipe. Three transects perpendicular to the pipe are indicated on the photo in Fig. 2. The drainpipe was expected to continue down grade to the right in the photo of Fig. 2. Instead, the pipe ran in the opposite direction.

To address the drainage/slope issue, profiling along transect (Line 4) was performed, which joined the pipe location determined from Lines 1, 2 and 3. This data provided a continuous profile along the axis of the pipe. From the GPR response, it was apparent that the depth of the pipe increased from right to left by about 3 ft (1 m). In addition, it was possible to see the reflections from the top and the bottom of the pipe. The storm drain appeared to be a concrete pipe with no apparent metallic structure associated with it. Using the depth of the top and bottom of the pipe and assuming the pipe was air filled, since there had been no rain for an extended period of time, it was possible to estimate the pipe diameter at about 36 inches (90 mm).

Fig. 2 Location of storm drain intake and GPR lines (Annan 2003)

This example illustrated the power of using GPR. The exercise of locating the pipe, marking the alignment and tracking the depth as well as estimating its diameter took about 10 min.

2.1.4 Utility Surveying in Hong Kong

Utility surveying was also performed at three sites in Hong Kong, which represented three different kinds of utilities. In this research, general procedures on how to design the grids and to conduct the radar surveys were reviewed. The main objective was to compare the radar results with the conventional survey methods in three trial sites. Finally, it was crucial to evaluate the use of antenna frequency and the accuracy of the obtained data. These utilities were:

- Drains along Kong Sin Wan Road, Cyberport, (Sites 1A and 1B, KSWR)
- Watermains at Nam Fung Path, Aberdeen (Site 2, NFP) and
- Electricity cables at Hong Tat Path, Tsim Sha Tsui East (Site 3, HTP) (Cheng 2013).

All are paved in flat surface without any irregularities. Site 1 was built for the purpose of accessing facilities nearby in 2009. Similarly, Site 2 was formed for the purpose of road facilities in 2008. Field inspection revealed that the study area was close to the construction site of the Mass Transit Railway south island line. With regard to Site 3, it contained a brunch of cables lying under pavements. General survey information relevant to these sites is described in Table 1.

Each plot was surveyed in two directions, but only the longitudinal transects were used subsequently for analysis, because it was more difficult to distinguish the reflectors along short horizontal profiles. Simply, one or two reflectors could be possibly collected in every two or three-metre horizontal profile, whereas plenty of reflectors along the longitudinal profiles could be identified as sample points.

Table 1 Survey details of three sites

Site	Grid dimensions	Total number of transects (horizontal and longitudinal)	Types of utilities	Burial depths (m)
1. KSWR (A)	19 × 1 m, 1 m spacing	22	300 mm diameter concrete drains	1.93–1.96
KSWR (B)	24 × 3 m, 1 m spacing	29	225 mm diameter concrete drains	1.89–1.93
2. NFP	19 × 2 m, 1 m spacing	23	150 mm diameter water drains	0.49–0.64
3. HTP	26 × 2 m, 0.5 m spacing	58	25-58 mm diameter power cable	0.45

In the study three ground coupled centre frequency antennas—100, 270 and 400 MHz were used. However, the use of 100 MHz antenna was abandoned because of lack of precise horizontal distance measurement without the usage of survey wheel. To ensure transect lines easily and to maintain a complete coverage of the targets a grid was used for the survey. It was achieved that the 400 MHz antenna was good at distinguishing underlying objects less than 2 m, whereas 270 MHz one was 2 m below.

2.1.5 Utility Mapping in São Paulo City, Brazil

Another utility-mapping was performed in two areas located in downtown São Paulo City, Brazil (Porsani et al. 2012). The main objective of this work was to locate subsoil utilities, such as, piping, galleries, electric cables etc., as well as concrete columns supporting the Roosevelt Road tunnel–viaduct complex in advances of the construction of the Line 4 subway (yellow) tunnel in São Paulo.

GPR measurements were performed using 200 MHz shielded antennas. Lines were surveyed along both the north-south and east-west directions in a polygonal area in order to achieve a pseudo 3D grid by 2D data interpolation. Interpretations of GPR results, combined with lithological information available from boreholes and trenches opened in the study areas provided important information for accurate location of shallow utilities in the subsurface. The study showed that the GPR method was a very important and useful step to precede excavation of São Paulo subway tunnels, providing the precise location of utilities in subsoil, as well as an estimate of their depths of occurrence. As a final point, based on these results, geotechnical work was safely carried out, and risks of dangerous accidents were avoided (Porsani et al. 2012).

Through these few examples the benefit of performing several parallel profiles was clearly demonstrated. An object located in one profile only could easily be interpreted as a pipe, if interpretation were done in 2D only. By combining several profiles and load them into a 3D software, it's easy to check whether the object is extending linearly or not. Thus this kind of interpretation improves the security of the investigation, at the cost of time. It is especially valuable when the site conditions are more difficult e.g. when many pipes and boulders are present. Moreover these researches showed that the ability of GPR to detect not only metallic objects, but also non-metallic objects (e.g. plastic, concrete ceramics or fiber optic cable), makes it a powerful tool to complement traditional methods in utility locating. This is of considerable importance, since an increasing fraction of buried pipes are non-metallic.

2.1.6 Utility Mapping in Trondheim, Norway

Utility mapping using a single-antenna GPR is a time consuming operation especially when large areas are to be covered. Data acquisition can be performed much more efficient by using an electronically scanned antenna array. An ultra-wideband GPR system was developed at the Norwegian University of Science and Technology (Eide and Hjelmstad 2002). The radar can be programmed to operate in the frequency range from 10 MHz to 3.4 GHz using a stepped- frequency waveform.

The radar is designed to operate with an antenna array for efficient 3D data acquisition of GPR data as shown in Fig. 3. The system uses a 1 meter wide antenna array which consists of 31 pairs of transmit/receive bow-tie monopoles mounted on a common ground plane. The antenna array is mounted on a lightweight trailer that is pulled by a vehicle along the street being surveyed. In this way, data can be collected on a 3.2 × 3.2 cm grid at a velocity of 1–2 m/s depending on the integration time required at each antenna element. During operation, the radar uses one transmit/receive antenna pair at the time and records radar trace at each spatial point.

The radar has successfully been used for mapping of pipes, cables and old tramlines in Trondheim during 2001. The wide bandwidth gives high enough

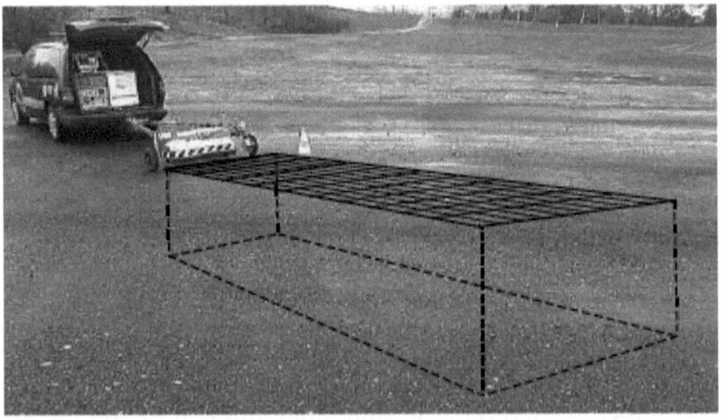

Fig. 3 3D data acquisition using the electronically scanned antenna array (Eide and Hjelmstad 2002)

resolution to map the asphalt thickness and the details of the base layers in addition to the utility lines. In this way the data from a survey can serve more than one customer. The results from the field test demonstrate the high user potential of 3D imaging compared to standard 2D GPR profiling (Eide and Hjelmstad 2002).

2.1.7 Utility Mapping in New York

In the framework of a research program performed by the New York State Department of Transportation (NYSDOT) utility-mapping at two sites was successful in that GPR was able to (a) identify buried utility infrastructure both at previously known and unknown locations, (b) successfully cover the entire pavement intersection at both sites in relatively short period of time, and (c) achieve superior 3D image quality because of the multi-channel (array system) GPR/GPS capability that is not available in typical single-channel and even two-channel GPR systems (Grivas 2006).

The system used covers 5.12 feet with each instrument pass obtaining 14 channels of GPR data within each swath. Multiple parallel and transverse instrument passes were done at each site in order to obtain sufficient coverage to produce the 3D images. Post processing and interpretation were performed to generate the 3D images, maps and CAD drawings that were delivered for the project. These 3D images produced from specialized proprietary multiple antenna hardware and processing software, are much more illustrative and accurate than the vertical slice provided by a single antenna. For instance, the images can be rotated to achieve the optimal view of a critical utility crossing, and can be superimposed over pavement edges, curbs and other surface features for ease of identification. The system produces utility locations in X, Y and Z coordinates to within a few centimeters, usually within one or two centimeters. The results of this study illustrated the ability of multi-channel GPR technology to detect and map buried utilities, indicating the utility depths, orientations, and proximities to other surrounding infrastructure.

2.1.8 Tree Root Mapping and Utilities, South Africa

Sometimes, a problem may be encountered in civil engineering/urban areas, where subsurface utilities such as water and drain pipes and electricity and telecommunication cables often occur in close proximity to growing trees with laterally extending root systems. Over time, these developing root systems may cause significant damage to utilities and to other types of infrastructure such as road surfaces and building foundations. GPR can be used as a monitoring tool in cases where tree root damage is suspected or anticipated (van Schoor 2009). A relevant case-study is presented below.

The survey site is a grassed area located on the campus of the Council for Scientific and Industrial Research (CSIR) in Pretoria, South Africa. A Burkea Africana (Wild Syringa) tree and an Ochna Pulchra (Lekkerbreek) tree are located in close proximity to a number of known subsurface utilities. Both tree types typically have lateral and fairly shallow root systems. A small 12 m × 12 m survey area, approximately the same size as the trees' combined canopy, was defined. This area is traversed by two separate subsurface electricity lines as well as two separate water pipe lines.

A total of 51 GPR profiles were acquired along two perpendicular directions. Data acquisition was done using a 500 MHz antenna. A profile length and spacing of 12 and 0.5 m, respectively, were used throughout the survey. The analysis showed that individual lateral tree roots that approach a linear geometry can be detected with GPR provided the property contrast with the background is sufficient. The GPR profile direction should ideally be perpendicular to the strike of the root. However, the variable geometry and unpredictable nature of tree roots often make them more difficult to pick out on individual 2D radargrams. Utilities are much easier GPR targets due to their typically distinct property contrast with the surrounding soil and their fairly predictable or known linear geometry and depth. Where discrimination between utilities and tree roots is required a detailed 3D survey approach is recommended (van Schoor 2009).

2.1.9 Water Leakage Detection, USA

Leaks not only waste precious natural resources, they create substantial damage to the transportation system and structure within urban and suburban environments. Surface geophysical methods are noninvasive, trenchless tools used to characterize the physical properties of the subsurface material. This characterization is then used to interpret the geologic and hydrogeologic conditions of the subsurface. Many geophysical techniques have been suggested as candidates for detecting water leakage, including GPR, acoustic devices, gas sampling devices and pressure wave detectors. In (Eyuboglu et al. 2003) a series of laboratory experiments were conducted to determine the validity and effectiveness of GPR technology in detecting water leakage in metal and plastic PVC pipes of an underground distribution system.

Initially, a prototype laboratory model was designed to simulate a pipe leak. Holes were drilled in the middle of the pipe to allow the water leak into a simulated soil (sand). The metal and PVC pipes were tested separately by burying them in sand to a depth of 18 and 20 cm, respectively. Water was then injected into the pipe from the surface through a plastic hose. A 1.5 GHz antenna was used to collect GPR data. Although the experiment was very well controlled, results obtained so far indicate that GPR is effective in detecting water leaks. Also, an outdoor test bed was under construction in collaboration with Central Arkansas Water (CAW) to simulate and detect water leaks in underground water systems using the GPR technique.

2.1.10 Water Leakage Detection from Plastic Pipes, Turkey

The effectiveness of GPR techniques in detecting and identifying water leaks from plastic pipes has been also investigated in (Demirci et al. 2011). For this purpose, experiments were conducted in a partly homogeneous outdoor soil environment. A PVC pipe with a diameter of 5 cm was located at a depth of 20 cm from the ground surface. Before embedding the pipe in soil, a small hole was drilled for the leakage of water. First, a reference scan measurement of the water free pipe was performed. Then, after starting injecting water into the pipe, three scan experiments at regular time intervals were performed. For each scan, a synthetic aperture of length 64 cm with 33 discrete spatial points and a frequency band from 0.8 to 5 GHz with 301 discrete frequency steps were used. Two double-ridged horn antennas with a length size of 0.5 m were utilized in a bistatic configuration during the measurements. After collecting data, the scan images were reconstructed by using a near-field back-projection algorithm.

From the analysis it can be interpreted that, as the water content of the soil around the leak become larger for an evolved time, the size of the region that causes strong EM wave reflections becomes also larger. Hence, EM wave could not penetrate beneath this water-saturated region. This phenomenon manifests itself as voids in the GPR images providing valuable information about the location of the leak. As the time passed, the reflecting area became larger and the leak position became more visible. Hence, it could be assessed that water leakages from plastic pipes can be successfully detected by GPR techniques.

2.1.11 Investigation at Road and Bridge Research Institute

In addition, Road and Bridge Research Institute (Internal Project PW.S 531) have investigated the implementation of 3D Radar into practical use in road measurements in order to assess its real abilities in road diagnostics (Krysiński and Sudyka 2012). This radar model is a step-frequency system working in range 300 MHz–3 GHz what correspond in practice to resolution of 700 MHz impulse antenna. The antenna chamber contains 15 transmitter-receiver pairs allowing simultaneous data acquisition along 15 parallel profiles on road lane of width 2.4 m. The system has odometer and GPS antenna that allow excellent synchronisation of data sets collected in different measurement passages and next construction of horizontal slices of some wider area. Using real-time GPS equipment, one decimetre precision of data positioning and synchronisation is available. Thus the system has excellent possibilities in identification and localisation of utilities, combining proper software that developed by the producer, for such large data sets.

The identification of linear objects is based on wide review of horizontal slices, where thick linear objects have fine manifestations, with the possibility to visualise sets of horizontal slices and sets of vertical cross-sections. The effective penetration depth in road environments can be estimated to about 1 up to 1.5 m where some ground horizons were observed in rare cases. Thus the system has some possibilities of sub-base thickness estimation, and can give also fine evidence of media having complicate, strongly irregular, chaotic structure. The equipment is very noise resistive giving fine results in difficult electromagnetic environment where typical impulse antennae fall.

This system has also some disadvantages, from intense multiplications which occur in the case of strongly reflective horizons, inducing reverberations inside the antenna chamber and then difficulties in echogram interpretation. Moreover, the echograms have low resolution and clarity in comparison to corresponding impulse antennae. Nevertheless, the evolution of such array systems tends to propose accurate approach due to high density of measurement, with precise positioning, enabling similar resolution at lower frequencies than single antenna systems.

2.2 GPR and Other Techniques

Among these experimental approaches estimating the performances of GPR systems, signal processing and imaging have been largely studied in numerous research laboratories. In the last few years, the ELEDIA Research Center (DISI—University of Trento, Italy) developed several techniques for real-time detection and classification of buried objects based on learning-by-examples (LBE), such as Support Vector Machines (SVM) (Massa et al. 2005; Lizzi et al. 2009). A lot of work has been also addressed towards the development of inversion schemes based on the integration of both global and local search algorithms with multi-focusing strategies. As a significant example, the integration of the Inexact Newton Method (INM) with a computationally efficient multi-focusing scheme has been successfully validated in (Salucci et al. 2013) when dealing with GPR measurements. A significant interest has been also addressed towards the application of innovative inversion procedures based on Bayesian Compressive Sensing (BCS) (Poli et al. 2013) and Interval Analysis (Manica et al. 2013) to the problem of subsurface prospecting. The localization of sparse metallic targets has been recently addressed in (Poli et al. 2013) by means of a new technique that models the targets through the local shape function (LSF) approach and solves the inversion problem in a BCS sense.

In (Lester and Bernold 2007) a data processing method, called Translation Invariant Wavelet Packet Detection (TIWPD) is presented. The method is applied to filtering GPR data collected on a university campus. The paper presents the results of experimental deployment of the system during a construction project that involved the extensive excavation trenches to lay chilled water pipes. The significance of this paper lies in its use of real-world GPR data to demonstrate the performance characteristics of the filtering process and its validation with the actual condition found during excavation. The encouraging results of this work can provide the basis for developing a near real time utility detection system that can be used by laborers in the field.

Moreover, another study has been performed at the Construction Automation and Robotics Laboratory (CARL) at North Carolina State University (NCSU). The investigation concerned a novel technology for detecting and locating buried utilities that attaches to the digging equipment and utilizes both EMI (Electromagnetic Induction) and GPR (Bernold et al. 2002). The newest effort involves the

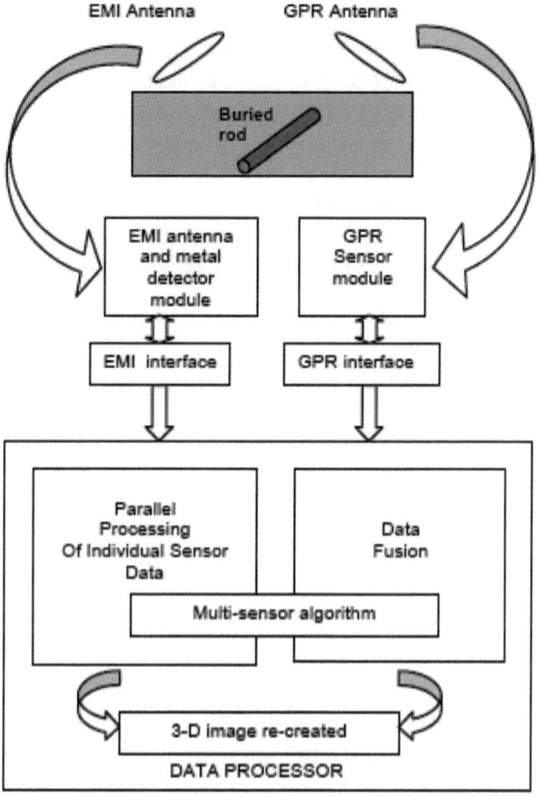

Fig. 4 Simplified data fusion model adapted to MS-BUDS (Bernold et al. 2002)

development and performance analysis of algorithms to detect and extract the features and characteristics of these utilities, such as their orientation, diameters etc. One prime focus is to minimize the percentage of false alarms. For that purpose, the two sensor systems are fused to create a multi-sensory approach to 3-D mapping of all the utilities without a priori knowledge of their location.

Figure 4 shows a data fusion model adapted to the Multi-Sensory Buried Utility Detection System (MS-BUDS). The data from the two sensors are initially conditioned by independent signal processing modules which later feed into a parallel processor running a multisensory algorithm.

3 Detection of Voids

The development of voids beneath roadways can lead to major pavement failures. Voids typically develop because of subsidence and erosion of the base and subgrade materials. Void-related roadway problems have often developed near water supply pipes or drainpipes. Leaks, pipe breaks or dislocated joints allow fines to be carried away, resulting in local base or foundation erosion and the formation of weak areas, which eventually become voids. Voids continue to increase in size until the load carrying capacity of the roadway is compromised (Holt and Eales 1997). Voids can be either air or water filled. Void detection beneath roadways using GPR is performed in many countries (Holt and Eales 1997; Morey 1998). Some related case-studies are presented below.

3.1 Case Studies

3.1.1 Detecting Voids Under Roadway Pavements by TxDOT and TTI

The Texas Department of Transportation (TxDOT) and Texas Transportation Institute (TTI) have successfully used the GPR technology to locate voids under roadway pavements (TxDOT 2010). The case-study presented here concerns the maintenance of a continuously reinforced concrete pavement (CRCP) section, which presented longitudinal cracks just after a patch repair. This has prompted district personnel to request an investigation to assess the safety of the structure and to determine if there were significant voids under the CRCP.

A 400 MHz ground-coupled antenna was used to survey and map the subsurface condition. GPR data was collected in the longitudinal direction parallel to the faulted joint and at selected transverse locations. A significant anomaly adjacent to the drainpipe was found which started directly under the CRCP. Based on the GPR image, the estimated size of the suspected void was significant (Fig. 5).

The GPR data indicated an anomaly over the transverse storm drain. This drain was separated and it was verified that this separation caused water to erode the area around the drain. Fortunately, the disjointed storm pipe was identified before severe roadway failure occurred.

3.1.2 Detecting Voids Under Roadway in Minnesota

A GPR survey of Interstate 94 at Reference Post (RP) 134 + 00 was performed on May 10, 2005 (Cao et al. 2007). The purpose of this survey was to investigate the potential of subsurface voids surrounding two steel-encased water and force mains that underlie the roadway at this location, at an approximate depth of 13 ft below surface. This potential manifests itself in a noticeable surface dip, especially in the EB driving lane. For the GPR survey of 30 ft length a 100 MHz Ground-Coupled antenna was used. Two surveys were taken, one in the eastbound passing lane and the other in the westbound passing lane at the location where three pipes have been. The GPR images indicated the potential presence of an underground void in the westbound driving lane and the eastbound passing lane at a depth ranging from 12.5 to 17 ft.

3.1.3 Detecting Voids Beneath Concrete Sections in UK

In terms of concrete investigation, GPR has an excellent reputation for being able to image voids beneath concrete sections (Cassidy et al. 2011). This kind of

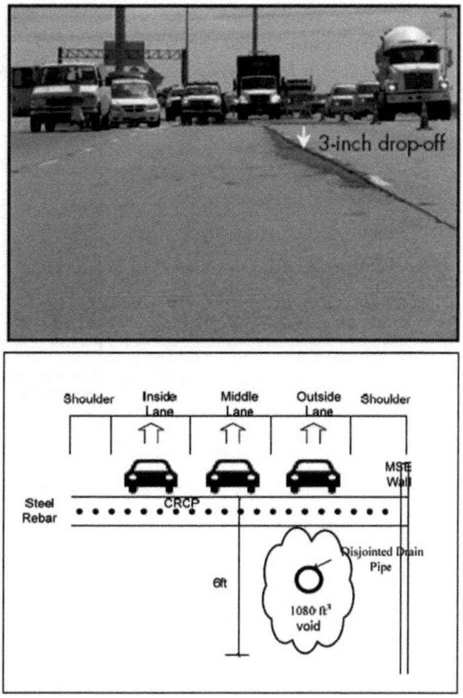

Fig. 5 Example of a drop-off of a continuously reinforced concrete pavement due to voids (TxDOT 2010)

application was performed in United Kingdom where a test slab was used. The test slab forms part of the NSGG (Near-Surface Geophysics Group of the Geological Society) shallow geophysics test facility at the University of Leicester. The focus of this research was the practical detection of sub-metre scale voids located under steel reinforced concrete sections in realistic survey conditions (a capped, relict mine shaft or vent). Figure 6 shows the general layout of the reinforced concrete test slab and images of data collection with the GPR.

A GPR unit with 450 and 900 MHz shielded antenna was used for all GPR surveys, in orthogonal grid, with each trace collected manually at defined intervals and processed using conventional methodologies. The average of thirty two trace stacks were recorded at each collection point with each manually triggered trace having a temporal increment of 0.1 ns and a total time window of 40 ns (approximately 2 m total depth of investigation). The choice of both 450 and 900 MHz antenna was based on the anticipated depth of investigation (<2 m), likely target size (a sub-metre isolated volume) and data collection time constraints. 1.2 GHz antenna surveys were initially considered, but the additional data collection time required for the more densely spaced data (0.1 m line spacing and 0.01 trace increments) was deemed prohibitive. The results of these surveys have shown that the selection of antenna frequency is important and that care must be taken with the mode and configuration of the survey geometries.

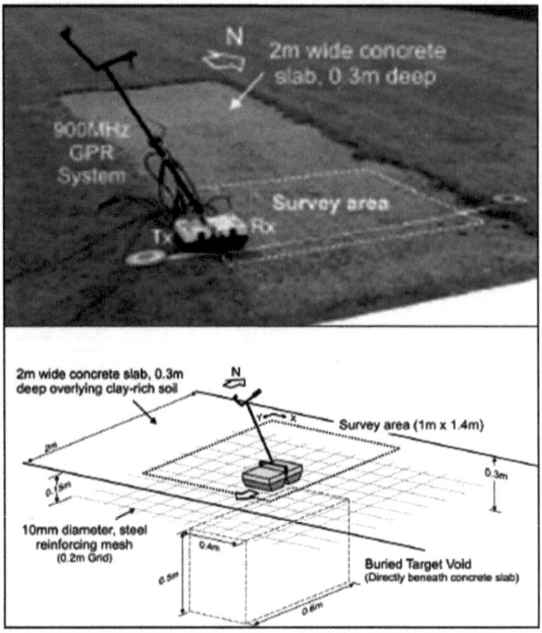

Fig. 6 Target void layout and data collection across the reinforced concrete test slab (Cassidy et al. 2011)

3.1.4 Detecting Voids in Malaysia

GPR surveys were executed in Shah Alam, Selangor, Malaysia with the aim of identifying a ground subsidence area and detecting voids. This region is located about 25 km west of the country's capital, Kuala Lumpur. GPR survey was conducted along the road of interest at main campus of public university in Shah Alam.

This study implemented a 250 MHz shielded antenna for data acquisition. There were total of 12 GPR survey lines conducted with different length as it is shown in Fig. 7. Survey line L1 and L2 were executed along brick pedestrian with total length of 16 and 33 m respectively. For survey line L3–L11, spacing between lines were 5 m with average length of 7 m for each lines. Orientations of these lines are across the road where cracks are obviously found. Line L12, with a length of 35 m, executed opposite L1 and L2. Site photos and sinkhole are also shown in Fig. 7.

From GPR profiles obtained, results show feature of soil subsidence observed at several depths. For L2, ground subsidence detected along the line with one suspected air-filled void identified at distance 31.3 m with depth of >2 m. For L3–L11, ground subsidence was also detected, resulting in cracks that were obvious on asphalt road. GPR profiles successfully detected areas of subsidence and void, generally with depth <5 m (Ismail et al. 2012).

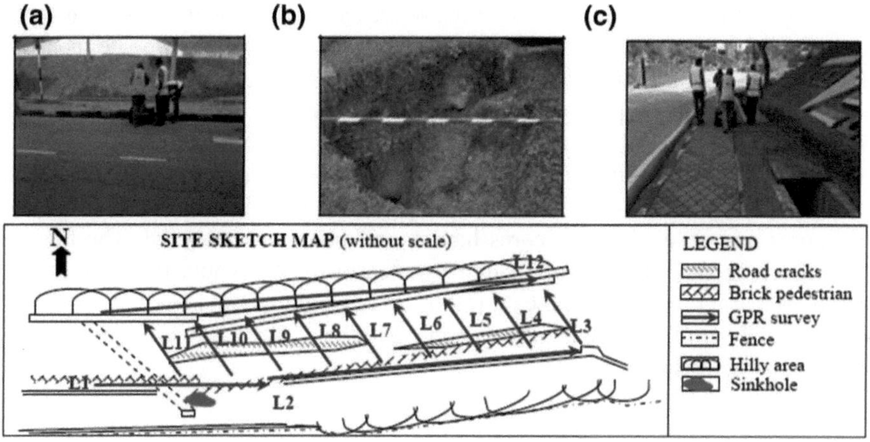

Fig. 7 Site photos **a** cracks on the road, **b** sinkhole with depth <7 m, **c** survey line L1 and L2 and site sketch map (Ismail and Saad 2012)

3.1.5 Estimation of Asphalt Air Void Content in Finland, Sweden and Norway

In 2011, Mara Nord, an international cooperative project financed by Interreg IV A Nord, has been initiated among Finland, Sweden and Norway. In this project, one goal was to produce common guidelines for the use of GPR in asphalt air voids content measurement that could be used as a reference in procurement processes in all three countries (Interreg IV A Nord Program 2011). In this framework, GPR technology was used to measure the dielectric value of the asphalt pavement, which was then used to calculate the air void content of the pavement. The method is suitable for measurement of air void content of new bituminous pavements only, regardless of the quality of the base course. It is suggested to be followed when doing survey design, data collection, analysis and reporting with a 2D GPR system prior to an asphalt air voids content measurement.

In pavement and top part of the pavement structure quality control surveys it is recommended that air coupled antennas be used especially when the interest is bituminous pavement and unbound base thickness and their quality. According to these guidelines, in normal cases, it is recommended that a 1 GHz antenna be used in asphalt air voids content surveys, but if the amount of new asphalt is 60 kg/m^2 or less then it is recommended that a 2 GHz antenna be used.

Concerning the number and location of survey lines Finland has required data collection only from the outer wheelpath and Swedish guidelines require the survey to be done from the outer wheelpath and between wheelpaths. Swedish guidelines can eliminate the effect of traffic compaction from the results. That is why these guidelines recommend that, if only one survey line is measured, it should be measured between the wheelpaths. However recent test results have shown that in addition to the two lines per lane it is also recommended that one line be measured along the pavement joint in lane centre (Interreg IV A Nord Program 2011).

When using air coupled antennas, after every survey session and before the GPR unit is switched off, a metal pulse reflection should be recorded. It is recommended that a metal reflection is also taken before the start of the measurement. Moreover, air voids content calculations using GPR technique requires calibration drill cores and at least 2 drill cores have to be taken and analyzed. The place of the drill cores should be selected first by measuring the whole road section and then by selecting places for the calibration cores from homogenous sections where dielectric value of the asphalt surface is close to the approximate average dielectric value of the new asphalt.

3.1.6 Detection of Air Voids During ICF Construction

GPR has been also used for detecting voids during the insulated concrete form construction (ICF) (Roger et al. 2011). Insulated concrete form construction consists of interlocking polystyrene forms and poured-in-place concrete. The forms are left in place after the concrete is poured to form an efficient insulation barrier. Consequently, it is difficult to ascertain the presence of flaws in the concrete, such as voids at the boundary between the form and the concrete, using traditional inspection methods without removing, and thus damaging the polystyrene forms.

In a research investigation, it was found that GPR utilizing a 1.6 GHz center-frequency antenna is useful for locating voids at the concrete-form boundary and voids buried in the concrete. Data were collected on the concrete over a number of days starting one day following pouring. The detection of voids buried 2.5–10.8 cm (1.0–4.3 in.) in the concrete was also achieved after 3 days of curing.

The detection of these voids could conceivably be performed during pouring to locate sections of concrete that require additional vibration. Alternatively, after the concrete is consolidated, GPR can be used as a quality control tool to assess the extent of voids flush to the back side of the form. The data collection methodology could consist of either a random set of vertical profile lines of data in which the percentage of voids detected per linear meter of data could be extrapolated to estimate the percentage of voids in the entire concrete structure. A second, more detailed, approach is to obtain data at a regular spacing and map the void extents in 3D coordinates so as to reconstruct the void surface area and provide an estimate of void depth (Roger et al. 2011).

3.1.7 Considerations

Several processing techniques, such as GA-based integrated strategy, have been proposed and assessed in (Benedetti et al. 2007) with unknown defects both in location and in size situated inside dielectric host mediums. Moreover, in

(Benedetti et al. 2010) a new approach based on the integration between a multi-scaling procedure and the level-set-based optimization has been proposed, aimed at the reconstruction of the shape of multiple and disconnected homogeneous scatters. Such an approach can be clearly applied to the problem of void detection, as well as to the retrieval of metallic objects, by exploiting measurements collected from GPR systems.

4 Conclusions

GPR has been used successfully with other technologies to identify and locate utilities—often previously unknown—prior to excavation, coring, or boring activities. Yet its full potential for augmenting subsurface utility-mapping has not been adequately researched, demonstrated, or determined. Part of this limitation has been due to the overwhelming use of single-channel GPR systems, as well as highly variable training and expertise in its use on an ad hoc basis.

Over the past several years, significant technological progress has been made both in the hardware and software imaging systems dedicated to simplifying utility-detection, particularly multi-channel 3D imaging and mapping systems integrating GPR with complementary technologies and survey-grade GPS. These systems are capable of collecting larger areas of data more efficiently and more accurately demonstrating the results. It is expected that these collective technological changes will benefit the inspection procedures for effective GPR sensing and mapping of underground utilities and voids.

Acknowledgments This work is a contribution to COST Action TU1208 "Civil Engineering Applications of Ground Penetrating Radar".

References

Annan, A.P.: Ground Penentrating Radar Principles, Procedures and Applications: Sensors and Software Incorporated (2003)

Benedetti, M., Franceschini, G., Azaro, R., Massa, A.: A numerical assessment of the reconstruction effectiveness of the integrated GA-based multicrack strategy. IEEE Antennas Wireless Propag. Lett. **6**, 271–274 (2007)

Benedetti, M., Lesselier, D., Lambert, M., Massa, A.: Multiple shapes reconstruction by means of multi-region level sets. IEEE Trans. Geosci. Remote Sens. **48**(5), 2330–2342 (2010)

Bernold, L., Venkatesan, L., Suvarna, S.: A multi-sensory approach to 3-D mapping of underground utilities. In: Proceedings of the 19th International Symposium on Automation and Robotics in Construction (ISARC), Washington, USA, 525–530 (2002)

Cao, Y., Dai, S., Labuz, J., Pantelis, J.: Implementation of ground penetrating radar. Minnesota Department of Transportation, Research Services Section, USA (2007)

Cassidy, N.J., Eddies, R., Dods, S.: Void detection beneath reinforced concrete sections: The practical application of ground-penetrating radar and ultrasonic techniques. J. Appl. Geophys. **74**, 263–276 (2011)

Cheng, Nga-Fong, Conrad Tang, Hong-Wai, Chan, Ching-To : Identification and positioning of underground utilities using ground penetrating radar (GPR)", Sustain. Environ. Res. **23**(2), 141–152 (2013)

Demirci, S., Yigit, E., Ozdemir, C.: Detection of movement and impedance changes behind surfaces using ground penetrating radar. Prog. Electromagnet. Res. Symp. (PIERS) **7**(1) (2011)

Eide, E.S., Hjelmstad, J.F.: 3D Utility mapping using electronically scanned antenna array. In: 9th International Conference on Ground Penetrating Radar, Santa Barbara, CA, 29 April–2 May 2002

Eyuboglu, S., Mahdi, H., Al-Shukri, H.J.: Detection of water leaks using ground penetrating radar. In: Proceedings of the Third International Conference on Applied Geophysics, Orlando-FL, 8–12 Dec 2003

French Standard NF S70-003-2 (2012): Travaux à proximité des réseaux. Partie 2: techniques de détection sans fouille/Works in the neighborhood of utilities. Part 2: Trenchless techniques of detection

Grivas, J.D.A.: Applications of ground penetrating radar for highway pavements, NYSERDA TIRC Project C-04-04. New York State Department of Transportation (2006)

Holt, F.B., Eales, J.W.: Nondestructive evaluation of pavements. Concr. Int. **9**, 41–45 (1997)

Interreg IV A Nord Program: Recommendations for guidelines for the use of GPR in asphalt air voids content measurement. Rovaniemi (2011)

Ismail, N.A., Saad, R., Muztaza, N.M., Ali, N. : Predictive mapping of underground utilities using ground penetrating radar. Caspian J. Appl. Sci. Res. 2, 104–108 (2013) (AICCE'12 & GIZ'12)

Ismail, NA., Saad, R.: A case study on ground subsidence using ground penetrating radar. In: International Conference on Environmental, Biomedical and Biotechnology, IPCBEE, vol. 41. IACSIT Press, Singapore (2012)

Jamil, H., Nomanbhoy, Z., Mohd Yusoff, M.Y.: "Underground utility mapping and its challenges in Malaysia", TS05 J—Mining and underground Engineering Surveying II, 5636, FIG Working Week 2012, Knowing to manage the territory, protect the environment, evaluate the cultural heritage, Rome, Italy, 6–10 May 2012

Jeng, Yih, Chen, Chih-Sung: Subsurface GPR imaging of a potential collapse area in urban environments. Eng. Geol. **147–148**, 57–67 (2012)

Krysiński, L., Sudyka, J.: "Abilities of Georadar 3D in Construction Identification of Roads" (in Polish), Drogi - Budownictwo infrastrukturalne, 4, 1–2 (styczeń/luty), 65–70. (Original title, Możliwości Georadaru 3D w rozpoznaniu konstrukcji dróg") (2012)

Lester, J., Bernold, L.E.: Innovative process to characterize buried utilities using ground penetrating radar. Autom. Constr. 16, 546–555 (2007)

Lizzi, L., Viani, F., Rocca, P., Oliveri, G., Benedetti, M., Massa, A.: Three-dimensional real-time localization of subsurface objects—From theory to experimental validation. In: IEEE International Geoscience and Remote Sensing Symposium, vol. 2, pp. II-121–II-124, 12–17 July 2009

Manica, L., Rocca, P., Salucci, M., Carlin, M., Massa, A.: Scattering data inversion through interval analysis under Rytov approximation. In: 7th European Conference on Antennas Propag. (EuCAP 2013), Gothenburg, Sweden, 8–12 Apr 2013

Massa, A., Boni, A., Donelli, M.: A classification approach based on SVM for electromagnetic subsurface sensing. IEEE Trans. Geosci. Remote Sens. **43**(9), 2084–2093 (2005)

Morey, R.M.: Ground Penetrating Radar for Evaluating Subsurface Conditions for Transportation Facilities, Synth. of Highway Practice 255, NCHRP. National Academy Press, Washington, D.C (1998)

NCHRP: Ground penetrating radar for evaluating subsurface conditions for transportation facilities. National cooperative highway research program, Synthesis of Highway practice 255. Transportation Research Board, Washington D.C. (1998)

Nissen, J., Johansson, B., Wolf, M.J., Skoog, L.: Ground Penetrating Radar—A Ground Investigation Method Applied to Utility Locating in No-Dig Technologies, pp. 1–6. Mala Geoscience Raycon, Stockholm (2001)

Poli, L., Oliveri, G., Rocca, P., Massa, A.: Bayesian compressive sensing approaches for the reconstruction of two-dimensional sparse scatterers under TE illumination. IEEE Trans. Geosci. Remote Sens. **51**(5), 2920–2936 (2013)

Poli, L., Oliveri, G., Massa, A.: Imaging sparse metallic cylinders through a local shape function bayesian compressive sensing approach. J. Opt. Soc. Am. A **30**(6), 1261–1272 (2013)

Porsani, L., Ruy, Y.B., Ramos, F.P., Yamanouth, G.R.B.: GPR applied to mapping utilities along the route of the Line 4 (yellow) subway tunnel construction in São Paulo City, Brazil. J. Appl. Geophys. **80**, 25–31 (2012)

Roger, R., Corcoran, K., Arvanitis, M., Schutz, A.: Insulated concrete form void detection using ground penetrating radar. In: Proceedings of Progress in Electromagnetics Research Symposium (PIERS), Marrakesh, 20–23 Mar 2011

Salucci, M., Sartori, D., Anselmi, N., Randazzo, A., Oliveri, G., Massa, A.: Imaging buried objects within the second-order born approximation through a multiresolution regularized inexact-newton method. In: International Symposium on Electromagnetic Theory (EMTS), Hiroshima, Japan, 20–24 May 2013 (invited)

TxDOT: Using Ground-Penetrating Radar (GPR) techniques to detect concealed subsurface voids. Texas Department of Transportation, USA (2010)

van Schoor, M., Colvin, C.: Tree root mapping with ground penetrating radar. In: 11th SAGA Biennial Technical Meeting and Exhibition, Swaziland, 16–18 Sep 2009

Effective GPR Inspection Procedures for Construction Materials and Structures

Lech Krysiński and Johannes Hugenschmidt

Abstract This paper is a review of methods related to assessment of construction details and material properties using of GPR. The focus is on recent research activities of the Project "Innovative procedures for effective GPR inspection of construction materials and structures" (project 2.4) in COST Action TU1208. The electromagnetic properties of investigated media are interesting because they reflect physical features of the materials (e.g. their composition), enabling a non-invasive inspection of their condition. Moreover, the assessment of electromagnetic properties (e.g. wave velocity) is an inherent part of any GPR structural study necessary for correct depth determination or amplitude interpretation. As a result of the review major directions of research are highlighted and some benefits and limits of different approaches are described.

1 Introduction

GPR datasets are characterised by the electromagnetic properties of the media and these properties play a crucial role in the further interpretation of the data. Structural interpretation of GPR datasets requires the wave velocity of the medium (Tosti et al. 2013a; Benedetto and Pensa 2007; Hugenschmidt 2002). On the other hand, the electromagnetic properties are closely related to other physical features of the investigated materials, for example to their composition and small-scale

L. Krysiński (✉)
Road and Bridge Research Institute,
ul. Instytutowa 1, 03-302 Warsaw, Poland
e-mail: lkrysinski@ibdim.edu.pl

J. Hugenschmidt
Rapperswil University of Applied Science,
Oberseestrasse 10, 8640 Rapperswil, Switzerland
e-mail: Johannes.hugenschmidt@hsr.ch

A. Benedetto and L. Pajewski (eds.), *Civil Engineering Applications of Ground Penetrating Radar*, Springer Transactions in Civil and Environmental Engineering,
DOI 10.1007/978-3-319-04813-0_6

structure (heterogeneous distribution of the components). Thus, the GPR data and the results of their interpretation provide, in many cases, important information on the structural details and the material properties of the inspected object. Data acquisition, processing and interpretation require adopted strategies depending on the type of object, the materials that are involved and the questions that have to be answered. In addition, advanced techniques, such as inversion may be considered. The choice of strategy requires experience and a thorough knowledge of the method. The investigation of particular types of civil engineering (CE) structures is associated with specific problems and different techniques were developed for an effective inspection and assessment of structural and material properties. The further development of these procedures and strategies has a large importance for the practical application of the method.

The issues related to GPR inspections of material properties and structures have several levels and relate to different types of research activities:

- Methodologies for efficient data collection, processing and visualization for different kinds of CE objects,
- A possible use of advanced techniques such as inversion,
- Methodologies for interpretation of GPR data aiming at an assessment of electromagnetic material properties (reflection amplitude determination, velocity analysis, travel time modelling, migration, hyperbola fitting, etc.),
- Methodologies for direct investigation of the electromagnetic properties of the materials,
- Combination with other methods,
- Investigations of the electromagnetic properties of specific categories of materials and their relation with other physical features.

The following sections present the basic notions and facts concerning real construction materials, the major research directions related to the most common construction materials and characterisation of the current research efforts in material and structural assessment of concrete constructions and bituminous pavements.

2 Electromagnetic Properties of Construction Materials

Most of phenomena observed in GPR practice can be described by the use of complex relative electric permittivity $\varepsilon_r^* = \varepsilon_r' + \mathrm{i} \cdot \varepsilon_r''$, characterising interaction of the homogeneous medium with electromagnetic field. The imaginary part of the permittivity $\varepsilon_r''(\omega)$ is responsible for wave absorption and it is closely related to frequency dependence of the real part $\varepsilon_r'(\omega)$, described by the Kramers-Kronig relations (Jackson 1975). The electrical conductivity of the medium, influenced for example by salt water or porous rocks filled with electrolytes, is a phenomenon belonging to the category described by the imaginary part of permittivity, but absorption is not necessarily associated with explicit conductivity (e.g. dry

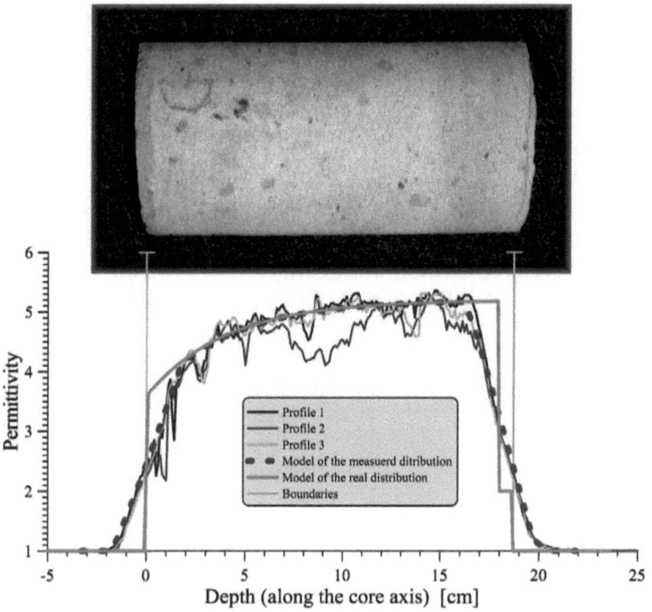

Fig. 1 Characteristic features of the permittivity distribution in cement stabilised sand determined by the use of capacimetric electrode (Percometer^{TM}) at about 50 MHz (after Krysiński 2013)

concrete, clay minerals) (Jackson 1975; Cassidy 2009). Secondary electromagnetic properties used for material characterization, such as wave velocity, reflection coefficients are determined by ε_r* (and permeability μ_r*).[1]

A special category of electromagnetic phenomena (reflection, refraction, diffraction etc.) is related to interaction of the GPR signal with inhomogeneities in media. These phenomena are referred to as scattering. These effects are particularly interesting when the characteristic dimensions of permittivity anomalies are comparable to the local wavelength of the GPR signal to cause significant response (scattering), but they are not strong enough for the generation of large scale manifestations on echograms which could be interpreted as a large-scale medium structure. Such effects are typical for real media and have important practical consequences which can be illustrated by the following examples.

The continuous depth stratification is an example for heterogeneity (Fig. 1). In the presented case, the permittivity in a core taken from a subbase slab made of cement stabilised sand increases significantly with increasing depth. It can be concluded that the slab has significantly different permittivity values at the top and at the bottom.

[1] In the case of materials containing significant amount of ferromagnetic minerals the magnetic permeability μ_r* can play important role (Cassidy 2009) in formation of GPR response, but it is very rare exception among construction materials. Thus usually the relative permeability is assumed to be equal to one.

Moreover, in this case the reflection amplitude (peak-to-peak amplitude) measured at the top of the plate (where gradient occurs) is frequency-dependent (Fig. 2) although the permittivity of the mediums is not frequency-dependent. In the recorded (in our case modelled) GPR data, the stratification is manifested only as a change in the shape of the surface reflection.

The second important type of heterogeneity is related to media of granular structure such as stone-asphalt (aggregate-binder) mixtures (Fig. 3). In the presented example the smoothed permittivity observed in centimetre space scales ranges from 6 to 8. This distribution illustrates the 'problem of scaling' in definition and determination physical properties (and their spatial distribution) of real

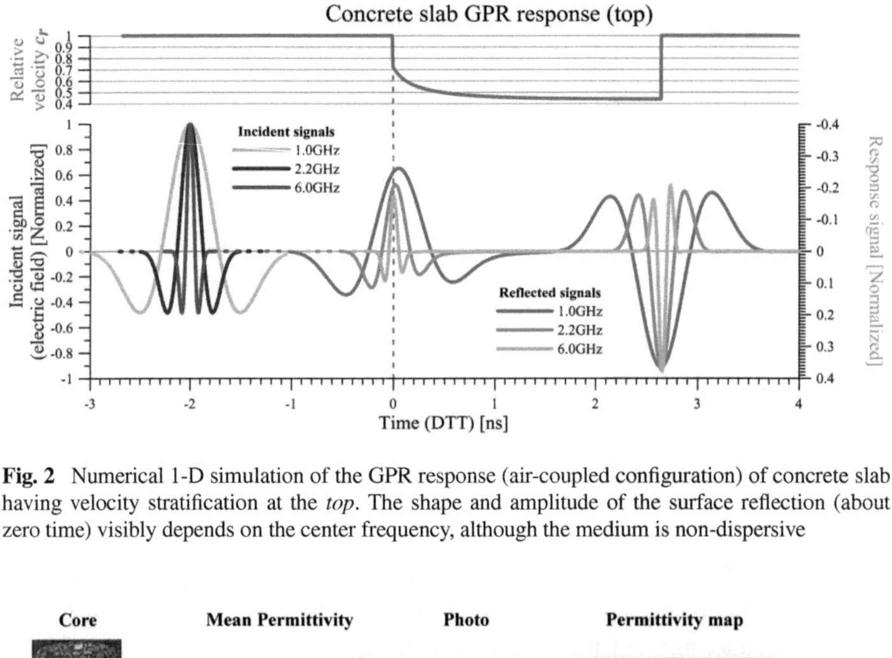

Fig. 2 Numerical 1-D simulation of the GPR response (air-coupled configuration) of concrete slab having velocity stratification at the *top*. The shape and amplitude of the surface reflection (about zero time) visibly depends on the center frequency, although the medium is non-dispersive

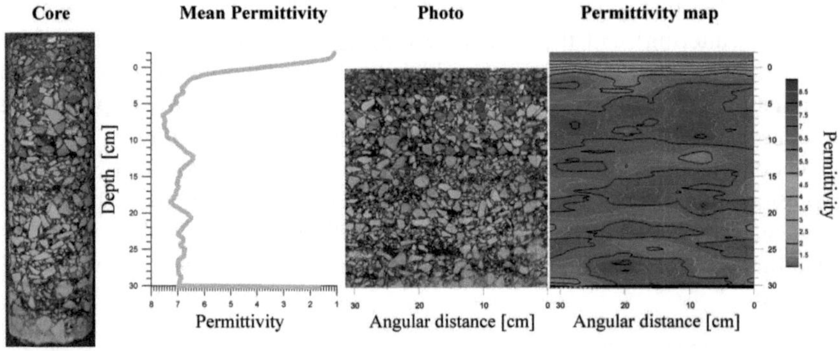

Fig. 3 Map of the permittivity distribution on the lateral surface of the drilling core (from the *right*), panoramic photo of the lateral surface and the depth distribution of the level average permittivity and photo of the core (Krysiński and Sudyka 2012b)

media. Any determination is dependent on resolution (smoothing scale) and results have large variation related to local inhomogeneity.

Electromagnetic properties are well determined material characteristics for homogeneous media only. But for inhomogeneous media the primary meaning of these notions falls down and these terms are used in practice as some provisional characterisation of the medium in the frame of the given method. Granular media also exhibit a special type of signal attenuation due to lateral scattering (non-dissipative attenuation). The absorption related to the imaginary part of permittivity should be distinguished from wave attenuation related to scattering. In practice, it is usually difficult to make this distinction.

3 Major Directions of the Research Related to Material Properties

Among research efforts, several directions related to different material types can be distinguished. Nowadays, concrete plays an important role among construction materials. Bituminous mixtures are very important in the road construction industry. We should note also shortly "improved (reworked) soils" which are treated also as construction materials, playing a fundamental role in engineering. These three examples will serve for a presentation (in following sections) of some major issues related to the GPR assessment of structural and material properties of object relevant in civil engineering. Obviously, further materials can be of interest but are not considered here. Despite of their shared methodological background, the inspection procedures have large technical differences with respect to data acquisition, properties being investigated, processing, imaging, and interpretation methods for different construction and material types.

Soils and subgrade materials have relatively long tradition of GPR investigations (Gołębiowski 2012). Field investigations are focused on the determination of water content (Ortuani et al. 2013; Benedetto et al. 2009, 2012, 2013a) and composition. The content of clay minerals (Benedetto and Tosti 2013a; Tosti et al. 2013b) responsible for important bearing properties of subbase (Benedetto and Tosti 2013b; Benedetto et al. 2013b) is of special interest. Among laboratory methods, one can find determinations of frequency-dependent complex permittivity of soils (Patriarca et al. 2013) and specific methods for aggregate permittivity determination based on 'frequency peak shift' (Benedetto and Tosti 2013a; Tosti et al. 2013b).

Investigations of concrete are focused on moisture content and composition manifested mostly in permittivity value, and on electrolytes (e.g. chlorides) content increasing the absorption. Chlorides are important because they are causing corrosion of the steel bars present in reinforced concrete structures. A simple approach (Hugenschmidt and Loser 2008) is based on the assumption that the combination of moisture and chlorides increases the electric conductivity and thus, the damping of the GPR signal. In a laboratory experiment nine specimens were built. The size was 0.90 m × 0.75 m × 0.08 m and the bottom of the specimens was a metal plate. Chlorides were added directly to the mixing water. The amount of chlorides added corresponds to 0, 0.4 and 1 % chloride content with respect to the cement mass. After the production, the specimens were stored at different relative humidities (35, 70, 90 % r. h.) for 97 days. Data were acquired using horn antennas (Fig. 4, left). The amplitudes of the recorded GPR signals were analysed. It was shown that the reflection amplitudes at the concrete surface increased both with increasing moisture and chlorides. In contrast, the amplitudes at the bottom of the specimen (at the metal plate) decreased with increasing moisture and chloride content. The quotient of the amplitudes (Fig. 4, right) at the concrete surface and the bottom of the specimen showed a clear relationship between moisture and chloride content on the one hand and the computed quotients on the other hand.

Several other approaches have been suggested by different authors, such as the analysis of the direct wave amplitude (Laurens et al. 2002; Sbartaï et al. 2006a) the analysis of amplitude, relative attenuation and spectrum of the slab bottom reflection (Laurens et al. 2005; Klysz and Balayssac 2007) full waveform inversion (Kalogeropoulos et al. 2013) and numerical modelling of the signal interaction with slab simulating the real geometry of the measuring system (Klysz et al. 2006; Ferrieres et al. 2009). These efforts allow for non-invasive determination of the slab permittivity and absorption in adequate (Ferrieres et al. 2009) and efficient way (Sbartaï et al. 2009) and some of them can be used in field investigations (Sbartaï et al. 2006b). Laboratory investigations of concrete properties focus on the determination of the complex permittivity as a function of frequency (Ihamouten et al. 2011, 2012), dependence of the permittivity on composition (Villain et al. 2010; Dérobert et al. 2009) and considerations related to perspectives of complex permittivity determination on the base of refracted wave analysis (Ihamouten et al. 2012). Investigations of concrete properties are of great interest in the scientific community and are likely to have an increasing significance for the construction industry.

Investigations of asphalt mixtures are focussed on composition, estimations of void content (Sebesta and Sculion 2002; Plati et al. 2013) and moisture

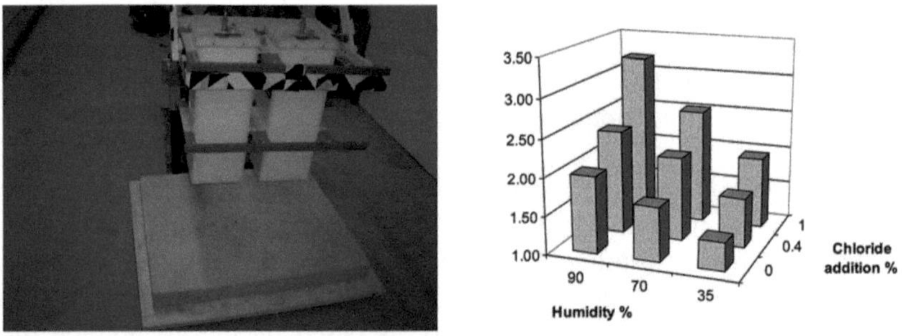

Fig. 4 Data acquisition with GPR antennas above specimen (*left*) and quotient of the concrete surface reflection and the metal plate reflection amplitudes (*right*)

content. Permittivity values are dominated by the properties of the stone fraction of mixtures but the degree of compaction may also have some significance. The laboratory determination of permittivity and its relation with composition is very difficult due to the coarse grain structure of this material. Thus, the investigations of fundamental properties are rare (Roimela 1999; Liu and Guo 2002; Krysiński and Sudyka 2012b; Fauchard et al. 2013) and some predictive relations (Cassidy 2009) are frequently used (Sebesta and Sculion 2002) instead of reliable estimations. The strongly heterogeneous structure of asphalt mixture (Krysiński and Sudyka 2012b) and its interaction with the high frequency GPR signal are likely to be an important field for future research.

For alternative determinations of permittivity and absorption some auxiliary non-GPR techniques are being used (capacimetry, resistivity measurements and imaging, TDR etc.). The combination of different methods can be very useful (Breysse 2012). Combinations can verify results obtained with a single method and can be used for limiting the width of ranges for results. In addition, combinations can be useful for the development of innovative methods and for the improvement of existing approaches. Comparative studies help also to identify problems which are not visible in the frame of one approach.

4 Advanced Data Processing and Inversion

In the following section, an example of advanced data processing and inversion is presented for a reinforced concrete structure. Data were acquired on a retaining wall using two antennas with a nominal center frequency of 1.5 GHz (Hugenschmidt et al. 2010). Both antennas were placed in an antenna box with orthogonal orientation. Data were recorded along vertical lines. The distance between the lines was 0.01 m. Data were processed in three different ways. First, data were processed in 2D using the dataset from one antenna only. In the second approach, the datasets from the two antennas were processed separately in 3D and merged afterwards using a fusion algorithm. The third approach used an inversion algorithm developed by Solimene et al. (2007). The results of the three different approaches are presented in Fig. 5. The vertical extent of the Fig. 5b, c is 2.6, 2.4 m for Fig. 5a. The horizontal extent is 1.40 m. The first approach (Fig. 5a) provides a mapping of the horizontal rebars. Vertical bars cannot be mapped properly because of the polarization (horizontal E-field) of the single antenna and the 2D data processing. The result of the 3D processing followed by the fusion of the two datasets (Fig. 5b) enables a complete mapping of both horizontal and vertical bars. In addition, increased reflection amplitudes can be spotted. The increased reflection amplitudes near the bottom and the top and left of the center of the time slice have a vertical orientation. The inverted result using both datasets (Fig. 5c) enables a complete mapping of both horizontal and vertical bars. In addition, the reason for the increased reflection amplitudes that are visible at the top and bottom of the Fig. 5b can be identified as caused by overlapping bars.

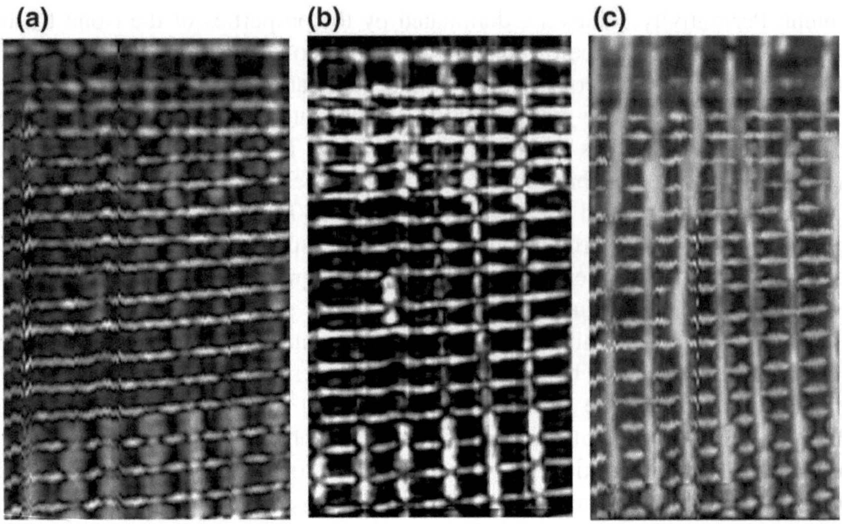

Fig. 5 **a** Time-slice, 1.45–1.95 ns, conventional 2D processing using one antenna only (Hugenschmidt et al. 2010). **b** Time-slice, 1.45–2.08 ns, 3D processing followed by data fusion. **c** Depth-slice, depth = 0.06 m, pre-processing followed by inversion using both datasets

5 Challenges in GPR Studies of Bituminous Mixtures and Pavements

Bituminous pavements have a relatively simple and flat planar structure. Usually horizontal layering is assumed in depth and thickness assessment (Plati and Loizos 2012). Despite this, there are numerous challenges related to the interpretation of the GPR data.

For inspections of large sections of road pavements high-frequency air-coupled antennas mounted on vehicles are used in reflection, zero-offset configurations. This equipment provides detailed (high resolution) insight into pavement structures (Hugenschmidt et al. 1998) based on a large amount of reflection data i.e. amplitude of surface reflection as well as arrival times and amplitudes of interlayer reflections if significant permittivity contrasts occur between contacting layers. These data are useful for non-invasive methods of layer thickness determination, but they are also an important source of information as to the material properties of construction materials (Roddis et al. 1992). The details of permittivity distribution in the vicinity of interlayer contact are also important for studies of the formation of the reflection signals (Krysiński and Sudyka 2012a) and conclusions have some consequences for precise localisation of the contact depth.

The field data are usually affected by significant EM noise. Thus, the processed data (e.g. bandpass filter, moving average, gain correction) are interpreted as a manifestation of the pavement permittivity. Some additional perturbations can also be caused by surfaces roughness in decimetre length scales or other effects, such as inhomogeneity of pavement material. The largest technical problems are related to instability of the GPR systems (Hugenschmidt 2000; Poikajärvi et al. 2012). It was observed that the GPR signal drifts both in time and amplitude even after a more than one hour after switching on the system (warm-up period). This phenomenon affects the resulting data changing amplitudes by up to 20 %. Thus the local relative changes of amplitudes are more reliable than its absolute values. Frequent calibrations (metal plate reference) and controlling repetitions of scanning are necessary to secure data quality, particularly if the reflection amplitudes are used for the calibration of signal velocities (Roddis et al. 1992). In this case, the interpretation of the surface reflection from the top layer of the asphalt layer package results in largest precision in permittivity and velocity determination. In the case of the deeper reflections, the transitions through upper layers must be accounted for (Plati and Loizos 2012; Tahmoressi et al. 1999) and the precision of the permittivity and velocity determination deteriorates. For newly constructed pavements, the tendency toward almost identical permittivities of subsequent layers is frequently observed in practice. In the case of the roads of long history a large variability of materials applied in layers occurs as a result of numerous repairs. In extreme situations there is not possible to distinguish precisely single horizons in their dense sequence (Hugenschmidt 1998). An example of a dataset with a length of 400 m from a Swiss pass road is presented in Fig. 6. Although the presented section has a length of only 400 m, at least 5 different sections (marked with yellow and red bars) can be distinguished. The problem of permittivity distribution determination in such densely layered system is a large challenge in practice.

5.1 Compositional Interpretations of Permittivity Value

The aggregates constitute usually more than 90 % of the stone asphalt mixture mass. Thus, the permittivity of the stone fraction has a decisive influence on effective

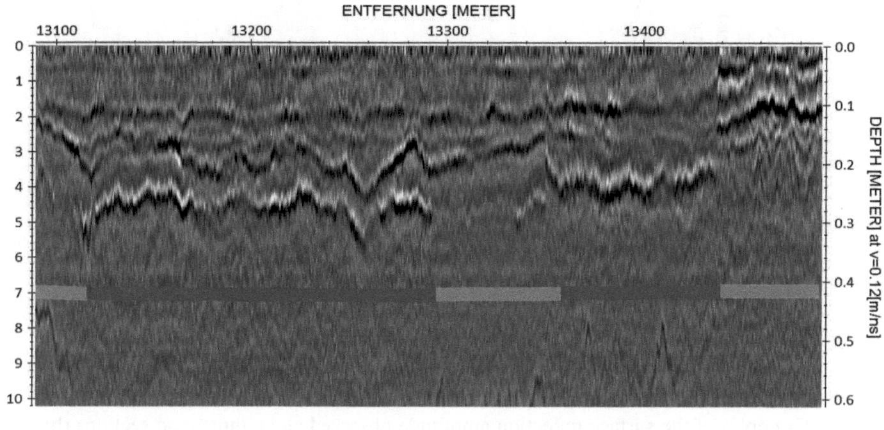

Fig. 6 Example from a dataset from a Swiss pass road, length = 400 m

permittivity of the mixture. Extensive surface reflection amplitude observations of bituminous pavements (Fig. 7) show that the range of its values corresponds well to the electrical properties of regional rocks. The change of lithology is usually

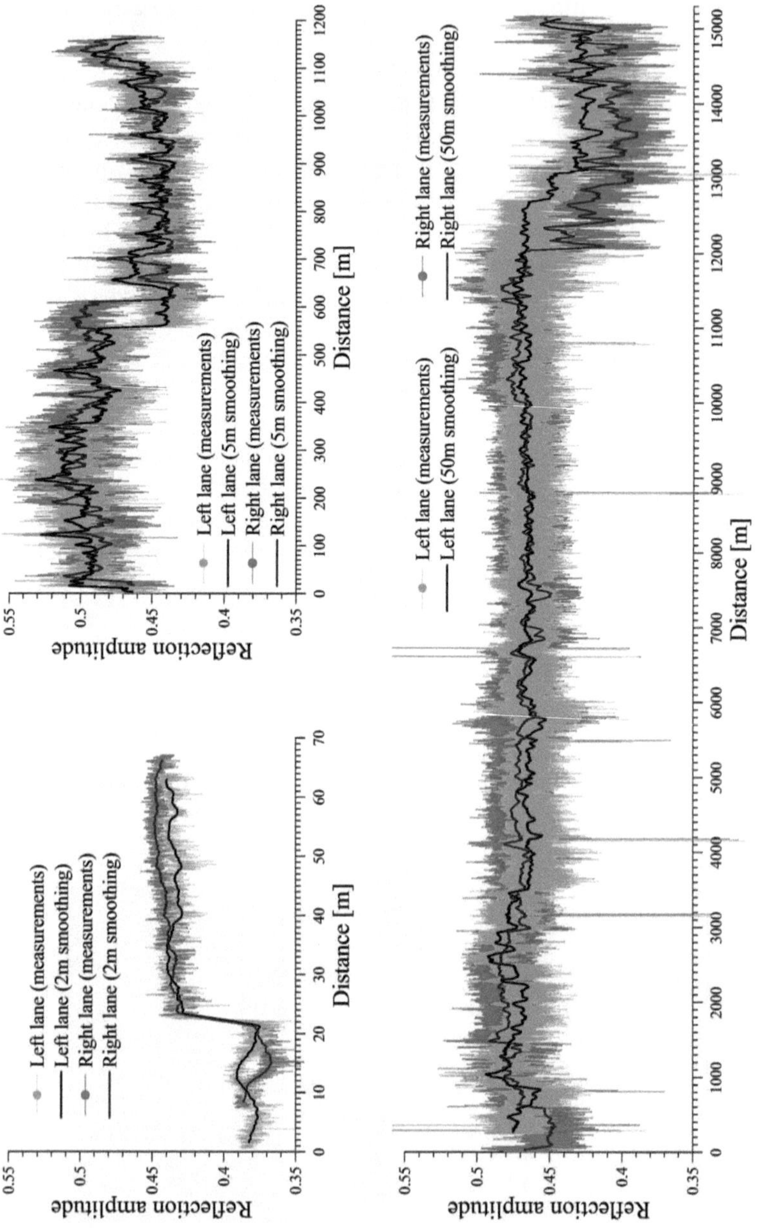

Fig. 7 Examples of the surface reflection amplitude observed along three road sections (bituminous pavements, 2.2 GHz)

associated with a change of reflection amplitude. The extreme values can be associated with some specific lithological types. The quantitative material interpretations of the reflection amplitude values need a database of characteristic permittivity values for regional rock materials (Ulaby et al. 1990; Fauchard et al. 2013). Some approximate values of permittivity for asphalt binders are also useful. Then the effective permittivity can be predicted by the use of mixing formulas. Some additional decrease of effective values can be related to unknown void content, but the variable lithology of stone fractions seems to be the major reason of local variations of the reflection amplitude (Fig. 7).

The method has some use as an auxiliary control tool, especially interesting in the case of airport pavements, where material specifications are rigorous. Some provisional assessment of the stone fraction type can be made by comparison with the table of characteristic values of regional rocks, but semi-quantitative discussion should use the recipe of the mixture assumed for the construction and mixing formulas for effective permittivity.

5.2 GPR Assessment of the Void Content

The compaction of the stone-asphalt mixture is one of the critical factors for the durability of bituminous pavements. Too large void content causes fast material deterioration. Thus, the possibility of void content estimation based on reflection amplitudes (Plati and Loizos 2013) is a very interesting area for material diagnostics. The permittivity of the asphalt mixture has an obvious relation with the void content. However, the variable mineral composition of the stone fraction has potentially much larger influence on the permittivity value. Thus, the variations of permittivity related to this compositional variability seem to be the major reason for the masking of the effect caused by the void content. The current diagnostic criteria put attention on distinct, local (several meters long) and significant anomalies of decreased permittivity value (more than 0.5 smaller than in neighbourhood, Saarenketo 2009; Tahmoressi et al. 1999). This form of anomalies can be associated with problems during paving (e.g. decreased temperature of the mixture).

The identification of the physical reasons of permittivity variations, the reduction of the noise level and the instability of the GPR system have to be addressed for a successful application of this technique. Another problem occurs in direct laboratory determination of the mixture permittivity dependence on void content. The inhomogeneity of the stone-asphalt mixtures (Fig. 3) and mineral variability of the stone fraction seem to be the main reasons for these difficulties. The results described in the literature show an extremely large variety (Roimela 1999; Liu and Guo 2002; Saarenketo 2009; Poikajärvi et al. 2012) and they differ when compared to results of combined laboratory-field approaches (Tahmoressi et al. 1999; Al-Qadi et al. 2011) and to successful laboratory tests well accordant with mixing formulas (Fauchard et al. 2013).

5.3 Characterisation of the Interlayer Interface Structure

The detection of internal defects in asphalt pavements is part of the characterisation of the construction. Insufficient or lacking bonding at interlayer interfaces of the asphalt pavement is an important problem in pavement technology. The predictions based on models of the mechanical behaviour of such system suggest that a lack of bonding leads to a significant decrease (several times) of the pavement durability (Sudyka et al. 2011). On the other hand it is very difficult to assess the actual decrease of the durability of real constructions. In some cases (e.g. grains interlocking at the interface) the loss of durability could not take place even if there is no bonding.

The GPR technique has some abilities in assessment of the interlayer connection problems. There is one important terminological difference between the two disciplines: in pavement technology interlayer connection refers to mechanical bonding, while in GRP technique it refers to permittivity contrasts at the contact (i.e. contrast between layers and thin contrasting lamina at the contact, Krysiński and Sudyka 2012a). Thus the result of GPR interpretations may have ambiguous material and mechanical interpretations. Delamination without gaps may not be visible in GPR data. Delaminations with air-filled gaps could be mistaken as normal interfaces with generously applied adhesives (Sudyka and Krysiński 2011). Delaminations with gaps filled with water or some high permittivity networks present at interface, produce strong positive double reflections well visible in GPR datasets. Although, the detection of delaminations or assessment of the interlayer connection type is not a well posed practical task, this kind of analysis usually provides interesting conjectures as to the properties of the construction. From the practical point of view, the development of diagnostic criteria is one of the major necessities in this area. In this particular case, these criteria have form of catalogue describing what kind of material phenomena at the interface (including some typical technical solution) can be expected for the given reflection type.

Delamination is only one example of different interlayer connection problems. The investigations of the physical phenomena occurring in the vicinity of the interface and its influence on the GPR signal has also more basic significance for GPR data interpretation. The interlayer boundaries in the bituminous pavement frequently are not thin laminas associated with simple single or double reflection in high-resolution echograms. But they are usually thicker, transitional zones manifested by elongated and deformed response signals. Such blurred zone would correspond to: undulated boundary surface, the belt where interlocking of the aggregate grains of the two contacting layers occurs, zone of changing porosity (e.g. insufficient compaction) or zone of intense material degradation, e.g. aggregate crumbling. These gradual zones cause significant changes in the resulting reflected signal shape and amplitude. Numerical simulations show that significant efficiency in backscattering has only these parts of the transitional zones, where the strongest velocity gradients occur. This effect can lead to improper interpretation of the boundary type and attribution of inadequate average velocity contrast between the two layers e.g. as a result of insufficient compaction in the base of the upper layer. Thus, the structure of the interlayer connection and the shape of its GPR response (Krysiński and Sudyka 2012a) have also some consequences for the boundary depth estimation when high precision is needed. Very wide zones of a smooth velocity change have weak backward scattering as a result of destructive interference, which leads to the disappearance of reflection signals.

5.4 Crack Detection and Characterisation

Cracks are common defects of the bituminous pavement and their detection and characterisation belongs to basic characterisation of the material quality. The detail search for GPR detection criteria shows (Krysiński and Sudyka 2013) that the cracks having clear manifestations on echogram are usually huge, well developed structures like several centimetre wide crevices filled with foreign material or widespread zones of material degradation or lithological changes. The initial, unopened cracks like new ones are not visible using GPR and the use of higher frequencies shouldn't make them visible due to masking signal generated in a strongly inhomogeneous, granulated medium like the asphalt mixture. Some parts of widespread cracking structures can have blurred character with vague boundaries. It was noted that the use of lower frequencies allow a better detection of large features of this structure shape, while higher frequencies can outline details. Therefore, using different frequencies (covering several octaves of the electromagnetic spectrum) is very helpful in crack diagnostics. Some other GPR manifestations are associated with dense cracks systems (e.g. alligator cracks) representing strong material degradation. Although the single crevices cannot be distinguished on echogram, the large amount of foreign material results in strong inhomogeneity and large change of the material effective permittivity.

GPR technique is useful in identification and localization of hidden (but well developed) cracks e.g. covered by newer layer, in determination of some geometrical features of cracks like depth range, width or shape, and sometimes it allows commenting on the reasons and course of cracking process. The method is more effective when a long term GPR observation (with constant equipment configuration) is performed; moreover, the availability of documentation describing the original state of the construction is very helpful. It should be noted that the pseudo 3D approaches with appropriate imaging (Benedetto 2013) has particular efficiency in cracks detection and studies with respect to possibility of tracing the lateral continuation of their manifestation.

6 Conclusions

The review of the research efforts related to construction materials GPR testing shows the large importance of comparative studies of the different methods. These comparisons are necessary for ability assessment and verification of these methods. This diagnostic discipline is an interesting and fruitful area for numerical modelling commenting on measuring procedures (with regarding the real geometry of the measuring systems). Investigations of heterogeneous materials and their GPR response constitute a large and relatively new research area.

Acknowledgments The authors acknowledge the COST Action TU1208 "Civil Engineering Applications of Ground Penetrating Radar", supporting this work.

References

Al-Qadi I.L., Leng Z., Larkin A.: In-Place Hot-Mix Asphalt Density Estimation Using Ground-Penetrating Radar. Technical Report of Research, Department of Civil and Environmental Engineering, University of Illinois at Urbana-Champaign, Urbana, Illinois, ICT Report No. 11-096, Dec 2011

Benedetto, A.: A three dimensional approach for tracking cracks in bridges using GPR. J. Appl. Geophys. **97**, 37–44 (2013)

Benedetto, A., Pensa, S.: Indirect diagnosis of pavement structural damages using surface GPR reflection techniques. J. Appl. Geophys. **62**, 107–123 (2007)

Benedetto, F., Tosti, F.: GPR spectral analysis for clay content evaluation by the frequency shift method. J. Appl. Geophys. **97**, 89–96 (2013a)

Benedetto, A., Tosti, F.: Inferring bearing ratio of unbound materials from dielectric properties using GPR: the case of runway safety areas. Airfield Highw. Pavement, pp. 1336–1347 (2013b). doi:10.1061/9780784413005.113

Benedetto, A., D'Amico, F., Fattorini F.: Measurement of moisture under road pavement: a new approach based on GPR signal processing in the frequency domain. International Workshop on Advanced Ground Penetrating Radar, Granada (ES) (2009)

Benedetto, A., Benedetto, F., Tosti, F.: GPR applications for geotechnical stability of transportation infrastructures. Nondestr. Testing Eval. **27**(3), 253–262 (2012)

Benedetto, A., Tosti, F., Ortuani, B., Giudici, M., Mele, M.: Soil moisture mapping using GPR for pavement applications. 7th International Workshop on Advanced Ground Penetrating Radar (IWAGPR), Nantes (2–5 July 2013), pp. 1–5 (2013a). ISBN 978-1-4799-0937-7, doi: 10.1109/IWAGPR.2013.6601550

Benedetto, A., D'Amico, F., Tosti, F.: GPR-based evaluation of strength properties of unbound pavement material from electrical characteristics. Geophysical Research Abstracts 15, EGU2013-3648, European Geosciences Union (EGU) General Assembly 2013, Wien (2013b)

Breysse D. (ed.): Non-destructive Assessment of Concrete Structures: Reliability and Limits of Single and Combined Techniques, Springer, pp. 63–71, (2012). ISBN 978-94-007-2735-9

Cassidy, N.J.: Electrical and magnetic properties of rocks, soils and fluids. In: Jol, H.M. (ed.) Ground Penetrating Radar: Theory and Applications, pp. 41–72. Elsevier, Amsterdam (2009)

Dérobert, X., Villain, G., Cortas, R., Chazelas, J.L.: EM characterization of hydraulic concretes in the GPR frequency-band using a quadratic experimental design. 7th International Symposium NDT-CE Proceedings, Nantes, July (2009)

Fauchard, C., Li, B., Laguerre, L., Héritier, B., Benjelloun, N., Kadi, M.: Determination of the compaction of hot mixasphalt using high-frequency electromagnetic methods. NDT&E Int. **60**, 40–51 (2013)

Ferrieres, X., Klysz, G., Mazet, P., Balayssac, J.-P.: Evaluation of the concrete electromagnetic properties by using radar measurements in a context of building sustainability. Comput. Phys. Commun. **180**(8), 1277–1281 (2009)

Gołębiowski, T.: Zastosowanie metody georadarowej do detekcji i monitoringu obiektów o stochastycznym rozkładzie w ośrodku geologicznym "Application of the GPR Method for Detection and Monitoring of Objects with Stochastical Distribution in the Geological Medium". AGH University of Science and Technology Press, Kraków, p. 257 (2012). ISBN 978-83-7464-449-5

Hugenschmidt, J.: GPR for road engineering. Mater. Struct. **31**(207), 192–194 (1998)

Hugenschmidt, J.: Concrete bridge inspection with a mobile GPR system. Constr. Build. Mater. **16**(3), 147–154 (2002)

Hugenschmidt, J.: Multi-Offset-Analysis for man-made structures. In: 8th International Conference on Ground Penetrating Radar, Gold Coast, Australia (2000)

Hugenschmidt, J., Loser, R.: Detection of chlorides and moisture in concrete structures with ground penetrating radar. Mater. Struct. **41**, 785–792 (2008)

Hugenschmidt, J., Kalogeropoulos, A., Soldovieri, F., Prisco, G.: Processing strategies for high-resolution GPR concrete inspections. NDT&E Int. **43**(4), 334–342 (2010)

Hugenschmidt, J., Partl, M., de Witte, H.: GPR inspection of a mountain motorway in Switzerland. J. Appl. Geophys. **40**, 95–104 (1998)

Ihamouten, A., Chahine, K., Baltazart, V., Villain, G., Dérobert, X.: On variants of the frequency power law for the electromagnetic characterization of hydraulic concrete. IEEE Trans. Inst. Meas. **60**(11), 3658–3668 (2011)

Ihamouten, A., Villain, G., Dérobert, X.: Complex permittivity frequency variations from multi-offset GPR data: hydraulic concrete characterization. IEEE Trans. Instrum. Meas. **61**(6), 1636–1648 (2012)

Jackson, J.D.: Classical Electrodynamics, 2nd edn. Wiley, Hoboken, New Jersey (1975)

Kalogeropoulos, A., Kruk, J., Hugenschmidt, J., Bikowski, J., Brühwiler, E.: Full-waveform GPR inversion to assess chloride gradients in concrete. NDT&E Int. **57**, 74–84 (2013)

Klysz, G., Balayssac, J.-P.: Determination of volumetric water content of concrete using ground-penetrating radar. Cem. Concr. Res. **37**(8), 1164–1171 (2007)

Klysz, G., Ferrières, X., Balayssac, J.-P., Laurens, S.: Simulation of direct wave propagation by numerical FDTD for a GPR coupled antenna. NDT&E Int. **39**, 338–347 (2006)

Krysiński L.: Use of impulse GPR for laboratory determination of road material permittivity in core samples. In: Geophysical Research Abstracts, Vol. 15, EGU2013-6779, European Geosciences Union (EGU) General Assembly 2013, Wien, (2013)

Krysiński, L., Sudyka, J.: Typology of reflections in the assessment of the interlayer bonding condition of the bituminous pavement by the use of an impulse high-frequency ground-penetrating radar. Nondestr. Testing Eval. **27**(3), 219–227 (2012a). doi:10.1080/10589759.2012.674525

Krysiński L., Sudyka J.: Ocena wpływu zagęszczenia warstwy asfaltowej na uzyskiwane wartości stałej dielektrycznej—research report, in Polish, Assessment of Asphalt Layer Compaction Influence on Resulting Values of Dielectric Constant, Pavement Diagnostic Division, Road and Bridge Research Institute, Warsaw, Nov (2012b)

Krysiński, L., Sudyka, J.: GPR abilities in investigation of the pavement transversal cracks. J. App. Geophys. **97**, 27–36 (2013)

Laurens, S., Balayssac, J.-P., Rhazi, J., Arliguie, G.: Influence of concrete relative humidity on the amplitude of Ground-Penetrating Radar (GPR) signal. Mater. Struct. **35**(248), 198–203 (2002)

Laurens, S., Balayssac, J.-P., Rhazi, J., Klysz, G., Arliguie, G.: Non destructive evaluation of concrete moisture by GPR: experimental study and direct modelling. Mater. Struct. **38**(283), 827–832 (2005)

Liu, L., Guo T.: Dielectric property of asphalt pavement specimens in dry, water-saturated, and frozen conditions. In: Proceedings of SPIE 4758, Ninth International Conference on Ground Penetrating Radar, p. 410, April 15, (2002). doi:10.1117/12.462222

Ortuani, B., Benedetto, A., Giudici, M., Mele, M., Tosti, F.: A Non-invasive approach to monitor variability of soil water content with electromagnetic methods. Procedia Environ. Sci. **19**, 446–455 (2013)

Patriarca, C., Tosti, F., Velds, C., Benedetto, A., Lambot, S., Slob, E.: Frequency dependent electric properties of homogeneous multi-phase lossy media in the, ground-penetrating radar frequency range. J. Appl. Geophys. **97**, 81–88 (2013)

Plati, C., Loizos, A.: Using ground-penetrating radar for assessing the structural needs of asphalt pavements. Nondestr. Testing Eval. **27**(3), 273–284 (2012)

Plati, C., Loizos, A.: Estimation of in-situ density and moisture content in HMA pavements based on GPR trace reflection amplitude using different frequencies. J. Appl. Geophys. **97**, 3–10 (2013)

Plati, C., Georgiou, P., Loizos, A. A comprehensive approach for the assessment of in-situ pavement density using GPR technique. In: Geophysical Research Abstracts 15, EGU2013-11094, European Geosciences Union (EGU) General Assembly 2013, Wien, (2013)

Poikajärvi J., Peisa K., Herronen T., Aursand P.O., Maijala P., Narbro A.: GPR in road investigations—equipment tests and quality assurance of new asphalt pavement. Nondestr. Testing and Eva. **27**(3), 293 (2012)

Roddis, W.M., Maser, K., Gisi, A.J.: Radar Pavement Thickness Evaluations for Varying Base Conditions, Transportation Research Record 1355. National Academy Press, Washington D. C., pp, 90–98 (1992)

Roimela, P.: Päällystetutka tiiviyden laadunvalvonnassa (in Finnish, The Use of Pavement Radar in Quality Control of Bituminous Pavement). Tielaitoksen selvityksiä, 6/1999, TIEL 3200499, Tielaitos Konsultointi, Rovaniemi, (1999). ISBN 951-726-496-8

Saarenketo, T.: NDT Transportation. In: Jol, H.M. (ed.) Ground Penetrating Radar: Theory and Applications, pp. 395–444. Elsevier, Amsterdam (2009)

Sbartaï, Z.M., Laurens, S., Balayssac, J.-P., Ballivy, G., Arliguie, G.: Effect of concrete moisture on radar signal amplitude. ACI Mater. J. **103**(6), 419–426 (2006a)

Sbartaï, M., Laurens, S., Balayssac, J.-P., Ballivy, G., Arliguie, G.: Ability of the direct wave of radar ground-coupled antenna for NDT of concrete structures. NDT&E Int. **39**(5), 400–407 (2006b)

Sbartaï, Z.M., Laurens, S., Viriyametanont, K., Balayssac, J.-P., Arliguie, G.: Non-destructive evaluation of concrete physical condition using radar and artificial neural networks. Constr. Build. Mater. **23**(2), 837–845 (2009)

Sebesta, S., Sculion, T.: Application of infrared imaging and ground penetrating radar for detecting segregation in hot-mix asphalt overlays. Transp. Res. Board: J. Transp. Res. Board **1861**, 37–43 (2002)

Solimene, R., Soldovieri, F., Prisco, G., Pierri, R.: Three-dimensional microwave tomography by a 2-D slice-based reconstruction algorithm. IEEE Geosci. Remote Sens. Lett. **4**(4), 556–560 (2007)

Sudyka, J., Krysiński, L.: Radar technique application in structural analysis and identification of interlayer bounding. Int. J. Pavement Res. Technol. **4**(3), 176–184 (2011)

Sudyka, J., Krysiński, L., Jaskuła, P., Mechowski, T., Harasim, P.: Radar technique in application of interlayer identification connections. In: 5th International Conference Bituminous Mixtures and Pavements, Thessaloniki, pp, 1449–1459 1–3 June (2011)

Tahmoressi, M., Head, D., Saenz, T., Rebala, S.: Material Transfer Device Showcase in El Paso, Texas. Research report DHT-47, Texas Department of Transportation, El Paso, Texas (1999)

Tosti, F., Benedetto, A., Calvi, A.: An Effective Approach for Road Maintenance through the Simulation of GPR-Based Pavements Damage Inspection. IJPC—International Journal of Pavements Conference, São Paulo, Brazil, (2013a), Paper 124–1

Tosti, F., Patriarca, C., Slob, E., Benedetto, A., Lambot, S.: Clay content evaluation in soils through GPR signal processing. J. Appl. Geophys. **97**, 69–80 (2013b)

Ulaby, F., Bengal, T., Dobson, M., East, J., Garvin, J., Evans, D.: Microwave dielectric properties of dry rocks. IEEE Trans. Geosci. Remote Sens. **28**(3), 325–336 (1990)

Villain, G., Ihamouten, A., Dérobert, X., Sedran, T., Burban, O., Coffec, O., Dauvergne, M., Alexandre, J., Cottineau, L.M., Thiéry, M.: Adapted mix design and characterization for non destructive assessment of concrete. International Conference on Marine Environmental Damage to Atl. Coastal and History Structure Proceedings, La Rochelle, May (2010)

Determination, by Using GPR, of the Volumetric Water Content in Structures, Substructures, Foundations and Soil

Fabio Tosti and Evert Slob

Abstract Volumetric water content evaluation in structures, substructures, soils, and subsurface in general is a crucial issue in a wide range of applications. The main weaknesses of subsurface moisture sensing techniques are usually related both to the lack of cost-effectiveness of measurements, and to unsuitable support scales with respect to the extension of the surface to be investigated. In this regard, ground-penetrating radar (GPR) is an increasingly used non-destructive tool specifically suited for characterization and imaging. Several GPR techniques have been developed for different application purposes. Moisture evaluation in concrete is important for diagnosing structures at early stages of deterioration, as water contributes to the transfer of degrading and corrosive agents e.g., chloride. Traditionally, research efforts have been focused on the processing of GPR signal in time domain, although more recent studies are being increasingly addressed towards frequency domain analysis, providing additional information on moisture content in concrete. Concerning the evaluation of subsurface soil water content, different models ranging from empirical to theoretical are used for converting permittivity values into moisture. In this regard, two main GPR approaches are commonly employed for permittivity evaluation in time-domain measurements, namely, the ground wave method and the reflection method. Furthermore, the use of borehole transmission measurements, traditional off-ground methods, and of an inverse modelling approach allowing for a full waveform inversion of radar signals have been developed in the past decade. More recently, a self-consistent approach based on the Rayleigh scattering theory has also allowed the direct evaluation of moisture content from frequency spectra analysis.

F. Tosti (✉)
Department of Engineering, Roma Tre University,
Via Vito Volterra 62, 00146 Rome, Italy
e-mail: fabio.tosti@uniroma3.it

E. Slob
Faculty of Civil Engineering and Geosciences, Section of Applied Geophysics and Petrophysics, Department of Geoscience and Engineering,
Delft University of Technology, Stevinweg 1, 2628 CN Delft, The Netherlands
e-mail: e.c.slob@tudelft.nl

A. Benedetto and L. Pajewski (eds.), *Civil Engineering Applications of Ground Penetrating Radar*, Springer Transactions in Civil and Environmental Engineering,
DOI 10.1007/978-3-319-04813-0_7

1 Introduction

Volumetric water content (VWC) θ dynamics in structures, substructures, soils, and subsurface in general is a key component in many fields of application such as agriculture, construction, geotechnical stability analyses, hydrology, and other environmental studies.

Traditional methods for moisture evaluation in concrete provide the use of destructive techniques, such as coring, drilling or otherwise removing part of the structure to allow inner visual inspections. Despite the high reliability of such techniques, they reveal as expensive and time-consuming and damages induced by their application usually become source points for further degradation processes.

More recently, non-destructive testing methods have been drawing more and more attention to the inspection of concrete structures by means of recent advances in both reliability and effectiveness of measurements. The capability to test directly in situ is being increasingly recognised so that a rising trend away from the traditional random sampling of concrete for material analysis or visual inspection is occurring to the use of enhanced non-destructive techniques (NDTs) on structures. Such methods may be direct or indirect, since information about size, depth or physical condition can be, respectively, directly measured or inferred from one or more measured parameters which in turn may be further used to evaluate risk of damage or on-going deterioration processes. Additionally, information related to strength and deformation properties of materials, their potential durability or bearing capacity can be inferred.

Amongst the most used NDTs for evaluating the condition of structures about the effects of moisture penetration into the concrete pore networks, half-cell potential is mainly used to determine the corrosion activity of steel in concrete (ASTM Standards 2009). Its apparatus consists of a voltmeter with one lead connected to a reference electrode placed on the surface of the concrete, and a second lead connecting the voltmeter to the reinforcing steel. Current passes from the reference electrode to the concrete surface through a sponge soaked with an electrolytic solution. The objective of such instrumentation is to measure the voltage difference between the rebar and the reference electrode.

Subsequently, a galvanostatic pulse measurement (GPM) technique has been developed and used combined with the above half-cell potential, which produces an enhanced potential map and enables the estimation of the corrosion rate (Gowers and Millard 1994).

The electrical resistivity (ER) of concrete is known to be sensitive to the degree of saturation (Lopez and Gonzalez 1993) and to chloride variations in concrete (Saleem et al. 1996). ER of concrete can be measured by several means (Polder et al. 2000) that can differ in the type of current applied e.g., alternating current (AC) and direct current (DC), and electrode configuration used. Typically, electrodes are placed on the surface and the resistance is measured. The resistivity can be therefore calculated from the cell geometry.

The use of ultrasonic impulse echo techniques in testing concrete has been widely analysed since the 1980s (Krause et al. 1995). Ohdaira and Masuzawa (2000) measured the ultrasound velocity and the frequency component on ultrasonic propagation in concrete as a function of water contents ranging from 0 to 8 % by weight, and dependence between those parameters was found.

According to a smaller scale of investigation, X-ray radiography has been used for on-site concrete inspections since the 1950s. Roels and Carmeliet (2006) analysed the moisture flow in microfractured cellular concrete specimen and assessed moisture content by logarithmic subtraction of a reference image (e.g., sample at dry state) from images of the wet sample. Nevertheless, its field of application can cover a very limited extent, partly due to the cost and immobility of X-ray equipment, which requires high voltages (Mitchell 1991).

Overall and despite the satisfactory levels of reliability, the above-mentioned methods are relatively slow, and require lane closures and traffic constraints for the sounding operations of bridge decks. In that respect, GPR has proved to be a powerful geophysical technique, relatively insensitive to environmental conditions, and suited to investigate large surfaces in a relatively short time. Many efforts have been devoted to the use of GPR since the 1960s, particularly in archaeology and civil engineering applications for detecting buried objects or investigating subsurface structures (Rodeick 1984; Vaughan 1986; Davis and Annan 1989; Bevan 1991; Maser and Scullion 1992; Annan 2002; Daniels 2004).

With regard to soil moisture sensing, the main physical properties investigated are traditionally the bulk electrical conductivity σ_a and the relative dielectric permittivity ε_r; alternatively, soil thermal properties as soil thermal conductivity λ, volumetric heat capacity C, and soil thermal diffusivity α are being increasingly considered alternatives to other techniques for assessing soil properties and parameters (Robinson et al. 2003).

Usually, water content measurement techniques at the field scale have proved to be invasive and time-consuming methods. In addition, a gap between small- (<0.01 m^2) and large- (>100 m^2) scale measurements is still encountered, with only a few amount of instruments being suitable for intermediate-scale (0.01–100 m^2) characterization of shallow subsurface soil properties.

Amongst the most widely used methods, thermogravimetric measurements from core drilling can be considered the major and more established destructive techniques for moisture sensing.

By lower levels of invasiveness, neutron probes have the advantage of providing a linear relationship between the count ratio and θ, thereby allowing for a smoother calibration. Nonetheless, the sampling volume is dependent on moisture content, so that it is generally within 0.15 m in wet soils and 0.50 m in dry soils.

Furthermore, electromagnetic (EM) sensors are widespread; these may include high-frequency impedance measurements i.e., capacitance probes (Wobschall 1978), and transmission line methods e.g., time domain reflectometry (TDR) (Fellner-Feldegg 1969). In particular, TDR can simultaneously determine θ and σ_a (Dalton et al. 1984), and operates at frequencies greater than 0.5 GHz such that a lower influence of σ_a is encountered. Nevertheless, capacitance probes generally operate in the frequency range between 5 MHz (Borhan and Parsons 2004) and 150 MHz (Gaskin and Miller 1996) and are they characterized by a geometry more suited to short electrodes and borehole applications.

In addition, heat pulse sensors (Campbell et al. 1991) allow to estimate both soil thermal properties and θ by inducing a short heat pulse from one sensor needle and by measuring the temperature response at a second sensor.

On a much broader scale of inspection, remote sensing instruments are employed for measuring θ without any contact with the ground, thereby enabling a high efficiency of measurements. The principle inspiring the evaluation of θ changes relies on the influence of soil on potential fields e.g., electric, magnetic, and gravitational fields. Three main remote sensing methods for measuring θ can be distinguished nowadays. In particular, passive (Njoku and Entekhabi 1996) and active (Ulaby et al. 1996) remote sensing methods account for, respectively, the electromagnetic radiation naturally emitted by the target, and the radiation scattered by the target once this is illuminated by a known source of radiation. Concerning this last method, detection is focused on changes in the gravity potential field above the soil which in turn are related to changes in the density of the soil and, thus, to θ variability (Tapley et al. 2004).

On the other hand, geophysical methods include a wide-ranging set of techniques for evaluating θ, with the main advantage to be minimally or fully non-invasive and to cover a huge spatial area for soil properties determination. Two main categories can be broadly identified, namely, those measuring ground electrical conductivity e.g., direct current (DC) resistivity (Samouelian et al. 2005) and electromagnetic induction (EMI) (Sheets and Hendrickx 1995), and those working through the transmission/reception of electromagnetic waves into the ground e.g., the above-mentioned GPR. With regard to the first group, DC resistivity is a direct current method for resistivity assessment providing electrodes inserted in the ground generally from 0.5 to 5 m distances between each other. It was first introduced by Briggs (1899) and has the twofold advantage to yield data rapidly and to monitor temporal changes associated with θ. Keller and Frischknecht (1966) gave an early description of EMI as a non-invasive borehole logging geophysical technique. The instrument is composed of a receiver at one end and a transmitter loop at the other, and enables to measure ground conductivity. Once the transmitter is energized, it creates magnetic field loops in the ground that produce electrical field loops, thereby creating in turn a secondary magnetic field. The ground conductivity depends non-linearly on the combined primary and secondary magnetic fields measured in the receiver. As a rule of thumb, the penetration into the ground is assumed to be directly proportional to the spacing of the loops and is also affected by their orientation, being 1.5 times the transmitter-receiver loop spacing for a vertical electromagnetic dipole configuration, and 0.75 times the spacing for a horizontal dipole.

In such a framework, GPR can reveal as a good compromise in near-surface moisture sensing for its intermediate scale capability of investigation and its higher time efficiency relative to other destructive and non-destructive sensing techniques. Several GPR-based methods focused on moisture sensing have been developed according to different applications (Nakashima et al. 2001; Mínguez Maturana 2012) and materials, concrete structures (Hugenschmidt and Loser 2008), natural soils (Lambot et al. 2004a; Benedetto and Pensa 2007) as well as hot-mix asphalt layers (Sybilski et al. 2012). They require specific approaches and model assumptions to reliably estimate the subsurface volumetric water content.

In this paper, a review on the most diffused techniques in VWC sensing of structures, substructures, and soils is presented. Some insights for further developments of the research are also provided.

2 GPR Principles

2.1 Overview

GPR principles took place from the electromagnetic theory wherein Maxwell's equations mathematically describe the physics of EM fields, while material properties are quantified by constitutive relationships. By combining such two elements, it is possible to quantitatively describe GPR signals.

2.2 Maxwell's Equations

EM fields and relationships can be expressed as follows:

$$\nabla \times \vec{E} = -\frac{\partial\left(\vec{B}\right)}{\partial t} \tag{1}$$

$$\nabla \times \vec{H} = \vec{J} + \frac{\partial\left(\vec{D}\right)}{\partial t} \tag{2}$$

$$\nabla \times \vec{D} = q \tag{3}$$

$$\nabla \times \vec{B} = 0 \tag{4}$$

where $\vec{E}$ is the electric field strength vector (V m^{-1}), q is the electric charge density (C m^{-3}), $\vec{B}$ is the magnetic flux density vector (T), $\vec{J}$ is the electric current density vector (A m^{-2}), $\vec{D}$ is the electric displacement vector (C m^{-2}), t is time (s), and $\vec{H}$ is the magnetic field intensity vector (A m^{-1}).

All classic EM fields of application (e.g., induction, radio waves, resistivity, circuit theory, etc.) can be derived from these equations when combined with constitutive relationships to characterize material electrical properties.

2.3 Constitutive Equations

Constitutive relationships allow to describe a material's response to EM fields. In the case of GPR, both electrical and magnetic properties are highly important. A macroscopic description on the response of a group of bow electrons, atoms, and molecules to the application of an EM field can be described by the following linear constitutive equations:

$$\vec{J} = \tilde{\sigma}\vec{E} \tag{5}$$

$$\vec{D} = \tilde{\varepsilon}\vec{E} \tag{6}$$

$$\vec{B} = \tilde{\mu}\vec{H} \tag{7}$$

When an electric field is present, free charge movement is characterized by electrical conductivity $\tilde{\sigma}$. Resistance to charge flow leads to energy dissipation. The displacement of charge constrained in a material structure to the presence of an electric field is characterized by dielectric permittivity $\tilde{\varepsilon}$. Charge displacement results in energy storage in the material. Magnetic permeability $\tilde{\mu}$ describes the response of intrinsic atomic and molecular magnetic moments to a magnetic field. For simple materials, distorting intrinsic magnetic moments store energy in the material.

Furthermore, the history of the incident field plays a crucial role in material properties. For GPR applications, the dielectric permittivity is an important quantity. It is defined as follows:

$$\varepsilon_r = \frac{\varepsilon}{\varepsilon_0} \tag{8}$$

where ε is the absolute permittivity (F m^{-1}), and ε_0 is the permittivity of vacuum, 8.89×10^{-12} F m^{-1}.

3 Volumetric Water Content in Concrete Structures

The detection of moisture is important for the diagnosis of concrete structures at early stages of deterioration as it determines most of the physicochemical pathologies, such as steel reinforcement corrosion, alkali-aggregate reaction, and freezing-and thawing-cycles.

Under normal conditions, the high pH-value of the pore solution protects reinforcement in concrete from corrosion, such that a stable film is formed on the steel surface, which renders the reinforcing steel chemically passive and prevents the electrochemical processes taking place during corrosion (Pourbaix 1966). Nevertheless, carbonation of concrete or penetrating chlorides can destroy such protective passivity (Neville 1995). In this respect, corrosion of the reinforcement can take place when exceeding a certain threshold value of chlorides and, in combination with water and oxygen, the protective passivity layer on the steel surface is locally destroyed (Montemor et al. 2003). Two main consequences of chloride inducing corrosion of steel can be encountered, namely, (i) cracking and spalling of the concrete cover since a volume several time larger than the original steel is occupied by the products of corrosion, and (ii) pitting of the steel reducing its cross-sectional area due to the highly localized chloride-induced corrosion at a small anode. Corrosion of the reinforcement is the main cause of structural concrete deterioration, thus it is one of the main factors affecting the rising of rehabilitation costs. In this review, techniques based on time domain and frequency domain analysis of radar signal are separately analysed.

3.1 Time Domain Analysis

3.1.1 Reflection Methods

The EM wave propagation is governed by the concrete permittivity, which is influenced by free water (Soutsos et al. 2001) and chlorides (Al-Qadi et al. 1997; Robert 1998). On the basis of such approach, changes of the wave reflected by asphalt–concrete or concrete–rebar interfaces have been used to assess corrosion probability. In this regard, Laurens et al. (2000) used a high-resolution ground-coupled 1.5 GHz antenna for detecting the possibility of steel corrosion in a bridge deck. By relating the corrosion probability to the reflectivity rate of a bridge area, the authors compared the amplitude of a reflective signal with the amplitude of the emitted pulse. High attenuation areas were considered as slightly reflective; on the contrary, low attenuation areas corresponded to highly conductive concrete. GPR and potential histogram analysis led to similar diagnosis concerning the general state of corrosion.

Hugenschmidt and Loser (2008) used an off-ground 2.5 GHz antenna to investigate the effects of moisture and chlorides in concrete on radar amplitudes through laboratory experiments. Nine concrete slabs from three concrete mixtures were built, with three different chloride contents being used for manufacturing the specimens. The base of the mould was covered with an aluminium sheet simulating the reflecting reinforcement. The change of the quotient of reflection amplitudes was evaluated as a possible method for mapping blackspots on concrete bridge decks covered with asphalt pavements.

Nevertheless, the presence of a reflector and a prior knowledge of its position is required for on-site applications of reflection methods. This issue can be considered one of the main drawbacks of the aforementioned laboratory-based approaches, since the exact position of rebars is difficult to determine in real conditions.

3.1.2 Direct Wave Methods

Direct wave methods rely on the use of the direct wave propagating in the first few centimeters of the concrete, such that an independence from the presence of any reflector can be taken into account. In addition, the distance between the transmitter and the receiver is fixed and known.

For reinforced concrete structures, the presence of steel bars in the near subsurface makes multi-offset radar measurements difficult to process and analyse. In particular, two main problems may be encountered, namely, the disturbance caused by reinforcements to the direct wave propagation and, moreover, an overlapping of direct and reflected signals.

For single offset measurements, when the receiving dipole is located in the near-field of the source (Roberts and Daniels 1997), it is possible to use the plane wave approximation for calculating wave velocity and dielectric permittivity. Some authors have shown that the amplitude of the near-field direct wave is affected by water content (Laurens et al. 2002; Klysz et al. 2004; Sbartaï et al. 2006a, b) and possible presence of chlorides (Sbartaï et al. 2006a). Laurens et al. (2005) developed an experimental study on the effect of concrete moisture on radar waves propagating through laboratory concrete slabs. Attention was particularly focused on the amplitude, velocity and frequency spectrum of the waveforms recorded. In particular, the behaviour of the transmitter-receiver direct wave was observed to be greatly influenced by moisture in concrete: according to theoretical expectations, the permittivity of the medium increases with increasing saturation degrees in concrete slabs. Moreover, a linear increase was observed in the amplitudes of the two those signals associated with the decrease in the degree of pore saturation.

Sbartaï et al. (2006a) demonstrated good correlation between direct wave and reflected wave attenuation concerning the physical state of the concrete. This correlation was validated on two bridges by comparison of direct wave attenuation maps with reflection attenuation maps. In addition, Sbartaï et al. (2007) used single offset radar measurements to compare information retrieved from direct wave measurement and those inferred from ER method. Results from laboratory tests showed that variations in concrete moisture and chloride contamination level strongly affected the radar direct signal. Tests performed in real road environment confirmed good correlation between radar direct wave attenuation and electrical resistivity and, thus, the possibility to detect concrete conditions leading to reinforcement corrosion through radar direct wave measurements. The attenuation A (dB) of the direct signal in concrete is given by:

$$A = -20 \times \log \left(\frac{A_c}{A_a} \right) \tag{9}$$

where A_c is the peak-to-peak amplitude of the direct wave signal recorded in concrete, and A_a is the peak-to-peak amplitude of the air wave signal (Fig. 1).

Overall, such promising results suggest that direct wave analyses of ground-coupled antenna systems can be applied for the rapid physical characterization of reinforced concrete in real conditions, thereby providing a great contribution in detecting the risk of steel bar corrosion.

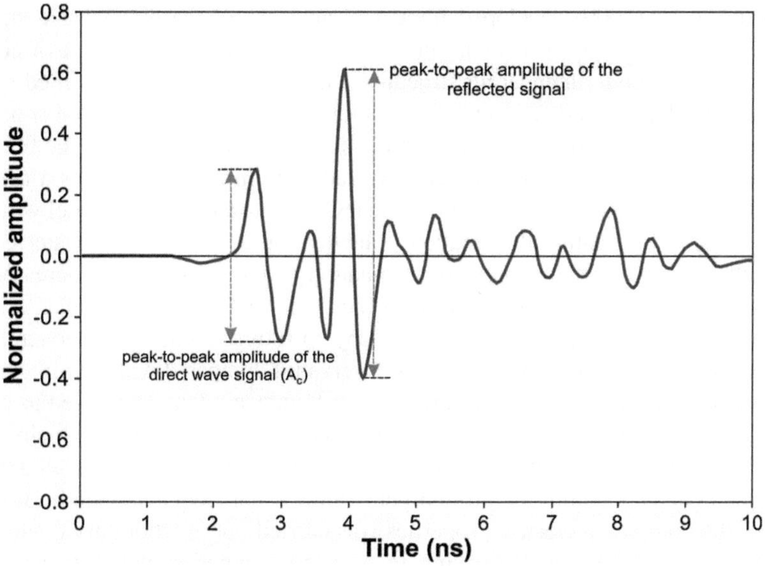

Fig. 1 Typical radar signal recorded on concrete slab

3.2 Frequency Domain Analysis

Research efforts on moisture content evaluation in concrete structures have been widely focused on time domain signal analysis, while there are still limited studies concerning signal processing in the frequency domain.

Klysz et al. (2004) studied concrete moisture changes according to the velocity analysis of the direct wave in frequency domain on the basis of Wide Angle Reflection Refraction (WARR) measurements, such that several profiles at the same place were taken using different distances between the two antennas. The authors demonstrated a difference between velocity values measured on dry and wet concretes in the 0.1–1.3 GHz frequency range. In addition, it was shown that the velocity of the radar direct wave is not affected by a low dispersive medium (e.g., concrete) at the considered frequency range, due to the slight dispersion observed on frequency-velocity curves.

Laurens et al. (2005) analysed the relation between the spectral content of the signal reflected by the bottom of laboratory slabs and the degree of saturation of the concrete by using a high-resolution ground-coupled 1.5 GHz antenna system. It was observed that the center frequency lies approximately between 1.18 GHz (saturated concrete) and 1.4 GHz (dry concrete). This behaviour can be explained by the presence of water causing electromagnetic dispersion of concrete. Specifically, high frequencies are generally more attenuated than low frequencies in concrete, and the presence of water contributes to amplify such behaviour.

Sbartaï et al. (2009) developed frequency analyses of the direct wave signal of a 1.5 GHz ground-coupled antenna in order to enhance the assessment of concrete water content in civil engineering structures. Several signals were recorded on concrete samples having various rates of saturation, from dry up to saturated conditions. A Fast Fourier Transform (FFT) algorithm was implemented to extract the frequency spectra of each signal, and the direct wave recorded in air was used as a reference signal for calculating the attenuation-frequency curve. An empirical model was therefore developed by relating the frequency attenuation to the concrete water content. The authors found that the spectrum attenuation of the direct signal increased with respect to frequency in the range analysed (i.e., 0.5–1.8 GHz); in particular, a change of the direct wave spectrum was induced by moisture variations in concrete, and more attenuation at high frequencies was observed if compared with low frequencies. Attenuation analyses on moisture variations in frequency domain seemed to provide more detailed information with respect to attenuation analyses in time domain.

Dérobert et al. (2009) have focused on the electromagnetic characterization of hydraulic concretes within the ground-penetrating radar frequency band. The evaluation of the complex dielectric properties was carried out in laboratory environment on mixtures manufactured according to a specific experimental design. In more details, various parameters of the mixing, namely, nature of aggregates, nature of cement, cement content, water to cement ratio (*w/c*), and chloride content were considered. A multi-linear polynomial model was used such that it was possible to evaluate the influence of each parameter and their interactions on the permittivity. It was demonstrated that the porosity has a relevant impact on the permittivity of the concretes, through their water content. Such effect was increased by chloride presence in the saturated mixing. In dry conditions, their effects remained negligible, with the nature of the aggregates being the most relevant. In this regard, completely different behaviours of chloride could be observed according to its ionic or crystallized state.

More recently, Ihamouten et al. (2012) studied the validity of a new technique for the complex permittivity extraction of several dispersive media, carried out by characterizing four concrete mixtures at different water contents on a large GPR bandwidth. Such approach was based on a coupling between the results of the wave–field transforms (transformations from time–displacement domain to temporal-frequency–spatial-frequency domain) and those of the *Q*-estimation methods. Promising results were obtained: the characterization of the hydric status of various concrete mixtures was possible by taking into account both the real and imaginary parts of permittivity over a large GPR bandwidth (0.05–3.0 GHz).

4 Volumetric Water Content in Unsaturated Soils

Over the past decades, GPR has been a widespread instrument in the areas of unsaturated zone hydrology and water resources. Amongst the various applications, it has been used to identify soil stratigraphy (Davis and Annan 1989), to assess subsurface hydraulic parameter (Hubbard et al. 1997), to locate water tables (Nakashima et al.

2001), and to measure soil water content (Hubbard et al. 2002; Huisman et al. 2003; Serbin and Or 2003). Different surveying techniques exist to estimate θ from GPR. Traditionally, an estimation of ε_r at the GPR measurement scale is carried out and a petrophysical relationship is used to convert ε_r to θ. More recently, research efforts have been devoted to find more self-consistent methods in order to directly evaluate near-surface moisture in soils without the need of petrophysical relationships and calibration of the GPR system (Pettinelli et al. 2007; Di Matteo et al. 2013).

4.1 Water Content Estimation Methods by Permittivity Measurements

4.1.1 Permittivity Estimation Methods

Overall, five main classes of GPR surveying techniques for permittivity estimation can be identified: (i) reflection methods (Fig. 2a, b), (ii) ground wave measurements (Fig. 2c), (iii) borehole transmission measurements (Fig. 2d), (iv) surface reflection methods (Fig. 2e), (v) inverse modelling of off-ground monostatic GPR systems (Fig. 2f). Herein, a review of the existing literature is aimed at grouping both traditional and recent research works under the above relevant categories.

4.1.2 Reflection Methods

The estimate of soil water content from reflected wave travel time data can be distinguished in two classes. In particular, methods that use a single antenna separation for soil water content estimation are included in the first class, while methods that require multiple measurements with different antenna separations can be grouped in a second one.

4.1.3 Single-Offset Reflection Methods

Estimation of θ by using single-offset reflection measurements has been mainly addressed on modelling the reflection hyperbolae originating at scattering objects in the subsurface (Grote et al. 2002; Loeffier and Bano 2004) or from zero-offset, two-way travel time to horizontal reflectors (Lunt et al. 2005).

It is worth noting that in both cases scattering objects and horizontal reflectors must be located at known depths. In fact, by using a single zero-offset measurement, prediction of reflector depth from GPR data is not possible from travel time information alone, and either true amplitude data must be available or an upstream knowledge of the target depth is needed for converting travel times measured into θ estimates.

Alongside the direction of scan x, a hyperbola is traced out by the reflected events in the radar section, as the anomaly below the GPR measurement is

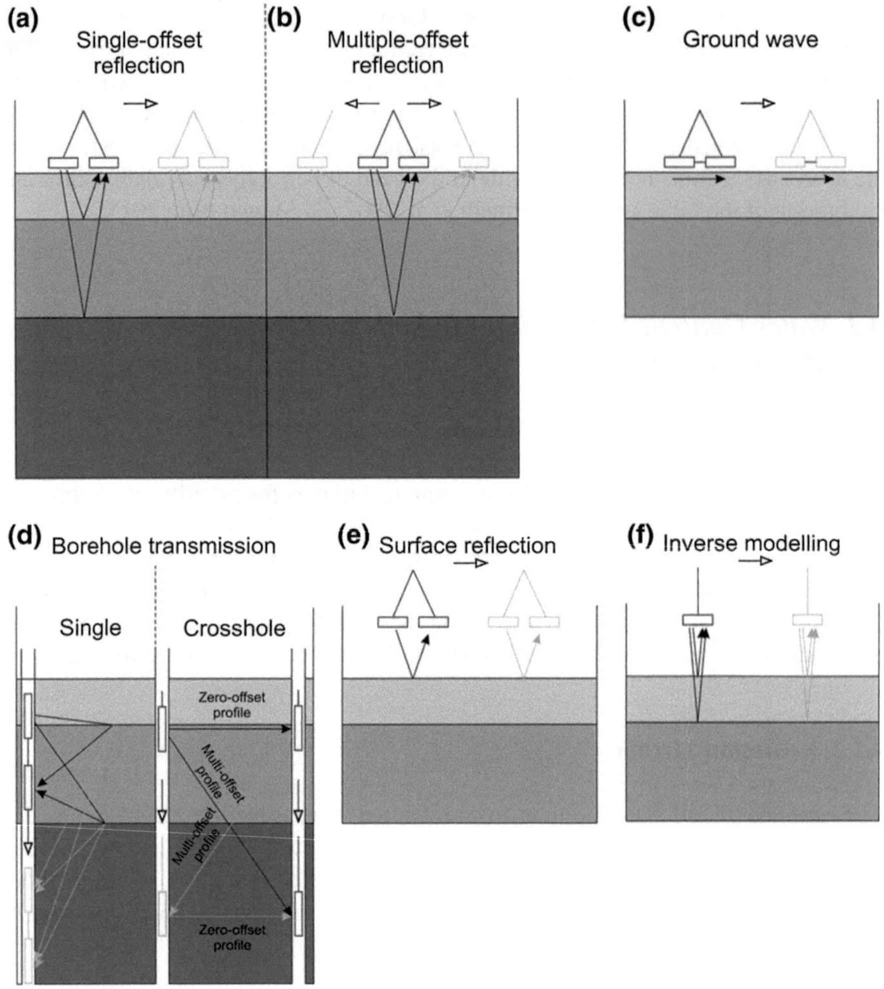

Fig. 2 Sketch of GPR inspection techniques for moisture evaluation through permittivity. **a** Single-offset reflection method. **b** Multi-offset reflection method. **c** Ground wave method. **d** Single- and cross-borehole transmission measurement. **e** Surface reflection method. **f** Inverse modelling

characterized by the shortest travel distance (i.e., time distance) and larger distances to travel will be encountered for all other waves. The average wave velocity in soil determines the convexity of the reflection hyperbola. By fitting the following hyperbola to measure arrival times at several positions *x*, it is possible to determine the average velocity between the ground surface and the anomaly, v_{soil}, as follows:

$$v_{soil} = \frac{2\sqrt{x^2 + h^2}}{t_x} \tag{10}$$

where x is the position of the scattering object at the apex of the hyperbola, h is the depth of the scattering object, and t_x is the travel time distance of the reflected wave at position x, filtered out by the additional travel time at the beginning of each measurement, that is mainly due to inner coupling to the antenna array.

For GPR systems with a significant antenna separation, s, (10) is expressed as follows:

$$v_{soil} = \frac{\sqrt{\left(x - \frac{1}{2}s\right)^2 + h^2} + \sqrt{\left(x + \frac{1}{2}s\right)^2 + h^2}}{t_x} \tag{11}$$

Concerning the zero time correction of arrival times, a standard procedure provides to align the arrival times of the air wave to correct possible shifts in the zero time, followed by the estimation of the average arrival time of the air wave and a calculation of the zero time correction from the average arrival time and the known antenna separation.

In the literature, scattering targets were embedded at known depths in test pits (Grote et al. 2002) such that a fit of the reflection hyperbolae, recorded while moving the single-offset GPR instrument, to a single average velocity representing the material between the target and the known scatterer depth was possible. Good agreement between θ estimates, assessed by converting permittivity values through a site-specific petrophysical relationship, and θ observed by gravimetric measurements was found. Furthermore, Loeffler and Bano (2004) obtained good correspondence between GPR-predicted θ and water volume added or removed from a tank at various known reflectors.

4.1.4 Multi-offset Reflection Methods

The average velocity between reflector and surface and depth can be estimated by a series of reflection hyperbola fitted by multi-offset GPR reflection datasets recorded over horizontal interfaces. Traditionally, two multi-offset GPR acquisition geometries, namely, Common MidPoint (CMP) and Wide Angle Reflection and Refraction (WARR) are used.

CMP acquisition provides an increasing distance between the antennas by fixing a common midpoint. Conversely, WARR acquisition is characterized by an increasing distance between the antennas with the transmitter at a fixed position. In case of multi-offset GPR measurement scenarios with consistent reflected waves, the average velocity to the reflecting layer, v_{soil}, can be directly estimated as follows:

$$v_{soil} = \frac{2\sqrt{h^2 + \left(\frac{1}{2}s\right)^2}}{t_x} \tag{12}$$

Soil water content can be therefore estimated by fitting v_{soil} to the zero time corrected arrival times of the reflected wave, t_x, for different antenna separations, s, at depth h (Tillard and Dubois 1995).

In order to provide reliable soil water content estimates and more time efficient analyses, many efforts have been devoted to develop approaches for velocity determination from reflected GPR waves, analogous to the velocity analysis approaches used in seismic data processing (Yilmaz 1987). Semblance analysis is a well-known method. The main goal of such approach is to evaluate velocity and travel time by considering a point wherein the reflection energy of a reflected wave in a multi-offset measurement collapses. In particular, the semblance plot is defined by evaluating again the arrival times of the CMP for a range of velocities (x axis), and by summing the normalized energy for each arrival time (y axis) for each velocity. Basically, reflected waves at certain arrival time are well described by the corresponding velocity when high values in semblance plot are obtained.

In order to convert the manually or semiautomatically determined average velocities to interval velocity $v_{int,n}$ of each layer n, the Dix (1955) formulation can be used:

$$v_{\text{int},n} = \sqrt{\frac{t_n v_{soil,n}^2 - t_{n-1} v_{soil,n-1}^2}{t_n - t_{n-1}}} \tag{13}$$

where $v_{soil,\,n}$ is the average velocity from the surface up to the bottom of layer n, $v_{soil,n-1}$ is the average velocity up to the bottom of layer $n-1$, t_n is the two-way travel time to the bottom of layer n, t_{n-1} is the two-way travel time to the bottom of layer $n-1$, being $n = 1$ the upper layer of the soil. Similarly to single-offset measurements, water content estimates with the multi-offset GPR reflection method require appropriate signal penetration and subsurface dielectric contrasts acting as GPR reflectors. Moreover, such method is usually time-consuming and do not allow for monitoring of soil water content variation. Examples of GPR-based θ estimation with multiple-offset methods can be found in several research works (Tillard and Dubois 1995; Greaves et al. 1996; van Overmeeren et al. 1997; Garambois et al. 2002).

4.1.5 Ground Wave Methods

Near-surface soil water content measurements can be performed using the direct ground wave travel time (Huisman et al. 2001). By definition, the direct ground wave propagates directly from the transmitter to the receiver through the top of the soil. Regardless from the presence of clearly reflecting soil layers, the ground wave is detected by the GPR receiver (Du and Rummel 1994; Berktold et al. 1998). One major requirement of ground wave methods concerns the need for a ground coupling of both transmitter and receiver, due to the intrinsic fading of the ground wave. A second one provides the ground wave to have a fixed velocity between transmitting and receiving antennas for single offset measurements.

Moisture content θ can be extracted using multiple-offset measurements from the linear relationship between travel time and offset for the direct wave. The direct dependence between slope of the ground wave in a multi-offset measurement and the ground wave velocity is a factor contributing to water content determination.

Nevertheless, it is also possible to use single-offset GPR measurements after knowing the approximate arrival time of the ground wave from a multi-offset measurement. In this respect, Du (1996) and Sperl (1999) proposed a procedure for mapping θ through ground wave methods providing, (i) identification of an approximate ground wave arrival time for different antenna separations in a multi-offset GPR measurement; (ii) selection of an antenna separation where a clear separation of the ground wave from the air and reflected waves occurs; (iii) use of this antenna separation for single-offset measurements and set relationships between changes in ground wave arrival and changes in soil permittivity.

Sperl (1999) provided a relationship between ground wave arrival time, t_G (s), antenna separation, s (m), and soil permittivity ε_r:

$$\varepsilon_r = \left(\frac{c}{v}\right)^2 = \left[\frac{c(t_G - t_A) + s}{s}\right]^2 \tag{14}$$

where t_A is the air wave arrival time, included as part of the zero time correction.

Overall, ground wave methods have proved to be promising techniques for mapping the spatial distribution of near-surface moisture (Huisman et al. 2002; Grote et al. 2003). Grote et al. (2003) suggested that this technique has the potential to detect θ values in the top 20 cm of soil depending on ground conditions and instrumentation settings. Notwithstanding the potential of the direct wave method, Huisman et al. (2003) suggested some drawbacks including (i) difficulties in separating the ground wave from refracted and reflected waves, (ii) difficulties in determining the proper antenna separation in field with varying soil water content, (iii) high attenuation of the ground wave causing constraints in the maximum antenna separation.

More recently, Laurens et al. (2005) used the ground wave technique for evaluating moisture content in pavement, while Weihermuller et al. (2007) demonstrated the unsuitability of such method for highly silty and clayey soils, due to poor data quality.

4.1.6 Borehole Transmission Measurements

The setup of borehole transmission measurements provides the transmitting and receiving antenna to be lowered into a pair of vertical access tubes. θ is evaluated by measuring the travel time of direct waves travelling across the distance separating the two nearby-placed boreholes. Two methods are commonly used namely, (i) the zero-offset profiling (ZOP) mode, where the midpoints of the lowered antennas are always at the same depth, and (ii) the multi-offset profile (MOP) where inversion of a large number of direct wave travel time between transmitter and receiver at multiple locations along the borehole lengths occurs for reconstructing two- or three-dimensional images of the soil water content distribution between the boreholes (Hubbard et al. 1997; Alumbaugh et al. 2002). In addition, vertical radar profiles (VRP) providing the measurement of the direct wave travel time between surface transmitter and receiver in a borehole are used to calculate vertical θ profiles.

Peterson (2001) obtained 2D tomograms by discretizing the area between the boreholes in rectangular cells of constant velocity and by minimizing an objective function for the estimate of each cell velocity.

One of the main issue related to the acquisition of 2D tomogram is the time required for obtaining the processed data that made the MOP mode particularly suited for steady-state θ conditions. Day-Lewis et al. (2003) studied the subsurface variations at the same radar data acquisition time scale and proposed an inversion of time-lapse tomographic data through a sequential approach aimed at providing more accurate imaging.

Another issue concerning borehole measurements is related to the correct identification of the path of the first arrival. In fact, in case of layered soils with a high degree of θ variation the first arrival may not configure as a direct wave. Rucker and Ferré (2003) and Rucker and Ferré (2004) showed that this may occur when antennas are in the low-velocity zone, since earlier arrivals of waves refracted can be observed at the layer interface.

4.1.7 Surface Reflection Methods

Surface reflection methods provide the measurement of the surface wave reflection through high-frequency off-ground GPR systems. θ is estimated by comparison between the reflection coefficient of the air-soil interface and the reflection coefficient of the air-perfect electric conductor (PEC) (Davis and Annan 2002; Serbin and Or 2004, 2005; Ghose and Slob 2006; Benedetto et al. 2013).

Theoretically, the reflection coefficient can be expressed as follows:

$$R = \frac{1 - \sqrt{\varepsilon_r}}{1 + \sqrt{\varepsilon_r}} \tag{15}$$

where ε_r is the soil permittivity that can be determined from the measured amplitude A_r and the amplitude A_m of the wave reflected from a PEC e.g., a metal plate larger than the footprint of the radar (Davis and Annan 2002; Redman et al. 2002) following the relationship given by:

$$\varepsilon_r = \left(\frac{1 + \frac{A_r}{A_m}}{1 - \frac{A_r}{A_m}} \right)^2 \tag{16}$$

The footprint of the radar can be calculated as the diameter of the first Fresnel zone (FZD) as follows:

$$FZD = \left(\frac{\lambda}{4} + 2h\lambda \right)^{\frac{1}{2}} \tag{17}$$

where λ represents the wavelength calculated at the center frequency of the GPR antenna, and h is the height of the antenna above the surface.

Evaluation of θ by this technique therefore occurs as a function of wave amplitudes, and it is not related to measurement of travel time through the medium e.g., in case of the aforementioned permittivity evaluation methods. Huisman et al. (2003) showed that the depth domain of such technique can cover the top 20 cm of soil. On the contrary, Serbin and Or (2004) estimated that such influence is concerned to only the top 1 cm of soil. Overall, the main drawbacks of this technique concern (i) the high dependence on surface roughness, such as in case of vegetation, as the reflection coefficient R is reduced by scattering from the combination of both varying θ profiles and roughness. Moreover, (ii) in case of lower θ values, this technique is very sensitive to θ variation.

4.1.8 Inverse Modelling

Over the last decade, inverse modelling of off-ground monostatic GPR systems has proved to be a very effective technique for achieving a better accuracy in the estimation of the surface dielectric permittivity and relevant water content (Lambot et al. 2004a, b, 2006a, b).

Overall, amongst the main advantages that can be reached, antenna effects can be filtered out through the antenna model so that a better accuracy in the estimation of the surface dielectric permittivity and correlated θ can be obtained. Such benefit increases when a need for a better horizontal resolution occurs, and lower distances of the antenna from the ground may be required. In addition, information on antenna height above the ground or measurements above a PEC are not necessary. This method can also take into account near-surface layering effects (Minet 2010), and various laboratory and field applications have recently confirmed its high reliability (Lambot et al. 2008, 2009).

4.1.9 Antenna Equation in the Frequency Domain

In the intrinsic far-field antenna model of Lambot et al. (2004a), which applies to plane layered media, the backscattered field over the antenna aperture requires a local plane wave field distribution such that the antenna radiation properties can be described by an equivalent single dielectric dipole.

On the basis of the linearity of Maxwell's equations, complex, frequency-dependent global reflection, and transmission coefficients account for wave propagation between the point source or field point and the radar transmission line reference plane. Antenna-medium interactions are accounted for as the above-mentioned reflection and transmission coefficients determine the antenna and transmission line internal transmissions and reflections. In the frequency domain, the radar equation expressing the relation between the radar-measured field and the 3-D layered medium Green's functions is given by Lambot et al. (2004a):

$$S_{11}(\omega) = \frac{b(\omega)}{a(\omega)} = R_i(\omega) + \frac{T(\omega)G_{xx}^{\uparrow}(\omega)}{1 - R_s(\omega)G_{xx}^{\uparrow}(\omega)} \tag{18}$$

where $S_{11}(\omega)$ is the raw radar signal expressed here as the ratio between the backscattered field $b(\omega)$ and incident field $a(\omega)$ at the radar transmission line reference plane, being ω the angular frequency; $R_i(\omega)$ is the global reflection coefficient of the antenna in free space, $T(\omega) = T_i(\omega) \cdot T_s(\omega)$ with $T_i(\omega)$ being the global transmission coefficient for fields incident from the radar reference plane onto the point source and $T_s(\omega)$ being the global transmission coefficient for fields incident from the field point onto the radar reference plane, $R_s(\omega)$ is the global reflection coefficient for the field incident from the layered medium onto the field point, and $G_{xx}^{\uparrow}(\omega)$ is the layered medium Green's function.

4.1.10 Green's Function

The theoretical basis for GPR wave propagation is given by Maxwell's equations. Green's function (i.e., the solution of the three-dimensional Maxwell's equations) for electromagnetic waves propagating in multilayered media is well-known (Michalski and Mosig 1997), and it is defined as the scattered x-directed electric field $E_x(\omega)$ at the field point for a unit-strength x-directed electric source $J_x(\omega)$ at the source point. A recursive scheme to compute the global reflection coefficients of the multilayered medium in the spectral domain can be used to derive the Green's function. The evaluation of the semi-infinite integral with a fast numerical procedure (Lambot et al. 2007) allows to perform the transformation back to space domain. In order to determine the antenna characteristic coefficients, a system of equations similar to (18) with different known Green's functions must be solved. Such functions correspond to different configurations for which the radar measurements are performed, such as measurements with the antenna at different heights over a PEC.

It has to be noted that when the antenna characteristic functions are known, it is possible to fully filter out the antenna effects from the raw radar data $S_{11}(\omega)$ to derive the measured medium Green's function $G_{xx}^{\uparrow}(\omega)$.

4.1.11 Model Inversion and Objective Function

Subsurface parameter identification by inverse modelling is a problem of non-linear optimization with the main goal to find the parameter vector $\mathbf{b} = [\varepsilon_n, \sigma_n, h_n]$, $n = 1, \ldots, N$, where N is the number of medium layers, such that an objective function $\varphi(\mathbf{b})$ is minimized.

The parameter vector $\mathbf{b}$ can be estimated for the nth layer with ε_n being the dielectric permittivity, σ_n [S m^{-1}] being the electrical conductivity, and h_n [m] being the thickness of the layer. In particular, the electrical conductivity σ is frequency-dependent assuming a linear relationship, valid if the bandwidth is not too large (Lambot et al. 2005):

$$\sigma(f) = \sigma_{f\min} + a(f - f_{\min}) \tag{19}$$

where $\sigma_{f\min}$ is the reference apparent electrical conductivity at the minimum frequency, and the medium-specific constant a is the linear variation rate of $\sigma(f)$.

The objective function φ(b) to minimize is given by:

$$\varphi(\mathbf{b}) = \left(\frac{\sum_{f\min}^{f\max} \left| G_{xx}^{*\uparrow} - G_{xx}^{\uparrow} \right|^2}{\sum_{f\min}^{f\max} \left| G_{xx}^{*\uparrow} \right|^2} \right)^{\frac{1}{2}} \tag{20}$$

where $\boldsymbol{G}_{\boldsymbol{xx}}^{*\uparrow} = G_{xx}^{*\uparrow}(\omega)$ and $\boldsymbol{G}_{\boldsymbol{xx}}^{\uparrow} = G_{xx}^{*\uparrow}(\omega\,\mathbf{b})$ are the vectors which take into account, respectively, the observed and simulated Green's functions. The solution of this minimization is found by using the global multilevel coordinate search (GMCS, (Huyer and Neumaier 1999)) sequentially combined with the Nelder-Mead Simplex algorithm (Lambot et al. 2002), since the non-linearity of the problem usually leads to multiple local minima.

4.1.12 Petrophysical Relationships

It is well-known from the literature that the permittivity of liquid water ranges from 78 to 88 as a function of several parameters, such as temperature and salinity (Cassidy 2009; Steelman and Endres 2011), while considerably lower values are encountered for the permittivity of air, equal to 1, and the permittivity of the solid grains of sand and loamy soils, that typically ranges between 4 and 6 (Cassidy 2009). Such large dielectric contrast between these three phases, mostly due to the high polarizability of water molecules, has led to the use of different types of approach for predicting volumetric soil water content. Overall, three classes of petrophysical relationships are commonly used, namely, the empirically-based relationships, the volumetric mixing formulae, and the effective medium approximations. It is worth noting that the use of an appropriate petrophysical relationship between the permittivity of the bulk material and its volumetric soil water content needs to be properly adopted for a reliable estimate of water content. In that respect, practitioners and end-users can exploit a wide range of research studies on various soils with different textural properties that may effectively lead to the best method to be adopted. A comprehensive comparison of petrophysical relationships for soil moisture estimation can be found in Steelman and Endres (2011).

Empirical Relationships

Empirical relationships are based on field and laboratory measurements. The main advantage from their use consists of a lower complexity than physical models, thus a higher ease of implementation is achieved according to specific soil textures and water content conditions. In this regard, very limited information about soil textural properties (e.g., porosity, bulk density, pore structure) are required.

Over the last decades, many relationships between θ and ε_r have been developed by numerous researchers using different range of soil textures.

In line with this, Topp et al. (1980) proposed the most used empirical relationship in the form of a third-order polynomial:

$$\theta = -5.3 \times 10^{-2} + 2.92 \times 10^{-2}\varepsilon_r - 5.5 \times 10^{-4}\varepsilon_r^2 + 4.3 \times 10^{-6}\varepsilon_r^3 \quad (21)$$

Such relationship was obtained by manufacturing four different fine-grained soil textures ranging from sandy loam to clay with a range of fresh and saltwater mixtures and saturation levels. The dielectric properties of soil samples placed in a coaxial transmission line were then evaluated through a TDR probe. Results demonstrated that this petrophysical relationship was suited to predict moisture conditions ranging from air dry to water saturated, with an error of estimate of 0.013 m^3 m^{-3}. In addition, the authors found the independence of permittivity measurement from soil density, texture, dissolved salt content, and temperature within the range of soil conditions tested.

The effects of water content layering (e.g., dry over wet, wet over dry) in silt loam soil through different TDR probe designs were investigated by Nadler et al. (1991). Similarly to the Topp relationship, a third-order polynomial equation was obtained. The authors demonstrated the accuracy of soil water content estimates from the TDR method regardless from both probe type and soil layering, with the exception of very dry soil overlying very wet soil, wherein difficult interpretations of traces were encountered.

More recently, the frequency dependent effects of measurements on water estimation in the GPR frequency range were tackled by Curtis (2001) and Patriarca et al. (2013).

More enhanced empirical relationships incorporating the effects of soil texture, density, and organic matter on permittivity measurements were developed by Jacobsen and Schjonning (1993) and Malicki et al. (1996), finding prediction errors of the same order of magnitude as Topp equation.

Volumetric Mixing Formulae

A more theoretical approach for relating θ to ε_r is based on volumetric mixing formulae, which use the volume fraction and the dielectric permittivity of each soil constituent to derive a relationship (e.g., Dobson et al. 1985; Roth et al. 1990; Jones and Friedman 2000; Tosti et al. 2013).

Generally, the form of these relationships in a system accounting for n dielectric components can be expressed as follows:

$$(\varepsilon_r)^\alpha = \sum_{i=1}^{n} f_i \left(\varepsilon_{r,i}\right)^\alpha \quad (22)$$

where α is a geometrical fitting parameter describing the structure within the heterogeneous medium (Lichtenecker and Rother 1931). f_i and $\varepsilon_{r,i}$ represent, respectively, the volume fraction and the permittivity of the ith constituent.

As regards near-surface soil moisture evaluation, it can be modeled as a typical three-phase system of air, solid, and water constituents as follows:

$$\theta = \frac{\varepsilon_r^\alpha - (1-\phi)\varepsilon_s^\alpha - \phi\varepsilon_a^\alpha}{\varepsilon_w^\alpha - \varepsilon_a^\alpha} \tag{23}$$

where Ø is the soil porosity, ε_r is the permittivity of the bulk material, ε_a, ε_s, ε_w are the permittivity values of air, soil matrix, and water components, respectively. Enhanced versions of volumetric mixing formulae considered a four-phase mixing model incorporating the contribution given by bound water on the solid mineral grain to the bulk permittivity (Dobson et al. 1985). It is worth mentioning that the geometric fitting parameter ranges between −1 and 1 for electric fields oriented perpendicular and parallel to a layered medium, respectively. The geometric factor is equal to 0.5 by assuming the travel time through the mixture equivalent to the sum of the volume-weighted travel times through the individual components. In addition, if ε_a is assumed to be equal to 1, (23) becomes the complex refractive index model (CRIM) (Birchak et al. 1974; Dobson et al. 1985; Roth et al. 1990; Heimovaara et al. 1994; Gorriti and Slob 2005) given by:

$$\theta = \frac{1}{\sqrt{\varepsilon_w} - 1}\sqrt{\varepsilon_r} - \frac{(1-\phi)\sqrt{\varepsilon_s} - n}{\sqrt{\varepsilon_w} - 1} \tag{24}$$

The CRIM model has been found to produce adequate results in most cases. Over the past years, different values of the geometric parameter α have been proposed, ranging from 0.46 for a three-phase medium (Roth et al. 1990), or 0.65 for a four-phase medium (Dobson et al. 1985). More recently, Brovelli and Cassiani (2008) showed α values spanning from 0.25 to 0.8 on the basis of geometry of mineral grains and pore fluids. Furthermore, Patriarca et al. (2013) presented a method for evaluating the optimal α values for two saturated sandy-textured soils at different clay rates from 0 to 25 % by weight. Such method revealed to be relatively helpful in three-phase systems (i.e., sand, air, and clay), and in four-phase systems (i.e., sand, clay, air, and water) in the dry end-member cases with 10 % clay content. The authors demonstrated that the application of a mixing model may fail for clayey soils in wet conditions, because of the swelling properties of clay. In particular, the use of such model in clay-rich soils may lead to under-predict electrical properties.

Effective Medium Approximations

This approach takes into account the textural and structural contribution of each component to the permittivity of a composite material.

Effective medium approximations are based on microscale geometric models incorporating textural and structural information about the heterogeneous system into the prediction of dielectric properties.

Research works based on such methods have proved that the assumption of considering the dielectric properties of one medium as a function of porosity and water content is not sufficient for a comprehensive modelling, thus unreliable moisture content may be estimated.

Many researchers have used effective medium approximations (Fiori et al. 2005) by incorporating pore structure or geometry (Endres and Bertrand 2006), grain shape (Sen 1984; Jones and Friedman 2000), and fluid distribution at the pore scale (Endres and Redman 1996; Chen and Or 2006).

Amongst the various effective medium approximations in literature, the EMA approach proposed by Fiori et al. (2005) provides an approximate water content-permittivity relationship obtained by employing the effective medium approximation in a random mixture of phases. The latter is modelled through the Multi-Indicator model with spherical elements of variable radii. The following equation was proposed:

$$\theta = \frac{(\varepsilon_w + 2\varepsilon_r)[(\varepsilon_a + 2\varepsilon_r)(\varepsilon_r - \varepsilon_s) + 3\varepsilon_r\phi(\varepsilon_s - \varepsilon_a)]}{2\varepsilon_r(2\varepsilon_r + \varepsilon_s)(\varepsilon_w - \varepsilon_a)} \tag{25}$$

where $\emptyset$ is the soil porosity, ε_r is the permittivity of the bulk material,, ε_a, ε_s, ε_w are the permittivity of air, soil matrix, and water components, respectively. The relative volume fractions are $n_s = 1 - \emptyset$ (solid volumetric fraction), $n_w = \theta$ (volumetric water content), $n_a = \emptyset - \theta$ (air volumetric fraction). The permittivity values of air and water are kept fixed, and once the bulk porosity is given, the proposed formula depends on one parameter, namely, the permittivity of the solid phase ε_s.

Furthermore, another approach known as the differential effective medium (DEM) approach by Norris et al. (1985) has been widely used to examine the dielectric properties of soils and rocks (Sen 1984; Endres and Bertrand 2006). Such method provides infinitesimally small volumes of the inclusions sequentially embedded into the heterogeneous system according to an iterative inclusion embedding process. The electrical property of the effective medium resulting from the aforementioned embedding process accounts for the electrical property of the background material for the further embedding step. From a mathematical point of view, Endres and Bertrand (2006) used a homogenization variable p that monotonically increases from 0 to 1 at the beginning and at the end of the embedding process, respectively.

4.2 Water Content Estimation by the Frequency Shift Method

A recent method based on signal processing in the frequency domain was proposed by Benedetto (2010). The main point of strength relies on the not need for destructive core sampling to calibrate the system, such that the volume fractions of the three phases in the medium are not accounted for. In that regard, θ can be directly estimated by frequency analysis without evaluating the dielectric permittivity. The main assumption concerns that in unsaturated soils electromagnetic waves are scattered by water droplets (Drude 1902), thereby a shifting of the frequency of the waves occurs (Bekefi and Barrett 1987; Bohren and Huffman 1983). It is worth noting that such an approach can be only focused on water content estimation purposes.

Rayleigh scattering is used to explain the shifting of the frequency of the scattered signals. A shift in the frequency distribution of the reflected signals has been observed in the past, although the cause of the shift was not identified or investigated. Narayana and Ophir (1983) analyzed ultrasonic waves reflected by normal and fatty livers, thereby noticing that different frequencies were attenuated in different media. The authors identified the Rayleigh scattering as the primary cause of the observed non-linear behaviour of reflected waves in fatty livers. In particular, Rayleigh scattering was originated by the presence of fat globules, which exhibited fourth-order frequency dependence. More recently, Ho et al. (2004) showed that a frequency shift also occurs when detecting buried mines with different shapes and sizes.

4.2.1 Rayleigh Scattering

It is well-known that scattering is generated by singularities or non-homogeneities in electromagnetic impedance. The process is described as Rayleigh scattering whether the dimensions of these non-uniformities are much smaller than the wavelength of the EM wave. The size of a scattering particle can be defined analytically by the ratio $x = 2\pi r/\lambda$, with r being the radius of the particle, and λ being the wavelength of the signal. By definition, Rayleigh scattering occurs in the small size parameter regime $x \ll 1$. Scattering from larger spherical particles is explained by Mie (1908) for an arbitrary size parameter x. In line with this, the Mie theory reduces to the Rayleigh approximation when small values of x are considered.

4.2.2 Theoretical Model

By means of both several assumptions on the three-phase porous medium properties and simplifications of the physics, Benedetto (2010) defined the following formulation:

$$I(\theta,\nu) = I_0(\nu)\frac{1+\cos^2\theta}{2R^2}\left[\frac{2\pi\nu}{c_0}\sqrt{\mu_r\left(\varepsilon_\infty + \frac{\Delta\varepsilon}{1+\nu^2\tau^2}\right)}\right]^4 \left[\frac{\mu_r\left(\varepsilon_\infty + \frac{\Delta\varepsilon}{1+\nu^2\tau^2}\right)-1}{\mu_r\left(\varepsilon_\infty + \frac{\Delta\varepsilon}{1+\nu^2\tau^2}\right)+2}\right]^2 \left(\frac{d}{2}\right)^6 \tag{26}$$

where R is the distance between the observer and the particle, θ is the angle of scattering, v is the frequency of the electromagnetic signal, c_0 is the velocity of free space, μ_r is the magnetic permeability, ε_∞ is the dielectric constant of the full-polarized medium at an infinite frequency electromagnetic field, $\Delta\varepsilon = \varepsilon_{static} - \varepsilon_\infty$ is the difference between the permittivity values of a steady and an infinite frequency electromagnetic field, τ is the relaxation time, and d is the diameter of the particle.

A non-linear modulation of the electromagnetic signal is produced by scattering, as a function of the moisture content. In this regard, the peak of frequency is assumed to be a comprehensive indicator, negatively related to moisture. According to (26), the various frequency components of the frequency spectra are differently scattered, depending on the soil type and water content.

A regression law, based on experimental evidences, was proposed to predict moisture content θ, expressed in %, from the value of the peak of frequency f_P, expressed in Hz $\times$ 10^8:

$$\theta = (A - f_P)/B \tag{27}$$

where A and B are regression coefficients. Table 1 lists the values of A and B and regression coefficients for five different soils (Benedetto 2010):

Further enhancements demonstrated the potential of this method to detect clay in soils, as the presence of clay is strictly related to moisture, by virtue of its strong hygroscopic capacity. In this regard, Tosti et al. (2013) proposed a semi-empirical model for assessing clay content [%] in a bi-phase configuration (dry condition) of the medium for three types of soil. Benedetto and Tosti (2013) presented super resolution methods, namely, parabolic, triangular, and sinc-based interpolators, to further refine the location of the frequency peak, thereby allowing for more accurate prediction of clay content, even for small amounts of clay.

In addition, recent applications of the frequency shift method showed the high potential of such technique in the field of civil engineering. Benedetto et al. (2012a) developed an effective survey method for efficiently monitoring the geotechnical stability of transportation infrastructures at the rural road network scale. Both off-ground (Tosti et al. 2014) and ground-coupled (Benedetto et al. 2013) radar systems with different center frequencies of investigation were used. A model for θ estimation from both traditional permittivity-based techniques (e.g., reflectivity and ground wave methods) and the frequency shift method was developed. In particular, θ variations along the longitudinal road axis that overcame an alarm threshold were considered to be critical to landslide occurrence. The authors provided a relocation of the former old dated landslide boundary, such that the most effective rehabilitation and maintenance actions were identified. Moreover, Benedetto

Table 1 Values of the parameters of the model from (27) (Benedetto 2010)

	Sandy soil	Alluvial soil	Subgrade material (Sm)	Sm + 5 % clay	Sm + 20 % clay	Average value
A $\times$ 10^8	5.7	7.0	6.2	5.6	5.3	5.7
B $\times$ 10^7	1.1	2.1	2.3	1.7	1.5	1.4

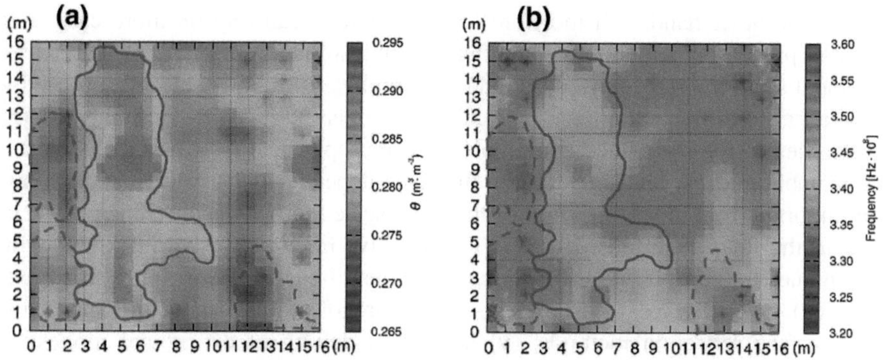

Fig. 3 Comparison between wet (*solid yellow line*) and dry (*grey dashed line*) areas using time domain and frequency domain methods for moisture evaluation. **a** Soil moisture map inferred by using a site-specific empirical relationship from Topp et al. (1980) after evaluating near-surface relative permittivity trough surface reflection method. **b** Map of frequency spectra peaks derived by the frequency shift method

et al. (2013) provided field measurement of a ground-coupled radar system over a 16 m × 16 m parcel within an agricultural field in order to map soil moisture variations. Very promising results were obtained by comparing the moisture map from the use of a site-specific empirical relationship by Topp et al. (1980), wherein near-surface permittivity was evaluated through the surface reflection technique, and the map of frequency peaks of spectra by using the frequency shift method, considered as a proxy of the near-surface moisture spatial field (Fig. 3).

5 Summary

In this chapter, it is demonstrated that the contribution of GPR in volumetric water content determination is being increasingly spread over a wide range of applications. In moisture content assessment of concrete structures, analysis of GPR signal in time domain have been traditionally carried out through reflection and direct wave methods, showing good results if compared to other common non-destructive techniques (Laurens et al. 2005; Sbartaï et al. 2006a; Hugenschmidt and Loser 2008). Nevertheless, the need for a reflector and a prior knowledge of its position i.e., reflection methods, and the weakly proportional linearity between signal parameters and water saturation levels, as resulted in some direct wave analyses, suggests the integration with other processing techniques, that could provide additional information on moisture content. In this regard, signal processing in the frequency domain may allow to assess the frequency dependence of water in concrete. Promising results were obtained over the last few years (Dérobert et al. 2009; Sbartaï et al. 2009) in laboratory environment, although further works is needed for validating results through in field tests.

On the other hand, soil moisture sensing in soils can rely on more case studies and processing techniques. Traditionally, an estimation of ε_r at the GPR measurement scale is firstly carried out and a petrophysical relationship is then used to convert ε_r to θ. In such a case, particular care must be taken to the mineralogical properties of the soil, since commonly used petrophysical relationships could not be suitable at every site. In addition, the scale-dependency between θ and ε_r plays an important role in the selection of the local-scale petrophysical relationship relative to the field-scale to investigate. Overall, five main classes of GPR surveying techniques for permittivity estimation can be identified, namely, (i) reflection methods, (ii) ground wave measurements, (iii) borehole transmission measurements, (iv) surface reflection methods, (v) inverse modelling of off-ground monostatic GPR systems. The first four listed above, are established techniques often revealing as time-consuming (i.e., reflection methods), minor invasive (borehole transmission measurements), dependent on soil surface roughness (i.e., surface reflection methods), and difficult in filtering out antenna effects from the raw radar data (i.e., ground wave measurements). In this regard, inverse modelling of off-ground monostatic GPR systems has proved high effectiveness in accurately predict surface dielectric permittivity and water content correlated (Lambot et al. 2004a). More recently, a self-consistent method has been developed based on signal processing in the frequency domain (Benedetto 2010). In particular, such method is focused on the Rayleigh scattering of the signal according to the Fresnel theory. Its main advantage consists in avoiding destructive core sampling for calibrating the system.

Concerning further insights, one of the major open issues is in determining hydraulic soil properties in a changing water content profile from time-lapse monitoring measurements. Another important open issue to be addressed in the future is related to the possibility of converting the measured ε_r to estimates of hydrogeological properties.

Acknowledgments The authors acknowledge the COST Action TU1208 “Civil Engineering Applications of Ground Penetrating Radar”, supporting this work.

References

Al-Qadi, I.L., Haddad, R.H., Riad, S.M.: Detection of chlorides in concrete using low radio frequencies. J. Mat. Civil Eng. **9**(1), 29–34 (1997)

Alumbaugh, D., Chang, P., Paprocki, L., Brainard, J., Glass, R.J., Rautman, C.A.: Estimating moisture contents in the vadose zone using cross-borehole ground penetrating radar: a study of accuracy and repeatability. Water Resour. Res. **38**, 1309 (2002)

Annan, A.: GPR-history, trends, and future developments. Subsurf. Sens. Technol. Appl. **3**, 253–270 (2002)

ASTM C 876-91: Standard test method for half-cell potentials of uncoated reinforcing steel in concrete. Ann. Book ASTM Stand. **04**(02) (2009)

Bekefi, G., Barrett, A.H.: Waves in dielectrics. Electromagnetic Vibrations Waves, and Radiation, pp. 426–440. MIT Press, Cambridge (1987)

Benedetto, A., Pensa, S.: Indirect diagnosis of pavement structural damages using surface GPR reflection techniques. J. Appl. Geophys. **62**, 107–123 (2007)

Benedetto, A.: Water content evaluation in unsaturated soil using GPR signal analysis in the frequency domain. J. Appl. Geophys. **71**, 26–35 (2010)

Benedetto, A., Benedetto, F., Tosti, F.: GPR applications for geotechnical stability of transportation infrastructures. Nondestr. Test. Eval. **27**(3), 253–262 (2012)

Benedetto, A., Tosti, F., Ortuani B., Giudici, M., Mele, M.: Soil moisture mapping using GPR for pavement applications. Paper presented at the 7th international workshop on advanced ground penetrating radar (IWAGPR), pp. 243–248. Nantes, 2–5 July 2013, doi:10.1109/IWAGPR.2013.6601550

Benedetto, F., Tosti, F.: GPR spectral analysis for clay content evaluation by the frequency shift method. J. Appl. Geophys. **1**(97), 89–96 (2013)

Berktold, A., Wollny, K.G., Alstetter, H.: Subsurface moisture determination with the ground wave of GPR. Paper presented at the 7th international conference on ground-penetrating radar, pp. 675–680. Lawrence, KS, May 1998

Bevan, B.W.: The search for graves. Geophysics **56**, 1310–1319 (1991)

Birchak, J.R., Gardner, C.G., Hipp, J.E., Victor, J.M.: High dielectric constant microwave probes for sensing soil moisture. Proc. IEEE **62**, 93–98 (1974)

Bohren, C.F., Huffman, D.: Absorption and scattering of light by small particles. Wiley, New York (1983)

Borhan, M.S., Parsons, L.R.: Monitoring of soil water content in a citrus grove using capacitance ECH2O probes. Paper no. 042110. Paper presented at the ASAE annual meeting, Ottowa, Canada, 1–4 August 2004

Briggs, L.J.: Electrical instruments for determining the moisture, temperature, and soluble salt content of soils. USDA Div. Soils Bull. **10**, pp. 52 (1899)

Brovelli, A., Cassiani, G.: Effective permittivity of porous media: a critical analysis of the complex refractive index model. Geophys. Prospecting. **56**, 715–727 (2008)

Campbell, G.S., Calissendorff, C., Williams, J.H.: Probe for measuring soil specific heat using a heat-pulse method. Soil Sci. Soc. Am. J. **55**, 291–293 (1991)

Cassidy, N.J.: Electrical and magnetic properties of rocks, soils and fluids. In: Jol, H.M. (ed.) Ground Penetrating Radar: Theory and Applications, pp. 41–72. Elsevier, Amsterdam (2009)

Chen, Y., Or, D.: Geometrical factors and interfacial processes affecting complex dielectric permittivity of partially saturated porous media. Water Resour. Res. **42**, W06423 (2006)

Curtis, J.O.: Moisture effects on the dielectric properties of soils. IEEE Trans. Geosci. Remote Sens. **39**, 125–128 (2001)

Dalton, F.N., Herkelrath, W.N., Rawlins, D.S., Rhoades, J.D.: Time domain reflectometry: simultaneous measurement of soil-water content and electrical-conductivity with a single probe. Science **224**, 989–990 (1984)

Daniels, D.J.: Ground Penetrating Radar, 2nd edn. The Institunial Electrical Engineering, London (2004)

Davis, J., Annan, A.P.: Ground penetrating radar for high resolution mapping of soil and rock stratigraphy. Geophys. Prospect. **37**, 531–551 (1989)

Davis, J.L., Annan, A.P.: Ground penetrating radar to measure soil water content. In: Dane, J.H., Topp, G.C. (eds.) Methods of Soil Analysis, Part 4, Soil Science Society of America (SSSA), pp. 446–463 (2002)

Day-Lewis, F.D., Lane, J.W., Harris, J.M., Gorelick, S.M.: Time-lapse imaging of saline-tracer transport in fractured rock using difference-attenuation radar tomography. Water Resour. Res., **39**(10) (2003). doi:10.1029/2002WR001722

Dérobert, X., Villain, G., Cortas, R., Chazelas, J.L.: EM characterization of hydraulic concretes in the GPR frequency band using a quadratic experimental design. Paper presented at the NDT conference on civil engineering, pp. 177–182. Nantes, France, 30 June–3 July 2009

Di Matteo, A., Pettinelli, E., Slob, E.: Early-time GPR signal attributes to estimate soil dielectric permittivity: a theoretical study. IEEE T Geosci. Remote **51**, 1643–1654 (2013)

Dix, C.H.: Seismic velocities from surface measurements. Geophysics **20**(1), 68–86 (1955)

Dobson, M.C., Ulaby, F.T., Hallikainen, M.T., El-Rayes, M.A.: Microwave dielectric behaviour of wet soil. Part II. Dielectric mixing models. IEEE Trans. Geosci. Remote **23**, 35–46 (1985)

Drude, P.: The Theory of Optics, pp. 268–396. Longmans, Green, and Co, New York (1902)
Du, S.: Determination of water content in the subsurface with the ground wave of ground penetrating radar. Ph.D. thesis. Ludwig-Maximilians-Universitat, Munich, Germany (1996)
Du, S., Rummel, P.: Reconnaissance studies of moisture in the subsurface with GPR. Paper presented at the 5th international conference on ground penetrating radar, pp. 1241–1248 (1994)
Endres, A.L., Bertrand, E.A.: A pore-size scale model for the dielectric properties of water-saturated clean rocks and soils. Geophysics **71**, F185–F193 (2006)
Endres, A.L., Redman, J.D.: Modelling the electrical properties of porous rocks and soil containing immiscible contaminants. J. Environ. Eng. Geophys. **1**, 105–112 (1996)
Fellner-Feldegg, H.: Measurement of dielectrics in time domain. J. Phys. Chem. **73**, 616–623 (1969)
Fiori, A., Benedetto, A., Romanelli, M.: Application of the effective medium approximation for determining water contents through GPR in coarse-grained soil materials. Geophys. Res. Lett. **32**, L09404 (2005)
Garambois, S., Senechal, P., Perroud, H.: On the use of combined geophysical methods to assess water content and water conductivity of near-surface formations. J. Hydrol. **259**, 32–48 (2002)
Gaskin, G.J., Miller, J.D.: Measurement of soil water content using a simplified impedance measuring technique. J. Agric. Eng. Res. **63**, 153–159 (1996)
Ghose, R., Slob, E.C.: Quantitative integration of seismic and GPR reflections to derive unique estimates for water saturation and porosity in subsoil. Geophys. Res. Lett. **33**(5) (2006). doi :10.1029/2005GL025376
Gorriti, A.G., Slob, E.C.: Comparison of the different reconstruction techniques of permittivity from S-parameters. IEEE Trans. Geosci. Remote **43**, 2051–2057 (2005)
Gowers, K.R., Millard, S.G.: Pulse mapping techniques for corrosion monitoring of reinforced concrete structures. In: Swamy, R.N. (ed.) Corrosion and Corrosion Protection of Steel in Concrete Sheffield Academic Press, pp. 1 86–199 (1994)
Greaves, R.J., Lesmes, D.P., Lee, J.M., Toksoz, M.N.: Velocity variations and water content estimated from multi-offset, ground-penetrating radar. Geophysics **61**, 683–695 (1996)
Grote, K., Hubbard, S.S., Rubin, Y.: GPR monitoring of volumetric water content in soils applied to highway construction and maintenance. Lead. Edge **21**, 482–485 (2002)
Grote, K., Hubbard, S.S., Rubin, Y.: Field-scale estimation of volumetric water content using GPR ground wave techniques. Wat Resour. Res. **39**(11) (2003)
Heimovaara, T.J., Bouten, W., Verstraten, J.M.: Frequency domain analysis of time-domain reflectometry waveforms: 2. A four component complex dielectric mixing model for soils. Water Resour. Res. **30**, 201–209 (1994)
Ho, K.C., Gader, P.D., Wilson, J.N.: Improving landmine detection using frequency domain features from ground penetrating radar. Presented at the 2004 IEEE international geoscience and remote sensing symposium, IGARSS '04, vol. 3, pp. 1617–1620 (2004)
Hubbard, S., Rubin, Y., Majer, E.: Ground-penetrating-radar-assisted saturation and permeability estimation in bimodal systems. Water Resour. Res. **33**, 971–990 (1997)
Hubbard, S., Grote, K., Rubin, Y.: Estimation of nearsubsurface water content using high frequency GPR ground wave. Lead. Edge Explor. Soc. Explor. Geophys. **21**(6), 552–559 (2002)
Hugenschmidt, J., Loser, R.: Detection of chlorides and moisture in concrete structures with ground penetrating radar. Mater. Struct. **41**, 785–792 (2008)
Huisman, J., Hubbard, S., Redman, J., Annan, A.P.: Measuring soil water content with ground penetrating radar: a review. Vadose Zone J. **2**, 476–491 (2003)
Huisman, J.A., Snepvangers, J.J.J.C., Bouten, W., Heuvelink, G.B.M.: Mapping spatial variation in surface soil water content: comparison of ground-penetrating radar and time domain reflectometry. J. Hydrol. **269**, 194–207 (2002)
Huisman, J.A., Sperl, C., Bouten, W., Verstraten, J.M.: Soil water content measurements at different scales: accuracy of time domain reflectometry and ground-penetrating radar. J. Hydrol. **245**(1–4), 48–58 (2001)
Huyer, W., Neumaier, A.: Global optimization by multilevel coordinate search. J. Global Optim. **14**, 331–355 (1999)

Ihamouten, A., Villain, G., Dérobert, X.: Complex permittivity frequency variation from multi-offset GPR data: hydraulic concrete characterization. IEEE Trans. Instrum. Measur. **61**(6), 1636–1648 (2012)

Jacobsen, O.H., Schjønning, P.: A laboratory calibration of time domain reflectometry for soil water measurements including effects of bulk density and texture. J. Hydrol. **151**, 147–157 (1993)

Jones, S.B., Friedman, S.P.: Particle shape effect on the effective permittivity of anisotropic or isotropic media consisting of aligned or randomly oriented ellipsoidal particles. Water Resour. Res. **36**, 2821–2833 (2000)

Keller, G.V., Frischknecht, F.C.: Electrical methods in geophysical prospecting. Pergamon Press, New York (1966)

Klysz, G., Balayssac, J.-P., Laurens, S.: Spectral analysis of radar surface waves for nondestructive evaluation of cover concrete. NDT E Int. **37**(3), 221–227 (2004)

Krause, M., Bämann, R., Frielinghaus, R., Fretzschmar, F., Kroggel, O., Langeberg, K., Maierhofer, C., Müller, W., Neisecke, J., Schickert M., Schmitz, V., Wiggenhauser, H., Wollbold, F.: Comparison of pulse-echo-methods for testing concrete. Presented at the international symposium non-destructive testing in civil engineering, pp. 281–295 (1995)

Lambot, S., Javaux, M., Hupet, F., Vanclooster, M.: A global multilevel coordinate search procedure for estimating the unsaturated soil hydraulic properties. Water Resour. Res. **38**(11), 1224 (2002). doi:http://dx.doi.org/10.1029/2001WR001224

Lambot, S., Slob, E.C., van den Bosch, I., Stockbroeckx, B., Vanclooster, M.: Modeling of ground-penetrating radar for accurate characterization of subsurface electric properties. IEEE Trans. Geosci. Remote Sens. **42**, 2555–2568 (2004a)

Lambot, S., Antoine, M., van den Bosch, I., Slob, E.C., Vanclooster, M.: Electromagnetic inversion of GPR signals and subsequent hydrodynamic inversion to estimate effective vadose zone hydraulic properties. Vadose Zone J. **3**(4), 1072–1081 (2004b)

Lambot, S., van den Bosch, I., Stockbroeckx, B., Druyts, P., Vanclooster, M., Slob, E.C.: Frequency dependence of the soil electromagnetic properties derived from ground-penetrating radar signal inversion. Subsurf. Sens. Technol. Appl. **6**, 73–87 (2005)

Lambot, S., Antoine, M., Vanclooster, M., Slob, E.C.: Effect of soil roughness on the inversion of off-ground monostatic GPR signal for noninvasive quantification of soil properties. Water Resour. Res. **42**(3) (2006a). doi:10.1029/2005WR004416

Lambot, S., Weihermuller, L., Huisman, J.A., Vereecken, H., Vanclooster, M., Slob, E.C.: Analysis of air-launched ground-penetrating radar techniques to measure the soil surface water content. Water Resour. Res. **42**(11) (2006b). doi:10.1029/2006WR005097

Lambot, S., Slob, E.C., Vereecken, H.: Fast evaluation of zero-offset Green's function for layered media with application to ground-penetrating radar. Geophys. Res. Lett. **34**(L21405) (2007). doi:http://dx.doi.org/10.1029/2007GL031459

Lambot, S., Slob, E.C., Chavarro, D., Lubczynski, M., Vereecken, H.: Measuring soil surface moisture in irrigated areas of southern Tunisia using full waveform inversion of proximal GPR data. Near Surf. Geophys. **6**(6), 403–410 (2008)

Lambot, S., Rhebergen, J., Slob, E.C., Lopera, O., Jadoon, K.Z., Vereecken, H.: Remote estimation of the hydraulic properties of a sandy soil using full-waveform integrated hydrogeopohysical inversion of timelapse off-ground GPR data. Vadose Zone J. **8**, 743–754 (2009)

Laurens, S., Rhazi, J., Balayssac, J.-P., Arliguie, G.: Assessment of corrosion in reinforced concrete by ground penetrating radar and half-cell potential tests. Presented at the RILEM workshop on life prediction and aging management of concrete structures, Cannes, France (2000)

Laurens, S., Balayssac, J.-P., Rhazi, J., Arliguie, G.: Influence of concrete moisture upon radar waveform. RILEM Mater. Struct. **35**(248), 198–203 (2002)

Laurens, S., Balayssac, J.-P., Rhazi, J., Klysz, G., Arliguie, G.: Non-destructive evaluation of concrete moisture by GPR: experimental study and direct modeling. Mater. Struct. **38**(283), 827–832 (2005)

Lichtenecker, K., Rother, K.: Die herleitung des logarithmischen mischungsgesetzes aus allgemeinen prinzipien der stätionaren strömung. Phys. Z. **32**, 255–260 (1931)

Loeffler, O., Bano, M.: Ground penetrating radar measurements in a controlled vadose zone: influence of the water content. Vadose Zone J. **3**(4), 1082–1092 (2004)

Lopez, W., Gonzalez, J.A.: Influence of the degree of pore saturation on the resistivity of concrete and the corrosion rate of steel reinforcement. Cem. Concr. Res. **23**(2), 368–376 (1993)

Lunt, I.A., Hubbard, S.S., Rubin, Y.: Soil moisture content estimation using ground penetrating radar reflection data. J. Hydrol. **307**(1–4), 254–269 (2005)

Malicki, M.A., Plagge, R., Roth, C.H.: Improving the calibration of dielectric TDR soil moisture determination taking into account the solid soil. Eur. J. Soil Sci. **47**, 357–366 (1996)

Maser, K., Scullion, T.: Automated pavement subsurface profiling using radar: case studies of four experimental field sites. Transp. Res. Rec. **1344**, 148–154 (1992)

Michalski, K.A., Mosig, J.R.: Multilayered media Green's functions in integral equation formulations. IEEE Trans. Antennas Propag. **45**, 508–519 (1997)

Mie, G.: Beiträge zur Optik trüber medien, speziell kolloidaler metallösungen. Ann. Phys. **330**, 377 (1908)

Minet, J., Lambot, S., Slob, E.C., Vanclooster, M.: Soil surface water content estimation by full-waveform GPR signal inversion in presence of thin layers. IEEE Trans. Geosci. Remote Sens. **48**(3), 1138–1150 (2010)

Mínguez Maturana R.: Aplicación del geo-radar 3D multifrecuencia como herramienta de alto rendimiento para la detección de zonas de acumulación de humedad en obras lineales. Rutas Técnica, vol. 150, 6 pp (2012)

Mitchell, T.M.: Radioactive/nuclear methods. In: CRC Handbook on Nondestructive Testing of Concrete. CRC Press, Florida, pp. 227–252 (1991)

Montemor, M.F., Simoes, A.M.P., Ferreira, M.G.S.: Chloride-induced corrosion on reinforcing steel: from the fundamentals to the monitoring techniques. Cem. Concr. Compos. **25**, 491–502 (2003)

Nadler, A., Dasberg, S., Lapid, I.: Time domain reflectometry measurements of water content and electrical conductivity of layered soil columns. Soil Sci. Soc. Am. J. **55**, 938–943 (1991)

Nakashima, Y., Zhou, H., Sato, M.: Estimation of groundwater level by GPR in an area with multiple ambiguous reflections. J. Appl. Geophys. **47**, 241–249 (2001)

Narayana, P.A., Ophir, J.: On the frequency dependence of attenuation in normal and fatty liver. IEEE Trans. Sonics Ultrason. **30**(6), 379–383 (1983)

Neville, A.M.: Properties of Concrete, 4th edn, pp. 277–284. Longman, Essex (1995)

Njoku, E.G., Entekhabi, D.: Passive microwave remote sensing of soil moisture. J. Hydrol. **184**, 101–129 (1996)

Norris, A.N., Callegari, A.J., Sheng, P.: A generalized differential effective medium theory. J. Mech. Phys. Solids **33**, 525–543 (1985)

Ohdaira, E., Masuzawa, N.: Water content and its effect on ultrasound propagation in concrete—the possibility of NDE. Ultrasonics **38**(1–8), 546–552 (2000)

Patriarca, C., Tosti, F., Velds, C., Benedetto, A., Lambot, S., Slob, E.C.: Frequency dependent electric properties of homogeneous multi-phase lossy media in the ground-penetrating radar frequency range. J. Appl. Geophys. **1**(97), 81–88 (2013)

Peterson, J.E.: Pre-inversion corrections and analysis of radar tomographic data. J. Environ. Eng. Geophys. **6**, 1–18 (2001)

Pettinelli, E., Vannaroni, G., Di Pasquo, B., Mattei, E., Di Matteo, A., De Santis, A., Annan, P.A.: Correlation between near-surface electromagnetic soil parameters and early-time GPR signals: An experimental study. Geophysics **72**(2), A25–A28 (2007)

Polder, R., Andrade, C., Elsener, B., Vennesland, O., Gulikers, J., Weidert, R., Raupach, M.: RILEM TC 154-EMC: electrochemical techniques for measuring metallic corrosion. Mater. Struct. **33**, 603–611 (2000)

Pourbaix, M.: Atlas of electrochemical equilibria in aqueous solutions. Pergamon Press, Oxford (1966)

Redman, J.D., Davis, J.L., Galagedara, L.W., Parkin, G.W.: Field studies of GPR air launched surface reflectivity measurements of soil water content. In: Proceesings of SPIE. Presented at the 9th Conference On Ground-Penetrating Radar, vol. 4758, pp. 156–161 (2002)

Robert, A.: Dielectric permittivity of concrete between 50 MHz and 1 GHz and GPR measurements for building materials evaluation. J. Appl. Geophys. **40**(1–3), 89–94 (1998)
Roberts, R.L., Daniels, J.J.: Modeling near-field GPR in three dimensions using the FDTD method. Geophysics **62**(4), 1114–1126 (1997)
Robinson, D.A., Jones, S.B., Wraith, J.M., Or, D., Friedman, S.P.: A review of advances in dielectric and electrical conductivity measurement in soils using time domain reflectometry. Vadose Zone J. **2**(4), 444–475 (2003)
Rodeick, C.A.: Roadbed void detection by ground penetrating radar. Highw. Heavy Constr. **127**, 60–61 (1984)
Roels, S., Carmeliet, J.: Analysis of moisture flow in porous materials using microfocus X-ray radiography. Int. J. Heat Mass Transfer **49**(25–26), 4762–4772 (2006)
Roth, K., Schulin, R., Fluhler, H., Attinger, W.: Calibration of time domain reflectometry for water content measurement using composite dielectric approach. Water Resour. Res. **26**, 2267–2273 (1990)
Rucker, D.F., Ferré, P.A.: Near-surface water content estimation with borehole ground penetrating radar using critically refracted waves. Vadoze Zone J. **2**, 247–252 (2003)
Rucker, D.F., Ferré, T.P.A.: Correcting water content measurement errors associated with critically refracted first arrivals on zero offset profiling borehole ground penetrating radar profiles. Vadose Zone J. **3**(1), 278–287 (2004)
Saleem, M., Shameem, M., Hussain, S.E., Maslehuddin, M.: Effect of moisture, chloride and sulphate contamination on the electrical resistivity of Portland cement concrete. Constr. Build. Mater. **10**(3), 209–214 (1996)
Samouelian, A., Cousin, I., Tabbagh, A., Bruand, A., Richard, G.: Electrical resistivity survey in soil science: a review. Soil Tillage Res. **83**, 173–193 (2005)
Sbartaï, Z.M., Laurens, S., Balayssac, J.-P., Arliguie, G., Ballivy, G.: Ability of the direct wave of radar ground-coupled antenna for NDT of concrete structures. NDT E Int. **39**(5), 400–407 (2006a)
Sbartaï, Z.M., Laurens, S., Balayssac, J.-P., Ballivy, G., Arliguie, G.: Effect of concrete moisture on radar signal amplitude. ACI Mater. **419**, 103–426 (2006b)
Sbartaï, Z.M., Laurens, S., Rhazi, J., Balayssac, J.-P., Arliguie, G.: Using radar direct wave for concrete condition assessment: Correlation with electrical resistivity. J. Appl. Geophys. **62**, 361–374 (2007)
Sbartaï, Z.M., Laurens, S., Breysse, D.: Concrete moisture assessment using radar NDT technique—comparison between time and frequency domain analysis. Presented at non-destructive testing in civil engineering (NDTCE '09), Nantes, France, June 30–July 3, 2009
Sen, P.N.: Grain shape eff ects on dielectric and electrical properties of rocks. Geophysics **49**, 586–587 (1984)
Serbin, G., Or, D.: Near-surface water content measurements using horn antenna radar: Methodology and overview. Vadose Zone J. **2**, 500–510 (2003)
Serbin, G., Or, D.: Ground-penetrating radar measurement of soil water content dynamics using a suspended horn antenna. IEEE Trans. Geosci. Remote Sens. **42**(8), 1695–1705 (2004)
Serbin, G., Or, D.: Ground-penetrating radar measurement of crop and surface water content dynamics. Remote Sens. Environ. **96**(1), 119–134 (2005)
Sheets, K.R., Hendrickx, J.M.H.: Noninvasive soil-water content measurement using electromagnetic induction. Water Resour. Res. **31**, 2401–2409 (1995)
Soutsos, M.N., Bungey, J.H., Miljard, S.G., Shaw, M.R., Patterson, A.: Dielectric properties of concrete and their influence on radar testing. NDT and E Int. **34**(6), 419–425 (2001)
Sperl, C.: Determination of spatial and temporal variation of the soil water content in an agroecosystem with ground-penetrating radar (In German). Ph.D. thesis. Technische Universitat Munchen, Munich, Germany (1999)
Steelman, C.M., Endres, A.L.: Comparison of petrophysical relationships for soil moisture estimation using GPR ground waves. Vadose Zone J. **10**, 270–285 (2011)
Sybilski, D., Bańkowski, W., Sudyka, J., Krysiński, L.: Reasons of premature cracking pavement deterioration—a case study. RILEM 2012–2012, pp. 31–169

Tapley, B.D., Bettadpur, S., Ries, J.C., Thompson, P.F., Watkins, M.M.: GRACE measurements of mass variability in the earth system. Science **305**, 503–505 (2004)

Tillard, S., Dubois, J.-C.: Analysis of GPR data: wave propagation velocity determination. J. Appl. Geophys. **33**, 77–91 (1995)

Topp, G., Davis, J.L., Annan, A.P.: Electromagnetic determination of soil water content: measurements in coaxial transmission lines. Water Resour. Res. **16**, 574–582 (1980)

Tosti, F., Patriarca, C., Slob, E.C., Benedetto, A., Lambot, S.: Clay content evaluation in soils through GPR signal processing. J. Appl. Geophys. **1**(97), 69–80 (2013)

Tosti, F., Benedetto, A., Calvi, A.: Efficient air-launched ground-penetrating radar inspections in a large-scale road network. Paper presented at the 3rd international conference on transportation infrastructure, pp. 703–709. Pisa, Italy, April 2014. doi:10.1201/b16730-103

Ulaby, F.T., Dubois, P.C., van Zyl, J.: Radar mapping of surface soil moisture. J. Hydrol. **184**, 57–84 (1996)

van Overmeeren, R.A., Sariowan, S.V., Gehrels, J.C.: Ground penetrating radar for determining volumetric soil water content; results of comparative measurements at two sites. J. Hydrol. **197**, 316–338 (1997)

Vaughan, C.J.: Ground penetrating radar surveys used in archaeological investigations. Geophysics **51**, 595–604 (1986)

Weihermuller, L., Huisman, J.A., Lambot, S., Herbst, M., Vereecken, H.: Mapping the spatial variation of soil water content at the field scale with different ground penetrating radar techniques. J. Hydrol. **340**(3–4), 205–216 (2007)

Wobschall, D.: Frequency-shift dielectric soil-moisture sensor. IEEE Trans. Geosci. Remote Sens. **16**, 112–118 (1978)

Yilmaz, O.: Seismic data processing. SEG Invest. Geophys. **62**, 1758–1773 (1987)

Part III
EM Methods for Near-Field Scattering Problems by Buried Structures; Data Processing Techniques

Methods for the Electromagnetic Forward Scattering by Buried Objects

Cristina Ponti

Abstract Methods developed to solve forward electromagnetic scattering by buried objects are useful tools for interpreting data from Ground Penetrating Radar responses. Time-domain methods, as Finite-Difference Time Domain or space-time integral equations, are well established tools in the modeling impulse Ground Penetrating Radar systems. Integral equation methods, when solved with Method of Moments discretization, lead to dense linear system. Therefore, the implementation of novel approaches approximating the integral equation via series expansions with lower computational complexity is called for. Analytical techniques have the advantage to be accurate and fast, as the geometry of the scattering problem is taken into account by an expansion of the fields in terms of suitable basis functions.

1 Introduction

The modeling of forward electromagnetic scattering by buried objects has a wide interest in the field of Ground Penetrating Radar (GPR) (Daniels 2004; Jol et al. 2008), for several aims. The analytical and numerical modeling of the GPR scenarios can improve the understanding of the complex interaction occurring between the electromagnetic field radiated by the radar antenna and the buried targets. The simulated data can be also used to test new inversion algorithms and data processing techniques, and in general as useful tools to rightly understand data from GPR surveys.

Several physical as well geometrical parameters must be taken into account in the scattering model to make it as realistic as possible. Such parameters deal with the target geometry, the nature of the background medium, the source field radiated by the radar antenna, and the general environment affecting the measurements.

C. Ponti (✉)
Department of Engineering, Roma Tre University, Via Vito Volterra 62, 00146 Rome, Italy
e-mail: cristina.ponti@uniroma3.it

A. Benedetto and L. Pajewski (eds.), *Civil Engineering Applications of Ground Penetrating Radar*, Springer Transactions in Civil and Environmental Engineering,
DOI 10.1007/978-3-319-04813-0_8

Many GPR applications, from the geophysical investigations, to the civil engineering surveys of roads, or the mapping of buried pipes, cables, conduits and tunnels, may be modeled by a two-dimensional (2D) scattering problem, as the transverse cross-section of the targets is small compared to the longitudinal size. In the simplest scenario, two dielectric half-spaces bounded by a planar interface, and a cylindrical target, dielectric or perfectly-conducting, buried in the lower medium may be used. Anyway, even with targets of canonical shape, as circular cross-section cylinders, the development of an analytical solution turns out to be not straightforward as in free space (Balanis 1989; Bowman et al. 1987), due to the presence of the planar discontinuity relevant to the interface. A very common technique for the modeling of 2D problems of scattering by buried objects is the integral-equation method (D'Yakonov 1959; Howard 1972; Ogunade 1981; Mahmoud et al. 1981; Budko and van den Berg 1999; Butler et al. 1985; Hongo and Hamamura 1986; Naqvi et al. 2000a, b; Ahmed and Naqvi 2008). The first analytical solution to the scattering of a line source by a cylinder in a homogeneous half-space has been proposed in the pioneering work by D'Yakonov (1959). Numerical evaluation of theory developed in D'Yakonov (1959) was done by by Howard (1972), who solved a two-dimensional integral equation by point-matching technique, in case of line source excitation. D'Yakonov's analytical work was extended by Ogunade (1981) in a for more suitable to obtain numerical results.

The analytical modeling of the 2D scattering problems by buried objects in a semi-infinite medium has been implemented with the Cylindrical Wave Approach (CWA) in Di Vico et al. (2005a, b); Frezza et al. (2012, 2013a, c).

The simple 2D model with two half-spaces has to be extended in many cases to model a multilayered medium as background of the scattering problem. Even the basic layered geometry with a dielectric slab between two half-spaces needs a cumbersome modeling, taking into account all the multiple reflections between the interfaces bounding the slab (Michalski and Zheng 1990a, b; Naqvi et al. 1998; Naqvi and Rizvi 2000; Jia and Yasumoto 2005; Lee 1999; Zhuck and Yarovoy 1994; Frezza et al. 2009, 2010). A multilayers model has important applications to characterize a soil with layers of different permittivities, a pavement, a or a wall of a building (Lambot 2014; Paknys 2003). The multilayered layout may also be useful to implement problems of Through-Wall scattering, when the targets are placed below the multilayer, as proposed with the CWA in Frezza et al. (2013b).

One or more rough interfaces may be also taken into account to deal with the typical unevenness of natural interfaces bounding the medium hosting buried targets. An analytical solution based on spectral plane-wave representation of the fields is employed by Lawrence and Sarabandi (2002); solution is given in the far-field as spectral integrals are evaluated with asymptotic techniques. Integral equation with numerical discretization approaches are also proposed in the literature. The particular case of a sinusoidal interface is solved by in Cottis and Kanellopoulos (1992), using Green's function theory combined with the extended boundary condition approach. In Altuncu et al. (2006), surfaces with arbitrary and localized roughnesses are considered, and the problem is solved with an integral equations/MOM approach. A geometry with a cylinder buried in layered media with rough interfaces is proposed in Kuo and Moghaddam (2002). Extended Boundary Condition Method and T-matrix algorithm are applied to construct reflection and transmission matrices of arbitrary rough interfaces as well as of an isolated single cylinder, respectively. A cylinder buried below a slightly rough interface is solved with CWA in Fiaz et al. (2012, 2013, 2014). In Fiaz et al. (2012, 2013), numerical solution is given for a sinusoidal surface profile, whereas an arbitrary profile is tackled in Fiaz et al. (2014).

The most common approach for the scattering by buried objects turns out to be the integral equation method. The resulting equations are solved in an approximated way, mainly with Method of Moments (MoM), as in (Michalski and Zheng 1990a, b), or asymptotic techniques, as in (Butler et al. 1985; Hongo and Hamamura 1986; Naqvi et al. 2000a, b; Ahmed and Naqvi 2008). Asymptotic techniques lead to limitation in the burial depth of the cylinder, or can lead to results only in the far-field region. As to MoM solution, the integral equations are solved in a discretized way, generally leading to a dense linear system, that turns out to be more and more costly the more complex is the scenario to be modeled. Another approach is the Contrast Source-Extended Born (CS-EB) method, i.e., a full-wave integral equation model proposed by Isernia et al. (2004) for free space scattering problems, and extended by Crocco et al. (2009) to the analysis of two-dimensional subsurface scattering problems. In the CS-EB method a rewriting of the traditional integral equation of two-dimensional scattering problems in terms of the so-called Contrast-Source equation is performed, exploiting the properties of Green's function in lossy media. Then, the obtained equation is linearized via series expansions, in a way very similar to the EB approximation. In this new approach, forward scattering problems can be conveniently solved by means of very simple series expansions, which allow a lower computational complexity and memory storage with respect to other iterative schemes.

In Bourlier et al. (2013), the integral equation solved with Method of Moments are addressed to evaluate the scattering by canonical objects, as a cylinder or a plate, by a randomly rough surface, and also by an object below or above a flat or rough surface. In the same book, the MoM is also combined with Extension-Propagation Inside Layer Method to efficiently solve the scattering by two scatterers, i.e. a coated cylinder, a cylinder below a flat or rough surface, or two rough surfaces.

Time-domain techniques, especially Finite-Difference Time-Domain (FDTD) method (Taflove and Hagness 2000; Kunz and Luebbers 1993), are the most suitable to simulate GPR responses. A free of charge FDTD software tool developed to simulate GPR responses is GprMax, developed by Giannopoulos (2005). Both two-dimensional (GprMax2D) and three-dimensional (GprMax3D) models can be analyzed with this FDTD software. GprMax2D is mainly used for GPR "signature" simulations, whereas GprMax3D is used for a more realistic modeling of GPR problems, especially when comparisons with measured GPR data must be performed. The main limitations in the FDTD method are in the staircase approximation of curved interfaces, and in the conditionally stable nature of the FDTD; moreover, the choice of small spatial steps in the simulation of structures of fine geometry in a large computation domain results in high computer memory requirements and long execution times. A solution to this problem has been proposed in Diamanti and Giannopoulos (2009, 2011), where subgrids with small spatial steps are introduced in a coarser FDTD grid. The implementation of the subgrids is particularly useful to model parts of the computation domain with finer detail, or when regions with high dielectric constants supporting waves propagating at very short wavelengths are included in the computational mesh.

The Chapter is organized as follows; in Sect. 2 the main issues relevant to definition of a simulation layout for a problem of forward electromagnetic scattering by buried objects are defined. Two techniques of simulation of forward scattering by buried objects are presented. In Sect. 3, Finite Difference Time Domain Method fundamentals are described. In Sect. 4, the theory of CWA is illustrated, through an example of application of scattering by buried circular cross-section cylinders with excitation from a line source.

2 The Simulation Layout

The scenarios measured in of the GPR surveys are often complex. The target is placed in a natural environment that can be difficultly modeled in a schematic way. The background medium, that may be a subsurface soil, a road, or a building is often non homogenous, and the target itself may have an irregular shape. The physical and geometrical parameters implemented in the forward solvers must be capable to represent the scattering scenario in a manner which is as realistic as possible. The basic elements to be modeled are the target, the source field radiated by the radar antenna, and the background medium. Such elements are represented at the bottom in the flow diagram in Fig. 1, showing the main issues relevant to the solution of a scattering problem: (1) Layout definition; (2) Method for the forward solution; (3) Computational issues. The final layout is decided on the basis of the modeling technique. On the side of numerical methods, quite general scenarios can be modeled, whereas on the side of analytical techniques it is often necessary to approximate the problem using canonical geometries. Therefore, the modeling technique determines the degree of complexity that can be achieved in the simulation of the scattering scenarios. The choice of the technique is made also on the basis of a time versus frequency analysis. The used technique entails three fundamental issues: accuracy, computational times, and memory requirements. High degree of complexity in the simulation of physical as well as geometrical details has a cost in terms of long execution times and high memory requirements.

In many cases, a simple scheme may be adopted for the characterization of the space under analysis. In particular, when the transverse dimensions of the target are much longer than the cross-section, a two dimensional model may be used. Moreover, according to the wavelength and the dissipative properties of the

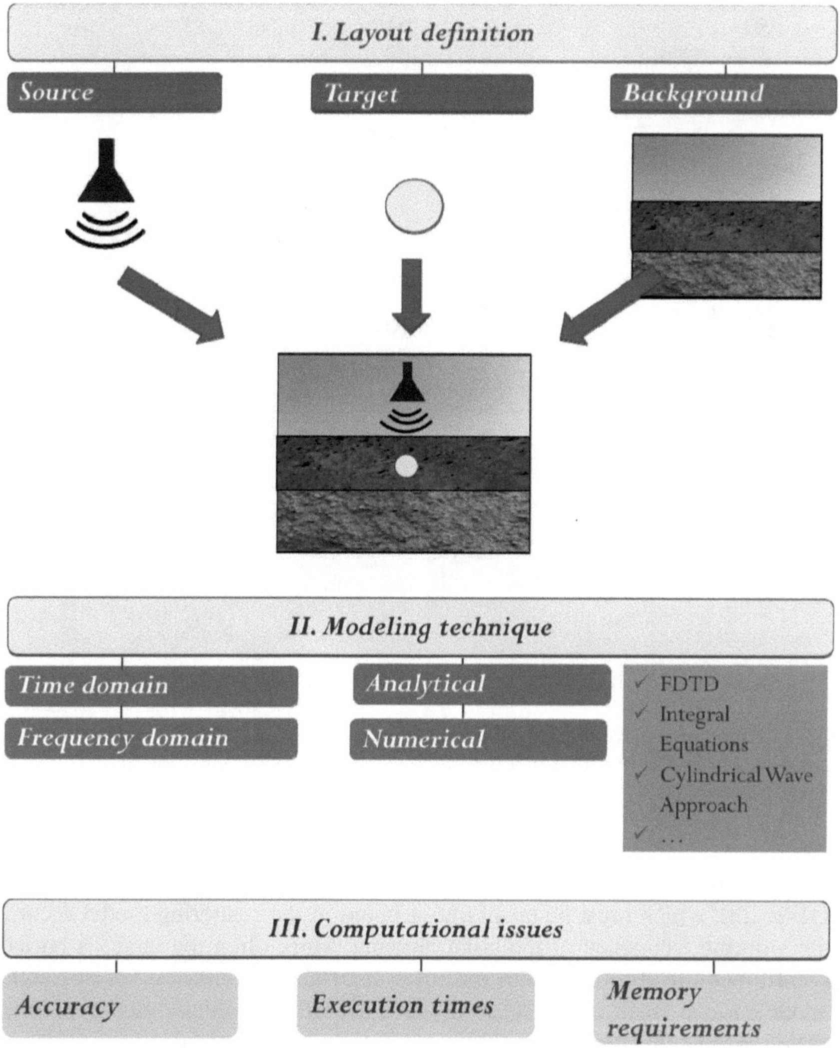

Fig. 1 Flow Diagram for the modeling of a scattering problem, from the definition of the geometrical and physical layout, that include models for the source, the target, and the background, to the choice of the modeling technique, and the computational issues deriving from the chosen technique

hosting material, a short penetration depth may justify a modeling of the problem in a limited domain. This is the simplest case reported in Fig. 2a, where the problem is represented by a 2D problem with two half-space separated by a flat interface. The upper half-space is filled with air (permittivity $\varepsilon_1 = \varepsilon_0$), and is the medium of propagation of the field radiated by the source mounted on the GPR equipment. The lower half-space (permittivity $\varepsilon_2 = \varepsilon_0\ \varepsilon_{r2}$) models the medium

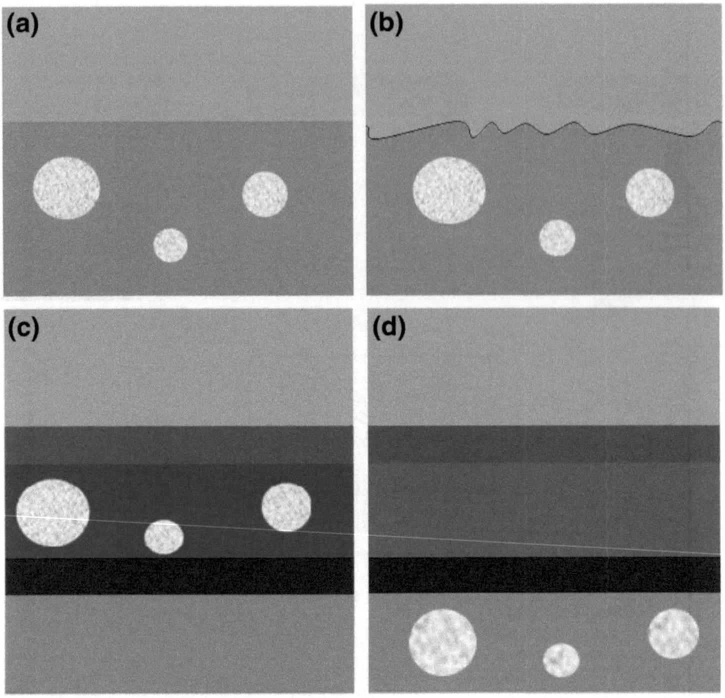

Fig. 2 Possible simple layout of two-dimensional modeling of a scattering problems, with circular cross-section targets buried: **a** in a semi-infinite medium, below a flat interface; **b** in a semi-infinite medium below a rough interface; **c** in a layered medium, **d** below a layered medium

hosting one or more buried scatterers. Real surfaces have unavoidable irregularities (Fig. 2b), which must be taken into account in the scattering model according to the working frequency of the radar antenna. More often the target is buried in a layered medium (Fig. 2c), with bounding interface that may be flat or rough, to be modeled as a stack of dielectric layers with different thickness and permittivity. The simplest layered layout is the one in Fig. 2c, that has to be modeled when penetration depth is such that it reaches the next medium. Otherwise, target itself may be buried below a multilayer, as in the basic case of Fig. 2d. A particular case of target may be also the multilayer itself, used in the modeling of roads' internal structure or of a general subsurface soil, in the frame of geophysical investigations.

The cross-section of target may be canonical (circular, square, …): this is the case of the buried services, as pipes and cables, or subsurface conduits and tunnels. Such targets may be accurately modeled through analytical techniques, as the involved geometries can be easily taken into account with suitable field expressions.

3 The Finite-Difference Time-Domain Method

Time-domain techniques are well suited in the frame of GPR applications, dealing with the scattering by a pulsed signal (Taflove and Hagness 2000; Kunz and Luebbers 1993). Among the possible approaches, FDTD method is a well-established one.

The starting point for the implementation of FDTD are the time-dependent Maxwell's equation:

$$\nabla \times \mathbf{E} = -\frac{\partial \mathbf{B}}{\partial t} \tag{1}$$

$$\nabla \times \mathbf{H} = -\frac{\partial \mathbf{D}}{\partial t} + \mathbf{J} \tag{2}$$

$$\nabla \cdot \mathbf{D} = \rho \tag{3}$$

$$\nabla \cdot \mathbf{B} = 0 \tag{4}$$

Equations (1)–(4) are written in a linear isotropic medium, with field and sources set to zero at the initial time, conveniently taken as time zero. The fields defined in the Maxwell equations are:

- **E**: electric field (V/m)
- **H**: magnetic field (A/m)
- **D**: electric induction field (A/m)
- **B**: magnetic induction field (A/m)
- **J**: electric current density (A/m^2)
- **ρ**: volumetric charge density (C/m^3)

where:

$$\mathbf{D}(\mathbf{r}) = \varepsilon_0 \varepsilon_r(\mathbf{r})\mathbf{E} \tag{5}$$

$$\mathbf{B}(\mathbf{r}) = \mu_0 \mu_r(\mathbf{r})\mathbf{H} \tag{6}$$

The two divergence equations turn out to be redundant, as already contained in the curl Eqs. (1) and (2). After replacing the (5) and (6) in (1) and (2), and the relation $\mathbf{J} = \sigma\mathbf{E}$, curl equations are cast in the form used in the FDTD analysis:

$$\frac{\partial \mathbf{H}}{\partial t} = -\frac{1}{\mu}\nabla \times \mathbf{E} - \frac{\sigma *}{\mu}\mathbf{H} \tag{7}$$

$$\frac{\partial \mathbf{E}}{\partial t} = -\frac{\sigma}{\varepsilon}\mathbf{E} - \frac{1}{\varepsilon}\nabla \times \mathbf{H} \tag{8}$$

The curl Eqs. (1) and (2) are discretized using a rectangular grid, that stores the field components and the material properties ε_r and μ_r. On the same grid, central differences for space and time derivatives are evaluated. The procedure returns a set of finite-difference equations, and updates the field components in time.

Finite differences replace derivatives as follows:

$$\frac{\partial f}{\partial t} = \lim \Delta t \to 0 \frac{f(x,t_2) - f(x,t_1)}{\Delta t} \approx \frac{f(x,t_2) - f(x,t_1)}{\Delta t} \quad (9)$$

$$\frac{\partial f}{\partial x} = \lim \Delta x \to 0 \frac{f(x_2,t) - f(x_1,t)}{\Delta x} \approx \frac{f(x_2,t) - f(x_1,t)}{\Delta x} \quad (10)$$

where in the approximation the time and space intervals have been turned into finite from infinitesimal.

The total volume must be discretized in a three-dimensional grid of unit 'field cells' of size Δx, Δy, and Δz. Cell size must much less than the smallest wavelength to get accurate results. Once the cell size has been fixed, the maximum time step follows from Courant condition.

For a 3D cell with size Δx, Δy, and Δz, the stability condition is the following:

$$\Delta t = \frac{1}{c\sqrt{\left(\frac{1}{\Delta x}\right)^2 + \left(\frac{1}{\Delta y}\right)^2 + \left(\frac{1}{\Delta z}\right)^2}} \quad (11)$$

Therefore, the temporal discretization step Δx and the spatial discretization Δx, Δy, and Δz cannot be assigned independently of each other, as FDTD is a conditionally stable numerical process. For a 2D case, the stability condition can be derived from (11) putting $\Delta z \to \infty$.

The electric and magnetic field are defined within each individual cell through a Cartesian component, i.e., $E(x, y, z)$ and $H(x, y, z)$, and they are staggered in such a way to form the Yee cell (Fig. 3).

Using the finite difference approximation of Maxwell equations, the electric field can be evaluated at any point in space, and at any time, from the knowledge of its neighboring magnetic field.

Let us examine how to build the Yee cell on the *xy* plane (Fig. 4). We solve the following scalar equation from the vector Eq. (8), supposing for the current density $\mathbf{J} = 0$:

$$\frac{\partial H_x}{\partial t} = -\frac{1}{\mu}\left(\frac{\partial E_y}{\partial z} - \frac{\partial E_z}{\partial x}\right) \quad (12)$$

The *x*-component of the magnetic field $E_y(x, y, z)$ on the Yee cell is defined as follows:

$$E_y^n(i\Delta x, j\Delta y, k\Delta z, n\Delta t) = E_y^n(i,j,k) \quad (13)$$

The field components and $E_z(x, y, z)$ and $H_x(x, y, z)$ are expressed in a similar manner within the Yee cell. Equation (12) must be turned to a finite difference form

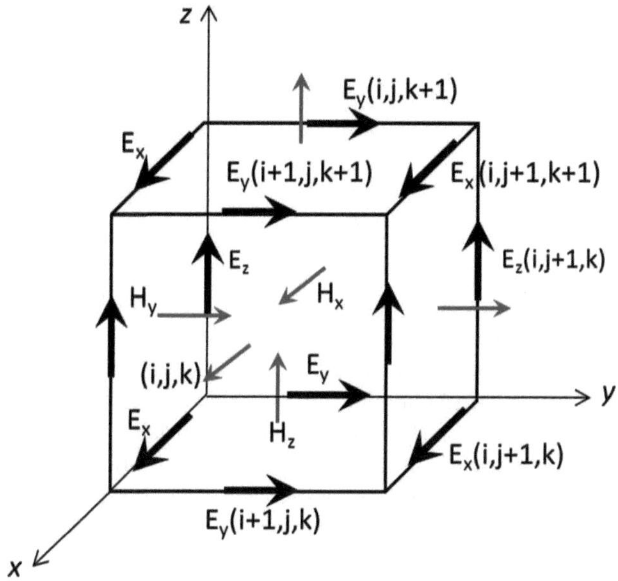

Fig. 3 The Yee cell geometry

Fig. 4 The Yee cell in the *xy* plane

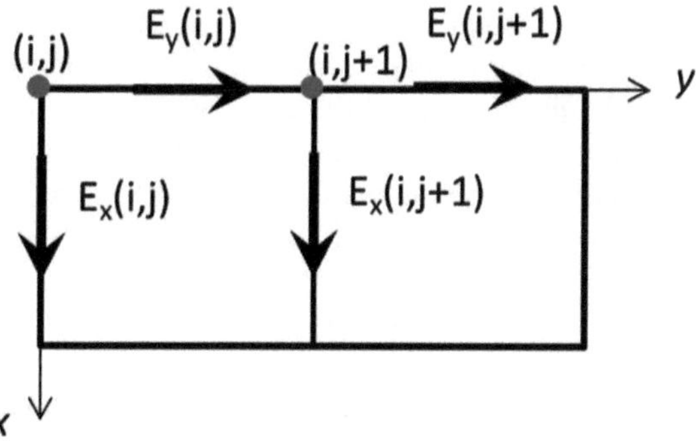

from a differential one, in order to remove space and time derivatives. Using approximations (9) and (10), Eq. (12) becomes:

$$\begin{aligned}&\frac{H_y^{n+1/2}(i,j+1/2,k+1/2)-H_y^{n-1/2}(i,j+1/2,k+1/2)}{\Delta t}\\&=\left[\frac{E_y^n(i,j+1/2,k+1)-E_y^n(i,j+1/2,k)}{\Delta z}\right.\\&\left.\quad-\frac{E_z^n(i,j+1,k+1/2)-E_z^n(i,j,k+1/2)}{\Delta x}\right]\end{aligned} \tag{14}$$

Equation (14) is cast in a form more suitable for FDTD computations:

$$\begin{aligned} H_y^{n+1/2}(i,j+1/2,k+1/2) = H_y^{n-1/2}(i,j+1/2,k+1/2) \\ + \frac{\Delta t}{\mu \Delta z}\left[E_y^n(i,j+1/2,k+1) - E_y^n(i,j+1/2,k)\right] \\ + \frac{\Delta t}{\mu \Delta y}\left[E_z^n(i,j,k+1/2) - E_z^n(i,j+1,k+1/2)\right] \end{aligned} \quad (15)$$

The equations for the field components E_x, E_y, E_z, H_x, and H_z are determined in a similar manner. In the Yee algorithm, the E-field components, calculated at times $n\Delta t$, are located at the edges of the grid cell, whereas the H-field components, evaluated at times $(n + 1/2)\Delta t$, are located at the face-centered points.

Using the finite-difference approximation of the Maxwell equations, the electric field can be evaluated at any point in space and any time from knowledge of the magnetic field in four neighboring points, and vice versa.

As to the source field, in the FDTD formulation it is specified analytically through any function of time $f(t)$. In the GPR applications, the source field is a pulse that may be given by a sine, Gaussian pulse, or a Ricker one.

The computational domain in the FDTD technique is necessarily bounded. Therefore, when modeling open boundary problems, as the GPR ones, the computational space is truncated at a finite distance from the source and the targets. Absorbing Boundary Conditions (ABC) are applied, that, while truncating the computational domain, absorb any waves impinging on them in order to simulate an unbounded space. ABC may be accomplished with several schemes. The most common are the Mur Absorbing Boundary conditions by Mur (1981), and the Perfectly Matched Layer (PML) by Berenger (1994). PMLs are special anisotropic and non-physical layers, that placed at the boundaries of the FDTD absorb any incident wave.

FDTD is a general-purpose electromagnetic technique. When used for the simulation of forward scattering by buried objects in the frame of GPR applications, FDTD has several advantages over other numerical methods. Being developed in the time-domain, the simulation of broadband GPR responses is straightforward. Dispersive materials can be also considered. Highly complex scenarios can be simulated, where fine details of the targets as well as of the surrounding environment can be taken into account through a suitable geometrical decomposition. Among the main limitations of such technique are its conditionally stable nature and the use of staircase approximation of curved geometries. When a small spatial step must be introduced for an adequate modeling of the objects in the computation domain, the allowable discretization time-step is small as well. Moreover, the choice of small spatial steps in the simulation of structures of fine geometry in a large computation domain results in high computer memory requirements and long execution times. A solution to this problem has been proposed by Diamanti and Giannopoulos in (2009), where subgrids with small spatial steps are introduced in a coarser FDTD grid. The implementation of the subgrids is particularly useful to model parts of the computation domain with finer detail, or when regions with high dielectric constants supporting waves propagating at very short wavelengths are included in the computational mesh.

4 The Cylindrical-Wave Approach

In this Section is described an example of analytical method for the analysis of the forward scattering by buried objects, the Cylindrical Wave Approach (CWA). CWA is a technique developed in the spectral domain to solve two-dimensional scattering problems by buried circular cross-section cylinders, with parallel radii and parallel to a planar interface. In its original formulation (Borghi et al. 1997) the technique was introduced to solve the scattering of a cylinder placed above a planar interface, and then extended to the modelling of scattering by buried objects, placing the cylinders below the interface, in a semi-infinite medium (Di Vico et al. 2005a, b).

Among the fundamentals of the method are:

- use of expansions into cylindrical waves to express the fields scattered by the cylinders;
- use of the plane-wave spectrum of a cylindrical function to evaluate the interaction of the cylindrical waves with the planar boundary, in terms of reflection and transmission;
- accurate evaluation of the obtained spectral integrals, to get results in far- as well near-field region.

The idea to expand the scattered fields by circular cross-section cylinder into cylindrical waves is also the basic principle of the analytical scattering by an isolated cylinder in free space (Balanis 1989). Anyway, the presence of at least one planar discontinuity to bound the medium hosting the buried cylinders turns the analytical solution to a higher degree of complexity, as the main geometry with circular boundaries is broken up. Due to the introduction of the interface, evaluation of transmission and reflection of the cylindrical waves used as basis functions of the scattered fields must be done. As Fresnel reflection and transmission coefficients are introduced only for plane-wave reflection and transmission, reflection and transmission of cylindrical waves is accomplished using plane-wave spectrum of cylindrical waves, presented in Cincotti et al. (1993).

A suitable use of the plane-wave spectrum of a cylindrical has been successfully applied in further extensions of the CWA, to deal with more compound geometries of the scattering problem. In Frezza et al. (2009, 2010), the method has been extended to scatterers buried in a dielectric layer, whereas in Frezza et al. (2013b) scatterers have been placed below a layer, to implement a through wall geometry. Scattering by a cylinder placed below a rough surface has been solved in Fiaz et al. (2012), with solution in the case of a surface with sinusoidal profile. The same surface profile has been proposed in Fiaz et al. (2013), but in this

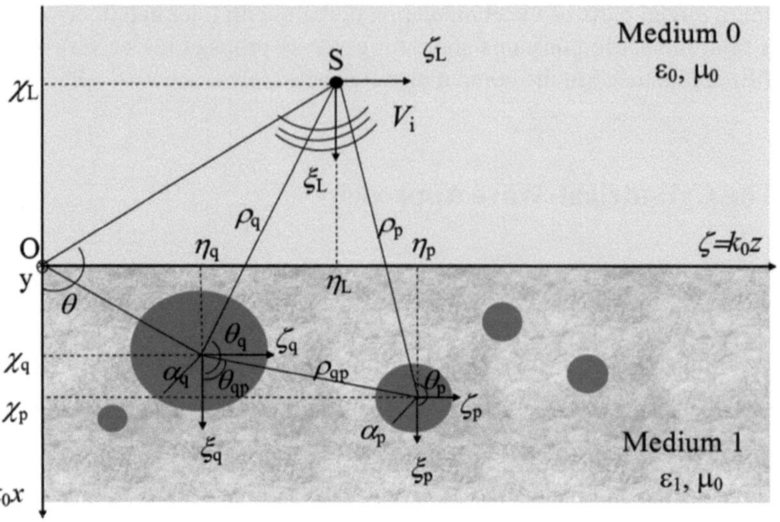

Fig. 5 Geometry of line-source scattering

work spectral integrals have been solved with asymptotic techniques. Solution to scattering by an arbitrarily rough surface has been developed in Fiaz et al. (2014), where a surface with Gaussian roughness spectrum has been implemented.

In the above-mentioned works, the involved dielectric media are linear, isotropic, homogeneous, and losses. Solution of CWA in case of a lossy medium has been presented in Frezza et al. (2013a, c). In Frezza et al. (2013a), an isolated cylinder is buried in a semi-infinite and dissipative half-space, having developed a solution for the spectrum of a cylindrical wave in a medium with complex permittivity. In Frezza et al. (2013c), CWA is solved for a set of cylinders buried in the same dissipative half-space.

The possibility to simulate a source field alternative to the plane wave excitation has been firstly investigated in Frezza et al. (2012), where the field radiated by a line-source excitation has been considered. In the frame of CWA, such a field is itself a cylindrical wave, and its interaction of a planar surface of separation between two media leads to a reflected and a transmitted cylindrical waves.

In order to introduce the CWA, the analytical solution of the method to a geometry of scattering of a line source by cylinders buried in a semi-infinite medium is here presented. The geometry is pictured in Fig. 5. Two semi-infinite media are considered, which are linear homogeneous, isotropic, and lossless. The upper medium is filled with air ($\varepsilon_1 = \varepsilon_0$, $\mu_1 = \mu_0$). A main reference frame of normalized coordinates (O, ξ, ζ) is used, being $\xi = k_0 x$ and $\zeta = k_0 z$, with k_0 the vacuum wavenumber. Cylinders have radii α_p, and centres in (χ_p, η_p), with p = 1, …, N, with axes parallel to the *y*-axis. Further N coordinates frame centred on each cylinder are introduced, i.e., (O, ξ_p, $\zeta_{p,}$), with $\xi_p = k_0 x_p$ and $\zeta = k_0 z_p$, and p = 1, …, N. In this spectral-domain analysis a time-dependence $e^{i\omega t}$ is omitted.

The incident field is a line-source with center in (χ_L, η_L); a further reference frame centred on the source is used: (O, ξ_L, ζ_L). The total field in each medium is described by a scalar function $V(\xi, \zeta)$, which stands for the electric field E_y, in the TM or E polarization state, or for the magnetic field H_y, in the TE or H one. Such a function is decomposed into several field contributions which are due to the interaction of the incident field, i.e. the scalar function V_i, with the interface and the cylinders. In particular, the reflected and transmitted fields V_r and V_t, respectively, are defined (Fig. 6a): they describe the interaction between the incident field and the interface, in the absence of the buried cylinders. Furthermore, as the transmitted field V_t impinges on the buried objects, scattered fields are also defined (Fig. 6b), i.e. the field V_s scattered by the cylinders in Medium 1, and the fields V_{sr} and V_{st}, i.e. the scattered-reflected and scattered-transmitted field, respectively, dealing with reflection and transmission of V_s by the interface (Fig. 6b).

The line source radiates a y-directed field expressed through a Hankel function of first kind and zero order, with argument proportional to the distance from the source to the observation point:

$$V_i(\xi,\zeta) = -V_0 H_0^{(1)}\left[\sqrt{(\xi-\chi_L)^2+(\zeta-\eta_L)^2}\right] \tag{16}$$

In the frame of CWA, expression (16) can be written in the following more compact form, highlighting that the first-kind Hankel function is cylindrical wave of zero-th order:

$$V_i(\xi,\zeta) = -V_0 CW_0(\xi_L,\zeta_L) \tag{17}$$

Reflection and transmission of the field $V_i(\xi, \zeta)$ through the planar interface in $\xi = 0$ can be evaluated expressing the cylindrical function CW_0 as a plane-wave spectrum (Cincotti et al. 1993):

$$CW_0(\xi_L,\zeta_L) = \frac{1}{2\pi}\int_{-\infty}^{+\infty} F_0(\xi_L,n_{||})e^{in_{||}\zeta_L}dn_{||} \tag{18}$$

with $n_{||}$ the parallel component with respect to the z-axis of the generic plane wave of the spectrum. The explicit expression of the spectrum F_0, evaluated in $\xi_L > 0$, is

$$F_0(\xi_L,n_{||}) = \frac{2e^{i\xi_L\sqrt{1-n_{||}^2}}}{\sqrt{1-n_{||}^2}} \tag{19}$$

To obtain the reflected field $V_r(\xi, \zeta)$, the generical plane wave (19) is evaluated on the plane $\xi_L = -\chi_L$ and multiplied by the Fresnel reflection coefficient $\Gamma_{01}(n_{||})$. A Reflected Cylindrical Function of zero-th order is derived

$$RW_0(\xi,\zeta) = \frac{1}{2\pi}\int_{-\infty}^{+\infty} \Gamma_{10}(n_{||})F_0(-\xi-\chi_L,n_{||})e^{in_{||}(\zeta-\eta_L)}dn_{||} \tag{20}$$

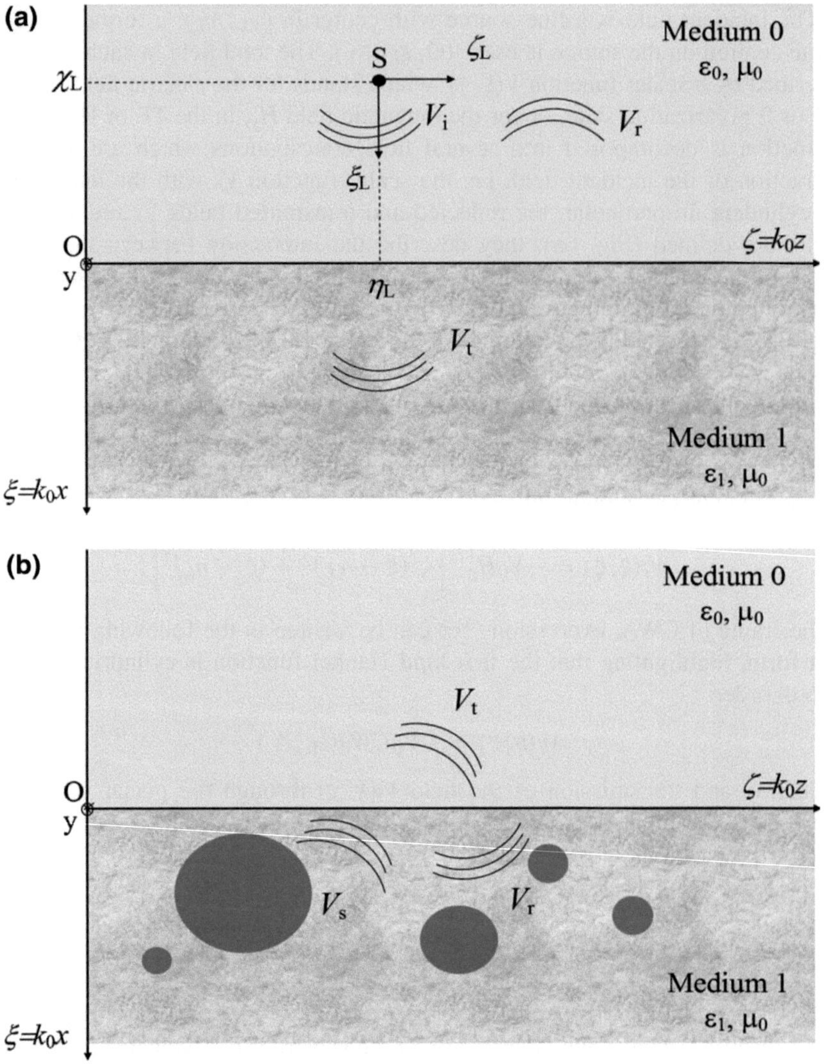

Fig. 6 Decomposition of the total field: **a** the incident file from the line-source, and the reflected and transmitted contributions, evaluated in the absence of any buried scatterer; **b** the scattered fields

The reflected field can be also written as follows

$$V_r(\xi,\zeta) = -V_0 R W_0(\xi,\zeta) \tag{21}$$

In a similar manner the transmitted field $V_t(\xi, \zeta)$ is expressed. The generical plane-wave of the spectrum (19), is evaluated on the plane $\xi_L = -\chi_L$ and multiplied by the Fresnel reflection coefficient $T_{01}(n_{\|})$. A Transmitted Cylindrical Function of zero-th order is defined

$$TW_0(\xi, \zeta; -\chi_L) = \frac{1}{2\pi} \int_{-\infty}^{+\infty} T_{10}(n_{||}) F_0(-\chi_L, n_{||}) e^{in_1\sqrt{1-(n_{||}/n_1)^2}\xi} e^{in_{||}(\zeta-\eta_L)} dn_{||} \tag{22}$$

thus obtaining

$$V_t(\xi, \zeta) = -V_0 TW_0(\xi, \zeta) \tag{23}$$

Since the final solution of the scattering problem is obtained imposing boundary conditions on the cylinders' surface, an expression of the transmitted field in (23) as a function of polar coordinates centered on the p-th cylinder's axis has to be employed. For this purpose, the cylindrical function (22) can be written in the following form

$$TW_0(\xi, \zeta; -\chi_L) = -V_0 \sum_{\ell=-\infty}^{+\infty} i^{\ell} J_{\ell}(n_1 \rho_p) e^{i\ell\theta_p} \frac{1}{2\pi} \int_{-\infty}^{+\infty} T_{10}(n_{||}) F_0(-\chi_L, n_{||})$$
$$\times e^{in_1\sqrt{1-(n_{||}/n_1)^2}\chi_p} e^{in_{||}(\eta_P-\eta_L)} e^{-i\ell \arctan\left[(n_1 n_{||})/\sqrt{1-(n_{||}/n_1)^2}\right]} dn_{||} \tag{24}$$

where the expansion of a plane-wave into Bessel functions has been used:

$$e^{in_1(n_\perp \xi_P + n_{||}\zeta_P)} = \sum_{\ell=-\infty}^{+\infty} i^{\ell} J_{\ell}(n_1 \rho_p) e^{i\ell\theta_p} e^{-i\ell\varphi_t} \tag{25}$$

Being φ_t the angle of transmission of the generic plane wave of the spectrum.

The transmitted field V_t impinging on the cylinders' surface in Medium 1 excites a scattered field $V_s(\xi, \zeta)$. It is expressed as an expansion into cylindrical functions with unknown coefficients c_{qm}:

$$V_s(\xi, \zeta) = V_0 \sum_{q=1}^{N} \sum_{m=-\infty}^{+\infty} c_{qm} CW_m(n_1\xi_q, n_1\zeta_q) \tag{26}$$

where the m-order cylindrical function is proportional to the first-kind Hankel function

$$CW_m(n_1\xi_q, n_1\zeta_q) = H_m^{(1)}(n_1\rho_q) e^{im\theta_q} \tag{27}$$

Making use of the Addition Theorem of Hankel functions, the scattered field (28) can be written in a reference frame centred on the q-th cylinder (q = 1, …, N), as follows

$$V_s(\xi_q, \zeta_q) = V_0 \sum_{\ell=-\infty}^{+\infty} i^{\ell} J_{\ell}(n_1\rho_p) e^{i\ell\theta_P}$$
$$\times \sum_{q=1}^{N} \sum_{m=-\infty}^{+\infty} i^m c_{qm} \left[CW_{m-\ell}(n_1\xi_{qp}, n_1\zeta_{qp})(1-\delta_{qp}) + \frac{H_{\ell}^{(1)}(n_1\rho_p)}{H_{\ell}(n_1\rho_p)} \delta_{qp}\delta_{\ell m} \right] \tag{28}$$

The scattered-reflected field $V_{sr}(\xi, \zeta)$ is derived expanding the cylindrical function CW_m into a Fourier spectrum:

$$CW_{\mathrm{m}}(\xi_L, \zeta_L) = \frac{1}{2\pi} \int_{-\infty}^{+\infty} F_m(\xi_L, n_{||}) e^{in_{||}\zeta_L} dn_{||} \tag{29}$$

The spectrum F_m is given by (Cincotti et al. 1993):

$$F_{\mathrm{m}}(\xi, \mathrm{n}_{||}) = \frac{2e^{i|\xi|\sqrt{1-\mathrm{n}_{||}^2}}}{\sqrt{1-\mathrm{n}_{||}^2}} \begin{cases} e^{-\mathrm{i}m \arccos \mathrm{n}_{||}}, & \xi \geq 0 \\ e^{\mathrm{i}m \arccos \mathrm{n}_{||}}, & \xi \leq 0 \end{cases} \tag{30}$$

Reflected Cylindrical Waves, which are the basis functions of the scattered-reflected field, are obtained evaluating the reflection on each plane wave of the spectrum (30):

$$RW_{\mathrm{m}}(n_1\xi, n_1\zeta) = \frac{1}{2\pi} \int_{-\infty}^{+\infty} \Gamma_{10}(n_{||}) F_{\mathrm{m}}(n_1\xi, n_{||}) e^{in_1 n_{||}\zeta} dn_{||} \tag{31}$$

An expression of the Cylindrical Wave (31), which is a function of polar coordinates centered on the p-th cylinder, is used when boundary conditions are imposed. The resulting expression of the scattered-reflected field is

$$\begin{aligned} V_{\mathrm{sr}}(\xi, \zeta) = V_0 \sum_{\ell=-\infty}^{+\infty} J_\ell(\mathrm{n}_1\rho_{\mathrm{p}}) e^{i\ell\theta_{\mathrm{p}}} \\ \times \sum_{\mathrm{q}=1}^{\mathrm{N}} \sum_{\mathrm{m}=-\infty}^{+\infty} c_{\mathrm{qm}} RW_{\mathrm{m}+\ell}\left[-\mathrm{n}_1(\chi_{\mathrm{q}} + \chi_{\mathrm{p}}), \mathrm{n}_1(\eta_{\mathrm{q}} - \eta_{\mathrm{p}})\right] \end{aligned} \tag{32}$$

In (32), the expansion of a plane wave into Bessel functions has been used.

In a similar way, Transmitted Cylindrical Waves are defined

$$\begin{aligned} TW_{\mathrm{m}}(\xi, \zeta, \chi) = \frac{1}{2\pi} \int_{-\infty}^{+\infty} T_{10}(n_{||}) F_{\mathrm{m}}(-n_1\chi, n_{||}) \\ \times e^{-i\sqrt{1-(n_1 n_{||})^2}(\xi+\chi)} e^{in_1 n_{||}\zeta} dn_{||} \end{aligned} \tag{33}$$

as basis functions of the scattered-transmitted field, which is defined as follows

$$V_{\mathrm{st}}(\xi, \zeta) = V_0 \sum_{\mathrm{q}=1}^{\mathrm{N}} \sum_{\mathrm{m}=-\infty}^{+\infty} c_{\mathrm{qm}} TW_{\mathrm{m}}(\xi_{\mathrm{q}}, \zeta_{\mathrm{q}}, \chi_{\mathrm{q}}) \tag{34}$$

If the scatterers are perfectly conducting cylinders, boundary conditions of zero electric field on the cylinders' surface are imposed, i.e.,

$$V_{\mathrm{t}} + V_{\mathrm{s}} + V_{\mathrm{sr}}|_{\rho_{\mathrm{p}}=\alpha_{\mathrm{p}}} = 0 \tag{35}$$

in TM polarization, and

$$\frac{\partial}{\partial\rho_p}(V_t + V_s + V_{sr})|_{\rho_p=\alpha_p} = 0 \tag{36}$$

in TE polarization.

From boundary conditions (35) and (36), after some algebra, a linear system in the unknown expansion coefficients c_{qm} is derived

$$\sum_{q=1}^{N}\sum_{m=-\infty}^{+\infty} A_{qp}^{m\ell(TM,TE)} c_{qm} = B_p^{\ell(TM,TE)} \qquad p = 1, \ldots, N \quad \ell = 0, \pm 1, \ldots, \pm\infty \tag{37}$$

being

$$A_{qp}^{m\ell(TM,TE)} = i^{-\ell} G_\ell^{(TM,TE)}(n_1\rho_p)\Big\{CW_{m-\ell}(n_1\xi_{qp}, n_1\zeta_{qp})(1-\delta_{qp}) + RW_{m+\ell}\big[-n_1(\chi_q+\chi_p), n_1(\eta_q-\eta_p)\big] + \frac{\delta_{qp}\delta_{\ell m}}{G_\ell^{(TM,TE)}(n_1\rho_p)}\Big\} \tag{38}$$

and

$$B_p^{\ell(TM,TE)} = -G_\ell^{(TM,TE)}(n_1\rho_p) TW_{0,\ell}(\chi_p, \eta_p - \eta_L, -\chi_L) \tag{39}$$

In (38), and (39), it is $G_\ell^{(TM)}(\cdot) = J_\ell(\cdot)/H_\ell^{(1)}(\cdot)$ and $G_\ell^{(TE)}(\cdot) = J'_\ell(\cdot)/H_\ell^{(1)\prime}(\cdot)$.

Dealing, instead, with dielectric scatterers, a further field contribution has to be taken into account, i.e., the field transmitted inside the q-th cylinder:

$$V_{cq} = V_0 \sum_{\ell=-\infty}^{+\infty} i^\ell d_{p\ell} J_\ell(n_{cp}\rho_p) e^{i\ell\theta_p} \tag{40}$$

As to boundary conditions, the continuity of tangential electric field to the cylinders' is imposed

$$V_t + V_s + V_{sr}|_{\rho_p=\alpha_p} = V_{cp}\big|_{\rho_p=\alpha_p} \tag{41}$$

in TM polarization, and

$$\frac{\partial}{\partial\rho_{cp}}(V_t + V_s + V_{sr})|_{\rho_p=\alpha_p} = \frac{\partial}{\partial\rho_{cp}}(V_{cp})\big|_{\rho_p=\alpha_p} \tag{42}$$

In the TE one.

From boundary conditions (42) and (43) two linear systems with two sets of unknowns, c_{qm} and $d_{p\ell}$, are derived. Such systems can be solved by eliminating the coefficients $d_{p\ell}$, thus obtaining a linear system with coefficients c_{qm} as unknowns:

$$\sum_{q=1}^{N} \sum_{m=-\infty}^{+\infty} D_{m\ell}^{pq(1,2)} c_{qm} = M_{\ell}^{p(1,2)} \tag{43}$$

with $D_{m\ell}^{pq(1,2)} = L_{\ell}^{p(2,1)} A_{m\ell}^{pq(1,2)} - L_{\ell}^{p(1,2)} A_{m\ell}^{pq(2,1)}$ and $M_{\ell}^{p(1,2)} = B_{\ell}^{p(1,2)} L_{\ell}^{p(2,1)} - B_{\ell}^{p(2,1)} L_{\ell}^{p(1,2)}$, and where

$$\begin{aligned} A_{qp}^{m\ell(1,2)} &= i^{-\ell} G_{\ell}^{(1,2)}(n_1\rho_p)\Big\{ CW_{m-\ell}(n_1\xi_{qp}, n_1\zeta_{qp})(1-\delta_{qp}) \\ &\quad + RW_{m+\ell}\left[-n_1(\chi_q+\chi_p), n_1(\eta_q-\eta_p)\right] + \frac{\delta_{qp}\delta_{\ell m}}{G_{\ell}^{(1,2)}(n_1\rho_p)} \Big\} \end{aligned} \tag{44}$$

$$B_p^{\ell(1,2)} = -G_{\ell}^{(1,2)}(n_1\rho_p) TW_{0,\ell}(\chi_p, \eta_p - \eta_L, -\chi_L) \tag{45}$$

$$L_{\ell}^{p(1)} = n_{cp}/n_1 [J_{\ell}(n_{cp}\alpha_p)/H_{\ell}^{(1)}(n_1\alpha_p)] \tag{46}$$

$$L_{\ell}^{p(2)} = n_{cp}/n_1 [J_{\ell}'(n_{cp}\alpha_p)/H_{\ell}'^{(1)}(n_1\alpha_p)] \tag{47}$$

Furthermore, it is $G_{\ell}^{(1)}(\cdot) = J_{\ell}(\cdot)/H_{\ell}^{(1)}(\cdot)$ and $G_{\ell}^{(2)}(\cdot) = J_{\ell}'(\cdot)/H_{\ell}^{(1)\prime}(\cdot)$, with the superscript (1) and (2) relevant to TM and TE polarization, respectively.

A key issue of the CWA is the numerical evaluation of spectral integrals introduced to define the cylindrical functions, that in the presented layout are the Eqs. (18), (20), (24), (31), and (33). Integrands are highly oscillating functions, definite on an infinite integration domain. An accurate integration algorithm has been developed for such integrals, the general rules of which are illustrated in Di Vico et al. (2005b). In the procedure the overall contributions of plane waves of the spectrum are evaluated, decomposing the total integration domain into sub-intervals. An evanescent, a homogeneous evanescent and a fully homogeneous interval are detected. Such sub-intervals are further decomposed having checked a regular behavior in the integrand, applying adaptive integration techniques. In the final integration subintervals, low order Gaussian rules, as Gauss-Legendre ones, are applied. The advantage of this integration technique is the possibility to obtain solution in any point of the spatial domain, i.e. in both near and far-field regions.

The analytical theory developed with CWA is restricted to cylindrical scatterers with circular cross-section. Anyway, in several works implementing the CWA it is proved that the technique may be extended to the simulation of arbitrary cross-section cylinders through a suitable arrangement of smaller cylinders. A buried perfectly conducting cylinder can be efficiently simulated in the frame of wire grid-approximation, as checked in Di Vico et al. (2005a), provide that the wires satisfy the same area rule

(Ludwig 1987). A buried dielectric cylinder of arbitrary cross-section may be also modeled, as described in Di Vico et al. (2005b), with smaller cylinders obeying in this case to the same volume rule (Elsherbeni and Kishk 1992).

5 Conclusions

Forward scattering by buried objects has been coped with in the literature with several techniques. The wide interest to this topic is especially due to the interesting applications in the frame of GPR technique.

The developed methods are mainly based on Integral Equation Method or analytical techniques, as Cylindrical Wave Approach. Direct solution with numerical techniques is also possible, and it is widely applied thanks to its extreme versatility in the simulation of arbitrarily complex scattering scenarios. Moreover, with a view to GPR applications, such a technique has the further advantage to be directly implemented in the time domain. Analytical techniques, as Cylindrical Wave Approach, show limitations in the arrangement of the scattering scenario. Anyway, several scattering problems, dealing with the scattering from pipes and tunnels, voids, as well as with rebars of reinforced concrete, may be successfully simulated with simple circular cross-section geometries relevant to the buried scatterers. Furthermore, approximating techniques may be used to implement the simulation of scatterers with arbitrary cross-section through suitable arrangements of smaller cylinders.

A further effort may be done on the side of purely numerical techniques, in order to reduce execution times in the simulation of complex shape targets and improve the accuracy in the approximating grid. On the side of analytical methods, as the Cylindrical Weave Approach, the technique may be further generalized to a wider class of scattering scenarios, due to its advantages in terms of accuracy of solution and reduced computation times.

References

Ahmed, S., Naqvi, Q.A.: Electromagnetic scattering from a perfect electromagnetic conductor cylinder buried in a dielectric half-space. Prog. Electromagn. Res. **78**, 25–38 (2008)

Altuncu, Y., Yapar, A., Akduman, I.: On the scattering of electromagnetic waves by bodies buried in a half-space with locally rough interface. IEEE Trans. Geosci. Remote Sens. **44**(6), 1435–1443 (2006)

Balanis, C.A.: Advanced Engineering Electromagnetics. Wiley, New York (1989)

Berenger, J.P.: A perfectly matched layer for the absorption of electromagnetic waves. IEEE Trans. Electromagn. Compat. **114**, 185–200 (1994)

Borghi, R., Santarsiero, M., Frezza, F., Schettini, G.: Plane-wave scattering by a dielectric circular cylinder parallel to a general reflecting flat surface. J. Opt. Soc. Am. A **14**, 1500–1504 (1997)

Bourlier, C., Pinel, N., Kubické, G.: Method of Moments for 2D Scattering Problems. Basic Concepts and Applications. UK: Focus Series in Waves, Ed. Wiley-ISTE, London (2013)

Bowman, J.J., Senior, T.B., Uslenghi, P.L.E.: Electromagnetic and acoustic scattering by simple shapes. Hemisphere, New York (1987)

Budko, N.V., van den Berg, P.M.: Characterization of a two-dimensional subsurface object with an effective scattering model. IEEE Trans. Geosci. Remote Sens. **37**, 2585–2896 (1999)

Butler, C.M., Xu, X.-B., Glisson, A.W.: Current induced on a conducting cylinder located near the planar interface between two semi-infinite half-spaces. IEEE Trans. Antennas Propag. **33**(6), 616–624 (1985)

Cincotti, G., Gori, F., Santarsiero, M., Frezza, F., Furnò, F., Schettini, G.: Plane wave expansion of cylindrical functions. Opt. Comm. **95**, 192–198 (1993)

Cottis, P.G., Kanellopoulos, J.D.: Scattering of electromagnetic waves from cylindrical inhomogeneities embedded inside a lossy medium with a sinusoidal surface. J. Electromagn. Waves Appl. **6**, 445–458 (1992)

Crocco, L., D'Urso, M., Isernia, T.: The contrast source-extended born model for 2D subsurface scattering problems. Prog. Electromagnet. Res. B **17**, 343–359 (2009)

D'Yakonov, B.P.: The diffraction of electromagnetic waves by a circular cylinder in a homogeneous half-space. Bull. Acad. U.S.S.R. **9**, 950–955 (1959)

Daniels, D.J.: Ground Penetrating Radar, 2nd edn. Institution of Engineering and Technology, London, UK (2004)

Di Vico, M., Frezza, F., Pajewski, L., Schettini, G.: Scattering by a finite set of perfectly conducting cylinders buried in a dielectric half-space. IEEE Trans. Antennas Propag. **53**(2), 719–727 (2005a)

Di Vico, M., Frezza, F., Pajewski, L., Schettini, G.: Scattering by buried dielectric cylindrical structures. Radio Sci. **40**, RS6S18 (2005b)

Diamanti, N., Giannopoulos, A.: Implementation of ADI-FDTD subgrids in ground penetrating radar FDTD models. J. Appl. Geophys. **67**, 309–317 (2009)

Diamanti, N., Giannopoulos, A.: Employing ADI-FDTD subgrids for GPR numerical modeling and their application to study ring separation in brick masonry arch bridges. Near Surf. Geophys. **9**, 245–256 (2011)

Elsherbeni, A.Z., Kishk, A.A.: Modeling of cylindrical objects by circular dielectric and conducting cylinders. IEEE Trans. Antennas Propag. **40**, 96–99 (1992)

Fiaz, M.A., Frezza, F., Pajewski, L., Ponti, C., Schettini, G.: Scattering by a circular cylinder buried under a rough surface: the cylindrical wave approach. IEEE Trans. Antennas Propag. **60**(6), 2834–2842 (2012)

Fiaz, M.A., Frezza, F., Pajewski, L., Ponti, C., Schettini, G.: Asymptotic solution for the scattered field by cylindrical objects buried beneath a slightly rough surface. Near Surf. Geophys. **11**(6), 177–183 (2013)

Fiaz, M.A., Frezza, F., Ponti, C., Schettini, G.: Electromagnetic scattering by a circular cylinder buried below a slightly rough Gaussian surface. J. Opt. Soc. Am. A **31**, 26–34 (2014)

Frezza, F., Pajewski, L., Ponti, C., Schettini, G.: Scattering by perfectly-conducting circular cylinders buried in a dielectric slab through the cylindrical wave approach. IEEE Trans. Antennas Propag. **57**(4), 1208–1217 (2009)

Frezza, F., Pajewski, L., Ponti, C., Schettini, G.: Scattering by dielectric circular cylinders in a dielectric slab. J. Opt. Soc. Am. A **27**(4), 687–695 (2010)

Frezza, F., Pajewski, L., Ponti, C., Schettini, G.: Line-source scattering by buried perfectly-conducting circular cylinders. Int. J. Antennas Propag. **2012**, 7 p (2012)

Frezza, F., Pajewski, L., Ponti, C., Schettini, G., Tedeschi, N.: Electromagnetic scattering by a metallic cylinder buried in a lossy medium with the cylindrical wave approach. Geosci. Remote Sens. Lett. **10**(1), 179–183 (2013a)

Frezza, F., Pajewski, L., Ponti, C., Schettini, G.: Through-wall electromagnetic scattering by N conducting cylinders. J. Opt. Soc. Am. A **30**, 1632–1639 (2013b)

Frezza, F., Pajewski, L., Ponti, C., Schettini, G., Tedeschi, N.: Electromagnetic scattering by a metallic cylinder buried in a lossy medium with the cylindrical wave approach. IEEE Geosci. Remote Sens. Lett. **10**(1), 179–183 (2013c)

Giannopoulos, A.: Modelling ground penetrating radar by GprMax. Constr. Build. Mater. **19**(10), 755–762 (2005)

Hongo, K., Hamamura, A.: Asympotitic solutions for the scattered field of a plane wave by a cylindrical obstacle buried in a dielectric half-space. IEEE Trans. Antennas Propag. **34**(11), 1306–1312 (1986)

Howard, A.Q.: The electromagnetic fields of a subterranean cylindrical inhomogeneity excited by a line source. Geophysics **37**, 975–984 (1972)

Isernia, T., Crocco, L., D'Urso, M.: New tools and series for forward and inverse scattering problems in lossy media. IEEE Geosci. Remote Sens. Lett. **1**, 331–337 (2004)

Jia, H., Yasumoto, K.: Scatteration and absorption characteristics of multilayered gratings embedded in a dielectric slab. Int. J. Infrared Millimeter Wave **26**(8), 1111–1126 (2005)

Jol, H.M.: Ground Penetrating Radar: Theory and Applications. Elsevier Science Ltd, Amsterdam (2008)

Kunz, K., Luebbers, L.J.: The Finite Difference Time Domain Method for Electromagnetics. CRC Press, London (1993)

Kuo, C.H., Moghaddam, M.: Electromagnetic scattering from a buried cylinder in layered media with rough interfaces. IEEE Trans. Antennas Propag. **54**(8), 2392–2401 (2002)

Lambot, S., André, F.: Full-wave modeling of near-field radar data for planar layered media reconstruction. Trans. Geosci. Remote Sens. **52**(5), 2295–2303 (2014)

Lawrence, D.E., Sarabandi, K.: Electromagnetic scattering from a dielectric cylinder buried beneath a slightly rough surface. IEEE Trans. Antennas Propag. **50**(10), 1368–1376 (2002)

Lee, S.C.: Light scattering by closely spaced parallel cylinders embedded in a finite dielectric slab. J. Opt. Soc. Am. A **16**(6), 1350–1361 (1999)

Ludwig, A.C.: Wire grid modeling of surfaces. IEEE Trans. Antennas Propag. **AP-35**(9), 1045–1048 (1987)

Mahmoud, S.F., Ali, S.M., Wait, J.R.: Electromagnetic scattering from a buried cylindrical inhomogeneity inside a lossy earth. Radio Sci. **16**(6), 1285–1298 (1981)

Michalski, K.A., Zheng, D.: Electromagnetic scattering and radiation by surfaces of arbitrary shape in layered media, part I: theory. IEEE Trans. Antennas Propag. **38**(3), 335–344 (1990a)

Michalski, K.A., Zheng, D.: Electromagnetic scattering and radiation by surfaces of arbitrary shape in layered media, part II: implementation and results for contiguous half-spaces. IEEE Trans. Antennas Propag. **38**(3), 345–352 (1990b)

Mur, G.: Absorbing boundary conditions for the finite-difference approximation of the time-domain electromagnetic-field equations. IEEE Trans. Electromagn. Compat. **23**(4), 377–383 (1981)

Naqvi, Q.A., Rizvi, A.A.: Scattering from a cylindrical object buried in a geometry with parallel plane interfaces. Prog. Electromagn. Res. **27**, 19–35 (2000)

Naqvi, Q.A., Rizvi, A.A., Ashraf, M.A.: Asymptotic solutions for the scattered field of a plane wave by a cylindrical obstacle buried in a grounded dielectric layer. Prog. Electromagn. Res. **20**, 249–262 (1998)

Naqvi, Q.A., Rizvi, A.A., Yaqoob, Z.: Asymptotic solutions for the scattered field of plane wave by a cylindrical obstacle buried in a dielectric halfspace. IEEE Trans. Antennas Propag. **48**(5), 846–848 (2000a)

Naqvi, Q.A., Rizvi, A.A., Yaqoob, Z.: Scattering of electromagnetic waves from a deeply buried circular cylinder. Prog. Electromagn. Res. **27**, 37–59 (2000b)

Ogunade, S.O.: Electromagnetic response of an embedded cylinder for line current excitation. Geophysics **46**, 45–52 (1981)

Paknys, R.: Reflection and transmission by reinforced concrete-numerical and asymptotic analysis. IEEE Trans. Antennas Propag. **51**(10), 2852–2861 (2003)

Taflove, A., Hagness, S.C.: Computational Electrodynamics: The Finite Difference Time Domain Method. Artech House, Boston (2000)

Zhuck, N.P., Yarovoy, A.G.: Two-dimensional scattering from an inhomogeneous dielectric cylinder embedded in a stratified medium: case of TM polarization. IEEE Trans. Antennas Propag. **42**(1), 16–21 (1994)

Development of Intrinsic Models for Describing Near-Field Antenna Effects, Including Antenna-Medium Coupling, for Improved Radar Data Processing Using Full-Wave Inversion

Anh Phuong Tran and Sébastien Lambot

Abstract Proper description of antenna effects on ground-penetrating radar (GPR) data generally relies on numerical methods such as the Method of Moments (MoM) or Finite-Difference Time-Domain (FDTD) modeling approaches. Yet, numerical methods are computationally expensive and accurate reproduction of real measurements has remained a major challenge for many years. Recently, intrinsic modeling approaches, through which radar antennas are effectively described using their fundamental features, have demonstrated great promise for near-field radar antenna modeling. Although such approaches are not suited for designing radar antennas, they are particularly powerful for fast and accurate modeling, which is a prerequisite when full-wave inversion is applied, e.g., for estimating medium electrical properties. These approaches are also of great interest for filtering out antenna effects from measured radar data for improved subsurface imaging.

1 Introduction

Ground-penetrating radar (GPR) for nondestructive imaging and characterization of the subsurface has been subject to intensive research for many years (Huisman et al. 2003; Giannopoulos 2005; Lambot et al. 2008; Slob et al. 2010). A major shortcoming in current knowledge is the modeling of the radar signal, which is necessary for quantitative reconstruction using full-wave inversion. Existing techniques usually rely strongly on simplifying assumptions regarding

A.P. Tran (✉) · S. Lambot
Earth and Life Institute, Environmental Sciences, Université Catholique de Louvain (UCL), Croix Du Sud 2 Box L7.05.02, 1348 Louvain-la-Neuve, Belgium
e-mail: aptran@lbl.gov

S. Lambot
e-mail: sebastien.lambot@uclouvain.be

A. Benedetto and L. Pajewski (eds.), *Civil Engineering Applications of Ground Penetrating Radar*, Springer Transactions in Civil and Environmental Engineering, DOI 10.1007/978-3-319-04813-0_9

electromagnetic wave propagation phenomena, and in particular, neglect antenna effects, which include frequency-dependent radiation pattern, gain, phase delay, mutual coupling, and coupling with the medium of interest.

Radar antennas can be modeled using numerical approaches, such as the finite-difference time-domain (FDTD) method (Hyun et al. 2009; Warren and Giannopoulos 2011; Meles et al. 2011; Millington et al. 2011), the finite element method (FEM) (Venkatarayalu et al. 2008; Ilic et al. 2009), or the method of moments (MoM) (Craeye et al. 2009; Chen et al. 2010). Yet, numerical approaches need significant computing resources to reproduce 3-D models and suffer from inherent differences between the real and conceptualized antenna models due to the discretization and sensitivity to small differences between the model and the reality: the computational domain needs to be modeled in detail (Diamanti and Giannopoulos 2011). For instance, Warren and Giannopoulos (2011) used a 3-D FDTD approach through which the different parts of transmitting and receiving bowtie antennas were reproduced in the discretized model. Although relatively good modeling results were obtained for data collected over different emulsions, still significant modeling errors could be observed. These issues can be addressed through the implementation of subgrids into the conventional FDTD mesh to simultaneously account for antenna details and economize on computational resources (Diamanti and Giannopoulos 2009). Pantoja et al. (2009) extended a MoM in the time domain for the transient analysis of thin-wire antennas located over a lossy half-space. Numerical analyses showed good results for near-field cases for which the antenna was not too close to a half-space medium.

More efficient techniques are based on electric field integral equation (EFIE) formulations (Sarkar and Taaghol 1999; Alvarez et al. 2007; Craeye and Gonzalez-Ovejero 2011; Crocco et al. 2005; Frezza et al. 2013a, b; Fiaz et al. 2013), through which, for the particular case of antennas, a set of infinitesimal electric dipoles and field points is used. The parameterization of these dipoles to properly describe real antenna radiation patterns is, however, not straightforward (Alvarez et al. 2007; Serhir et al. 2010). In addition, such formulations do not directly account for wave propagation between the source or field points and the radar transmission line reference plane, i.e., where the field is actually measured, and hence, antenna-medium interactions and mutual coupling are either not accounted for or not adequately described. For instance, Gentili and Spagnolini (2000) modeled a GPR horn antenna at some distance over a 3-D layered medium using an array of frequency-independent source dipoles and a feeding line characteristic impedance. Yet, with this approach the multiple reflections between the antenna terminal section and the medium were not accounted for. Slob and Fokkema (2002) used a thin-wire approximation to study the effects of two antennas placed close together on the Earth's surface, and in particular, investigated the coupling between the antennas. They observed that the coupling is not negligible for GPR applications. The influence of the half-space medium on antenna behavior is strong but limited to a fraction of the wavelength in depth.

For the particular case of far-field GPR with applications to planar layered media, Lambot et al. (2004) proposed a closed-form, frequency-domain, radar equation that simultaneously accounts for: (1) all antenna effects through frequency-dependent global reflection and transmission coefficients and (2) wave propagation in layered media through 3-D Green's functions. This intrinsic antenna-medium model relies on the assumption that the spatial distribution of the backscattered field locally tends to a plane wave over the antenna aperture, which is asymptotically valid in far-field conditions. The model has demonstrated a high level of accuracy for describing radar data and retrieving medium electrical properties, including frequency dependence, in a series of hydrogeophysical and engineering applications (Lambot et al. 2004; Minet et al. 2010; Soldovieri et al. 2011; Patriarca et al. 2011). In addition, the validity of that model being theoretically independent of frequency and antenna type, the approach also applies to electromagnetic induction (EMI), i.e., in the kHz frequency ranges where diffusive phenomena are dominant. In that respect, Moghadas et al. (2010) successfully applied this model to a loop antenna operating in the 30–60 kHz range for soil electrical conductivity determination. Whether for GPR or EMI, it was observed that the so-called far-field condition for the planar field approximation holds when the distance between the antenna and the medium is larger than the antenna aperture dimension. For subsurface characterization and imaging, the far-field condition however strongly limits resolution and penetration depth.

More recently, by resorting to the superposition principle, the far-field model of Lambot et al. (2004) was generalized to near field conditions (Lambot and Andre 2014), for which the backscattered field distribution over the antenna aperture depends on the medium distance and properties. With this approach, the radar antennas are described using an equivalent set of infinitesimal electric dipoles and characteristic, frequency-dependent, global reflection, and transmission coefficients. These coefficients determine through a planar field decomposition over the antenna aperture wave propagation between the radar reference plane, point sources, and field points. The fields are calculated using 3-D Green's functions for wave propagation in planar layered media. In this chapter, a general overview of this antenna modeling approach is presented and insights are provided for future research avenues.

2 Intrinsic Antenna Method

2.1 Far-Field Model

In the far-field radar model of Lambot et al. (2004), which particularly applies to planar layered media, a local plane wave field distribution is assumed for the backscattered field over the antenna aperture, and hence, an equivalent single electric dipole approximation holds for describing antenna radiation properties. Relying on the linearity of Maxwell's equations, wave propagation between the point source or field point and the radar transmission line reference plane is accounted for by means of complex, frequency-dependent global reflection and transmission

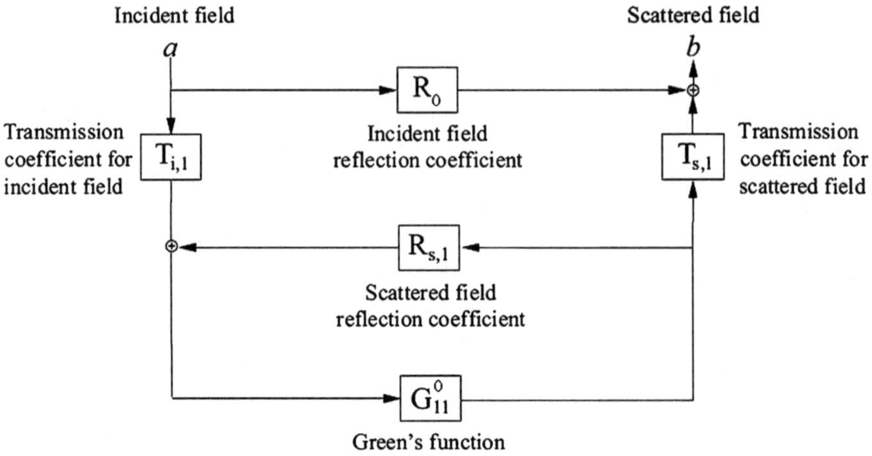

Fig. 1 Linear block diagram representing the far-field antenna model with global reflection and transmission coefficients that describe wave propagation between the radar reference plane and the source or field point

coefficients resulting from the impedances changes within the antenna (see Fig. 1). These characteristic coefficients determine the antenna and transmission line internal transmissions and reflections, and thereby antenna—medium interactions. In far-field conditions, the relation between the radar-measured field and the 3-D layered medium Green's function is described in the frequency domain as follows (Lambot et al. 2004):

$$S(\omega) = \frac{b(\omega)}{a(\omega)} = R_0(\omega) + \frac{T_{s,1} G_{11}^0 T_{i,1}}{1 - G_{11}^0 R_{s,1}} \tag{1}$$

where $S(\omega)$ is the radar signal expressed here as the ratio between the backscattered field $b(\omega)$ and incident field $a(\omega)$ at the radar transmission line reference plane, with ω being the angular frequency; $R_0(\omega)$ is the global reflection coefficient of the antenna for fields incident from the radar reference plane onto the source point, corresponding to the free-space antenna response; $T_{s,1}(\omega)$ is the global transmission coefficient for fields incident from the field point onto the radar reference plane; $T_{i,1}(\omega)$ is the global transmission coefficient for fields incident from the radar reference plane onto the source point; and $R_{s,1}(\omega)$ is the global reflection coefficient for fields incident from the layered medium onto the field point. Subscript 1 denotes the index of the source point or field point. The source-field point corresponds to the antenna phase center. The Green's function G_{11}^0 is defined as the scattered x-directed electric field $E_{x,.}(\omega)$ at the field point for a unit-strength x-directed electric source $J_{x,.}$ which is situated at the same location. The Green's function is derived using a recursive scheme to compute the global reflection coefficients of the 3-D planar layered medium in the spectral domain (Chew

1990; Michalski and Mosig 1997; Slob and Fokkema 2002). The transformation back to the spatial domain is performed by evaluating numerically a semi-infinite integral, for which a fast procedure is applied (Lambot et al. 2007). The spatial-domain Green's function is defined as:

$$G_{..} = \frac{1}{8\pi} \int_{0}^{+\infty} \tilde{G}_{..}(k_\rho) k_\rho dk_\rho \tag{2}$$

where the spectral Green's function is

$$\begin{aligned}\tilde{G}_{..}(k_\rho) = \Bigg[& J_0(k_\rho \rho)\left(\frac{\Gamma_0 R_0^{TM}}{\sigma_0 + j\omega\varepsilon_0} - \frac{j\omega\mu_0 R_0^{TE}}{\Gamma_0}\right) \\ & - J_2(k_\rho \rho)\cos(2\theta)\left(\frac{\Gamma_0 R_0^{TM}}{\sigma_0 + j\omega\varepsilon_0} + \frac{j\omega\mu_0 R_0^{TE}}{\Gamma_0}\right)\Bigg] \exp(-2\Gamma_0 h_0)\end{aligned} \tag{3}$$

In this expression, J_0 and J_2 are, respectively, the first kind zero- and second-order Bessel's functions; ρ and θ are, respectively, the distance and angle in the xy-plane between the field and source points (referred to by the dots in $\mathbf{G}_{..}$, respectively). In the far-field expression (1), we consider $\rho = 0$. Subscript 0 refers to the upper half-space (free space); h_0 is the distance between the source/receiver points and the first medium interface; R^{TM} and R^{TE} are, respectively, the transverse magnetic (TM) and transverse electric (TE) global reflection coefficients accounting for all reflections in the multilayered medium; Γ is the vertical wave number defined as $\Gamma = \sqrt{k_\rho^2 - k^2}$ while $k^2 = \omega^2\mu\left(\varepsilon - \frac{j\sigma}{\omega}\right)$ with the magnetic permeability μ, dielectric permittivity ε, and electrical conductivity σ. For the free-space layer 0 (upper half-space), we have $k_0^2 = \left(\frac{\omega}{c}\right)^2$ with c being the free-space wave velocity.

2.2 *Near-Field Model*

When the layered medium is situated in the near-field of the antenna, the planar field assumption over the antenna aperture does not hold anymore and the scattered field distribution depends on the antenna—medium distance and medium properties. An efficient approach for producing the correct antenna radiation field and capture the scattered field distribution is to consider an equivalent set of infinitesimal electric dipoles ($J_{x,n}$, $n = 1, \ldots, N$) for the source and a set of points ($E_{x,m}$, $m = 1, \ldots, N$) where the field is sampled for the receiver (Serhir et al. 2010; Gentili and Spagnolini 2000), thereby resorting to the superposition principle. The number of points to consider depends on the complexity of the scattered field distribution, which is intrinsically decomposed into a number of elementary distributions, i.e., local plane waves. As for the far-field model described above, wave propagation between the point sources or field points and the radar transmission line reference plane is accounted for by means of complex, frequency-dependent global

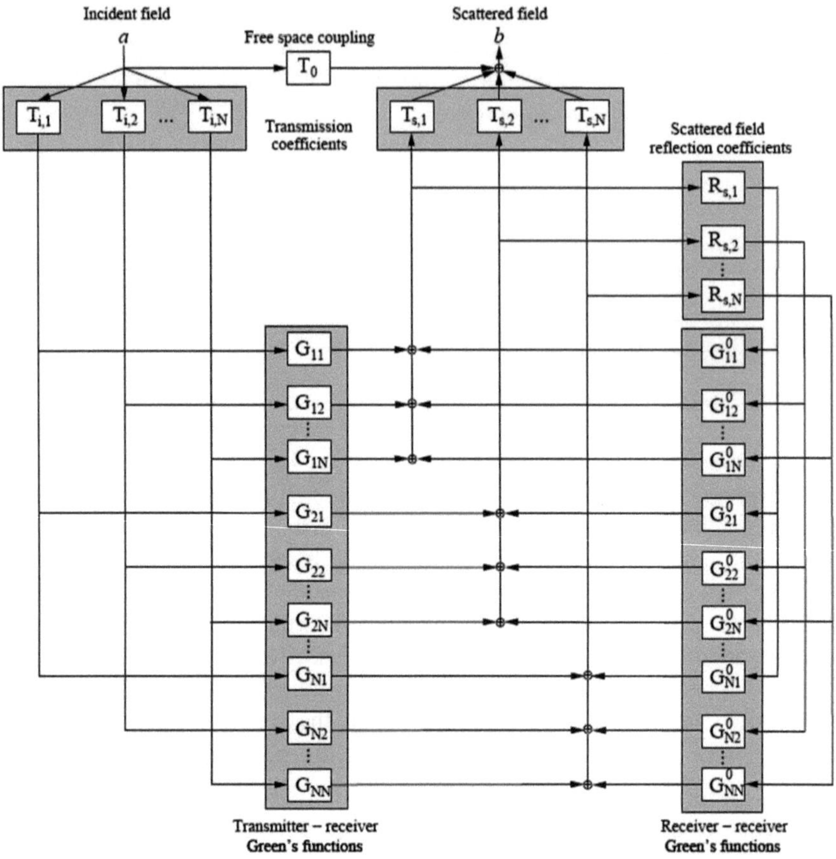

Fig. 2 Linear block diagram representing the transmitting and receiving antenna model where the fields are decomposed into plane waves to which correspond global transmission and reflection coefficients. This block diagram permits to describe wave propagation between the radar reference plane, point sources, and field points

reflection and transmission coefficients related to the variations of impedance along the field paths. The generalized model is depicted in Fig. 2. Because of the scattered field antenna reflection coefficients, the field points act as well as point sources within infinite reflection loops. Although the part of mutual antenna coupling that depends on the medium is not explicitly included in the proposed model (only free-space coupling is physically described through T_0), it can be shown that it is implicitly accounted for through the linearity of Maxwell's equations. The antenna block diagram shown in Fig. 3 can be solved in a closed form in the frequency domain as the solution of a system of linear equations as follows:

$$S(\omega) = \frac{b(\omega)}{a(\omega)} = T_0(\omega) + \mathbf{T_s}\left(\mathbf{I_N} - \mathbf{G^0 R_s}\right)\mathbf{G T_i} \tag{4}$$

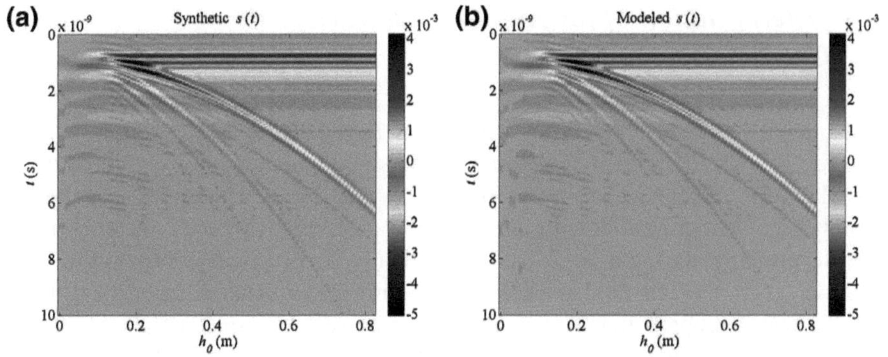

Fig. 3 GprMax3D synthetic (**a**) and near-field modeled GPR data (**b**) in the time domain for the calibration stage with the antenna located at different heights above a PEC [modified from Tran et al. (2013)]

with

$$\mathbf{T_s} = [\,T_{s,1}\;T_{s,2}\;\ldots\;T_{s,N}\,]$$
$$\mathbf{T_i} = [\,T_{i,1}\;T_{i,2}\;\ldots\;T_{i,N}\,]^T$$
$$\mathbf{R_s} = \mathrm{diag}(R_{s,1}, R_{s,1}, \ldots R_{s,N})$$
$$\mathbf{G^0} = \begin{bmatrix} G^0_{11} & G^0_{12} & \ldots & G^0_{1N} \\ G^0_{21} & G^0_{22} & \ldots & G^0_{2N} \\ \vdots & \vdots & \vdots & \vdots \\ G^0_{N1} & G^0_{N2} & \ldots & G^0_{NN} \end{bmatrix}$$

And

$$\mathbf{G} = \begin{bmatrix} G_{11} & G_{12} & \ldots & G_{1N} \\ G_{21} & G_{22} & \ldots & G_{2N} \\ \vdots & \vdots & \vdots & \vdots \\ G_{N1} & G_{N2} & \ldots & G_{NN} \end{bmatrix}$$

where the different quantities are defined as in (1), with $T_0(\omega)$ being the global transmission or reflection coefficient of the antenna in free space (referred to by subscript 0) for non-zero or zero-offset source-receiver, respectively; $\mathbf{G}^0_{..}$ and $\mathbf{G}_{..}$ being the layered medium Green's functions for fields incident from the field points onto the field points themselves and from field points to source points; $\mathbf{I_N}$ being the N-order identity matrix; and superscript T denoting transpose. When the number of source and receiver points limits to 1 (far-field, zero-offset configuration), (4) reduces to (1) and the block diagram shown in Fig. 2 reduces to the scheme of Fig. 1.

2.3 Determination of the Antenna Characteristic Coefficients

The antennas global transmission coefficients $T_0(\omega)$, $T_{i,.}(\omega)$ and $T_{s,.}(\omega)$, and reflection coefficients $R_{s,.}(\omega)$ can be determined by solving (4) for known layered media for which the Green's functions $\mathbf{G}_{..}(\omega)$ and $\mathbf{G}^0_{..}(\omega)$ can be calculated and the corresponding radar measurements $S(\omega)$ can be performed, thereby formulating an inverse problem that can be solved numerically. A practical configuration is to consider near- and far-field measurements with the antenna situated at different heights over a copper plane considered as a perfect electrical conductor (PEC). The size of the copper plane should be large enough so that it can be considered as an infinite plane. The antenna calibration inverse problem involves the optimization of a multidimensional objective function, the topography of which is relatively complex and thereby requires a specific optimization strategy that is outlined below. The proposed optimization scheme is solved independently frequency per frequency and, for one specific frequency, relies on a progressive calibration using radar subdatasets. First, far-field measurements are used to determine the antenna coefficients assuming a single point source and field point [see (1)]. This optimization problem is formulated analytically in a closed form in terms of a system of linear equations (Lambot et al. 2006). Subsequently, these coefficient values are used as initial guess for the near-field calibration procedure, and in particular, are uniformly distributed over the full set of antenna coefficients pertaining to the complete set of point sources and field points, thereby constituting an initial guess for the complete optimization problem. The initial guess for the transmission functions is $T_{.,.}(\omega) = T_{.,1}(\omega)/N$ where $T_{.,1}(\omega)$ comes from the far-field model calibration [see (1)] and N represents the number of source/field points. As the incident transmission coefficients $T_{i,n}$ ($n = 1,\ldots, N$) are mathematically fully correlated to their scattering counterpart $T_{s,n}$, only the products $T_{i,n}T_{s,n}$ can be estimated, which thereby reduces the number of unknown functions to determine. It is worth noting that although the $T_0(\omega)$ function is obtained from the far-field calibration, it can also be obtained from a free space measurement. This function is therefore known for the near-field calibration phase. The initial guess for the reflection functions is $R_{s,.}(\omega) = R_{s,1}(\omega)$ where $R_{s,1}(\omega)$ comes from the far-field model calibration. The symmetry of an antenna can be used to decrease the number of unknowns in the calibration inverse problem. Once the initial guesses from the far-field calibration are determined, a subset of the near-field radar data pertaining to the highest heights is added to the initial far-field subset and the antenna coefficients are updated using local optimization (e.g., using Levenberg–Marquardt algorithm) to fit (4) to the new radar dataset. The procedure is repeated until the full radar dataset (down to an antenna height of zero) is used. The selection of the radar data subsets should be made such that the optimization algorithm iteratively progresses in the correct direction in the parameter space. Thus, the height steps should be relatively small with respect to the wavelength to ensure proper convergence of the optimization procedure. The antenna heights should be known with a relatively high level of accuracy, especially relative to each other. When the frequency step in the radar data is sufficiently small, the optimization procedure can be accelerated by using the solution from a previous frequency as initial guess to directly invert the full radar dataset. It is noteworthy that the antenna characteristic coefficients fully describe the antenna, including antenna-medium coupling effects, and do not depend on the medium. Theoretically, they should only be determined once for a specific transmitting-receiving antenna system.

3 Model Validation

In this section, we summarize some studies that validated the intrinsic near-field antenna model presented in Sect. 2. These studies range from numerical evaluations to real field applications.

3.1 Numerical Evaluation

We evaluated the near-field, full-wave antenna model by comparing it with the GprMax3D model. The GprMax3D is part of a suite of electromagnetic simulation software, developed by Giannopoulos (2005) originally in 1996, which is particularly suited to modeling GPR. The model is based on the FDTD method and approximates Maxwell's equations to simulate the propagation of electromagnetic waves in different media. The near-field antenna model was compared with the GprMax3D by inverting synthetic GPR data generated from the GprMax3D to retrieve the electric properties of an artificial medium with different scenarios of layering, thickness, electrical properties and antenna height. The 1.5 GHz GSSI bowtie antenna from Geophysical Survey Systems, Inc. (GSSI, Salem, Massachusetts, USA) was modeled in details in GprMax3D (Warren and Giannopoulos 2011).

For calibrating the near-field antenna model, 98 numerical measurements corresponding to antenna heights ranging from 0 to 0.82 m above an artificial PEC were generated by the GprMax3D model. Figure 3 presents the GPR data generated by the FDTD antenna model and those obtained from calibrating the near-field antenna model described above. The figure shows that all main reflections are well reproduced by the intrinsic antenna model, with a correlation coefficient of 0.9995. The constant horizontal lines represent the direct coupling between the transmitter and the receiver. Below the antenna coupling signals is the PEC reflection whose travel time increases with increasing antenna height. The other reflections under the PEC reflection are multiple reflections between the antenna and the PEC. It should be noted that when the antenna height is less than 10 cm, the PEC reflection and the antenna coupling occur at almost the same time, and therefore, they cannot be readily distinguished.

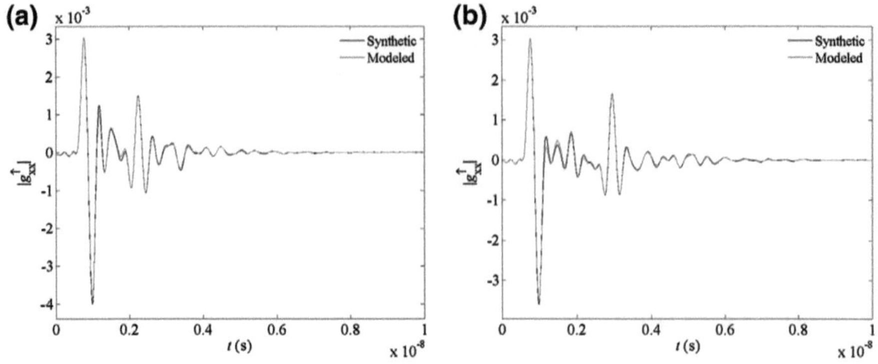

Fig. 4 Example of the synthetic and modeled GPR data for 1- (**a**) and 2-layered (**b**) medium scenarios with the antenna located at a height of 5 cm. **a** $\varepsilon_r = 11.2$, $\sigma = 0.095\,\mathrm{S/m}$, **b** $\varepsilon_{1r} = 4.5$, $\sigma_1 = 0.0015\,\mathrm{S/m}$, $\varepsilon_{2r} = 11.2$, $\sigma_2 = 0.095\,\mathrm{S/m}$

Based on the data obtained from the FDTD antenna model, we applied near-field full-wave inverse modeling to reproduce the GPR signals and predict the electric permittivity and conductivity of the 1- and 2-layered scenarios considering different antenna heights. To control the number of parameters, we also assumed that the thickness of the dielectric layer was already known. Accordingly, there were 3 unknown parameters, namely, the antenna height (h_0), electric permittivity (ε_r) and conductivity (σ) for the 1-layered scenarios, and 5 parameters (h_0, ε_{1r}, σ_1, ε_{2r}, σ_2) for the 2-layered scenarios.

Figure 4 shows an example of synthetic and modeled GPR data in the time domain for the scenarios 2 (1-layered medium) and 5 (2-layered medium) for an antenna height of 5 cm. The results indicate that the near-field full-wave GPR model can reproduce very well the GPR signals synthetically generated by the FDTD antenna model. The main reflections are accurately reproduced both in arrival time and amplitude. We can see that antenna-medium coupling is visible at around 0.9–1 ns in both figures. Reflections from the medium surface are difficult to detect as they are coupled with the antenna-medium coupling. The reflections from the PEC appear, respectively, at 2.3 ns for the 1-layered scenario, and at 3 ns for the 2-layered scenario. For the 2-layered scenario, we also observed the reflections from the upper–lower layer boundary at 2 ns.

Figure 5b–d presents the relative errors of the optimized permittivities and conductivities as a function of the antenna height. The optimized permittivities are in excellent agreement with the modeled values with relative errors lower than 5 %. However, the difference between model and optimization lessens with increasing antenna height. This indicates that the agreement between the GprMax3D and near-field antenna models reduces when the antenna is nearer the medium surface. Figure 5a compares the antenna heights used by the FDTD antenna model with those derived from GPR signal inversion. It shows that for each of the antenna heights the optimized antenna heights are slightly

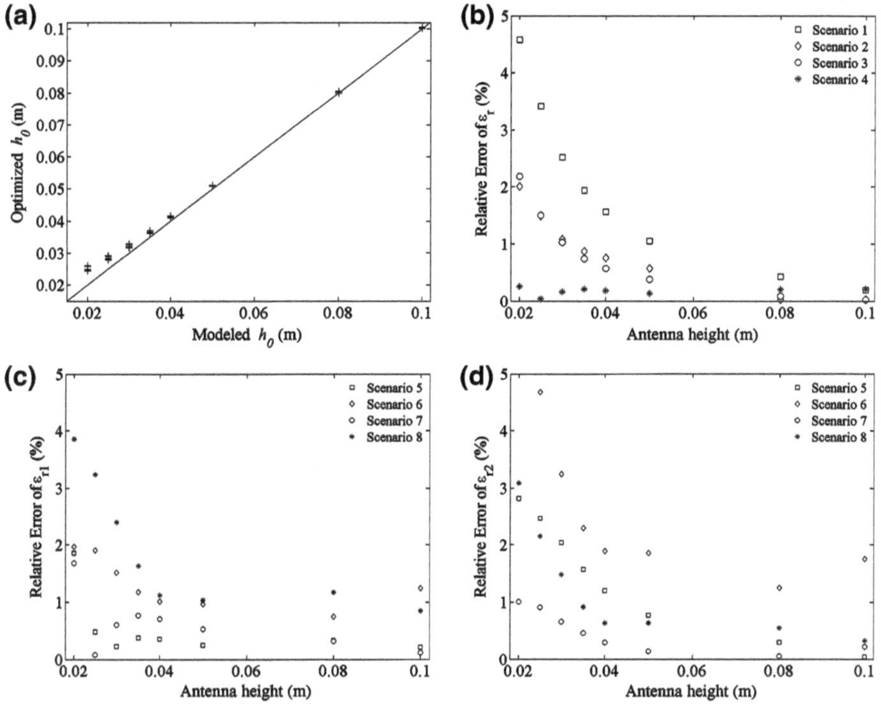

Fig. 5 **a** Comparison of the estimated and synthetic antenna heights; **b–d** Relative errors of the estimated permittivity corresponding to 1-layered scenarios, upper and lower layers of 2-layered scenarios

overestimated but approximately identical in all scenarios. The largest difference is found at the height of 2 cm with errors around 4–6 mm. However, when the antenna height is greater than 5 cm, the difference between model and optimization is negligible.

3.2 Laboratory Evaluation

For laboratory evaluation, we used two GPR systems, working in the frequency (FD-GPR) and time domain (TD-GPR). The FD-GPR system consists of a stepped-frequency continuous-wave vector network analyzer (VNA, ZNB8, Rohde & Schwarz, Munich, Germany) and a homemade single Vivaldi antenna, which simultaneously works as the transmitter and receiver. The antenna is made of aluminum and has an aperture of 24 cm and a height of 20 cm (Sarkis and Craeye 2010). The measurements were performed in the frequency range 1–3.0 GHz, with a step of 6 MHz. The TD-GPR system is the model SIR-20 from Geophysical Survey Systems, Inc. equipped with 900 MHz centre frequency transmitting and receiving bowtie antennas with antennas box dimensions of 33 cm × 18 cm × 8 cm. We recorded the GPR data with a time range of 50 ns and 512 samples per scan. The collected radar data in the time domain were transformed into the frequency domain using the fast Fourier transform. Only data between 0.46 and 1.34 GHz, with the highest signal-to-noise ratio, were considered, resulting in 45 observation frequencies (frequency step of 20 MHz).

To calibrate the two antennas, measurements were performed with the antenna at 100 different heights in both the near- and far-fields (h_0 = 0–0.72 m) over a 3 × 3 m^2 copper sheet assumed as a PEC. Once these coefficients are determined, (4) can be used to simulate the radar data for any layered medium configuration. Figure 6 shows the measured and modeled radar data, expressed in the time domain, over the copper sheet. The antenna coupling can be clearly observed at early times as well as the copper sheet reflection that appears at larger times for increasing heights (h_0). Antenna-medium coupling in particular results in a decrease of the waveform frequency spectrum (larger periods are observed in the time-domain), which is well reproduced by the intrinsic model. At later

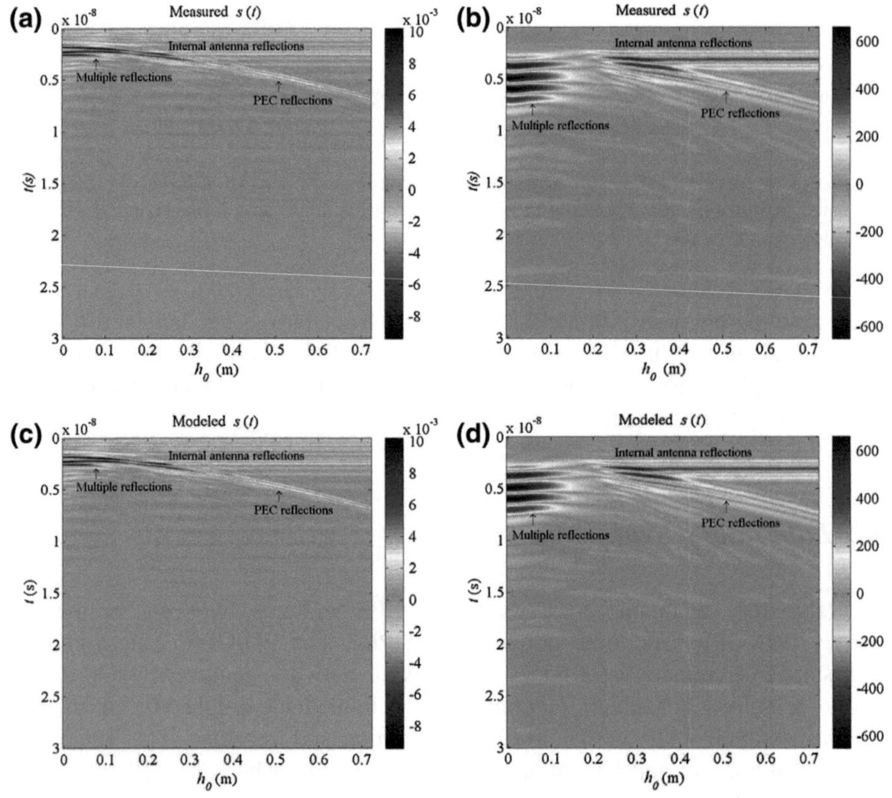

Fig. 6 Measured (**a**, **b**) and modeled (**c**, **d**) GPR data expressed in the time domain for the antennas at different heights (h_0) over the copper sheet [modified from Tran et al. (2014)]

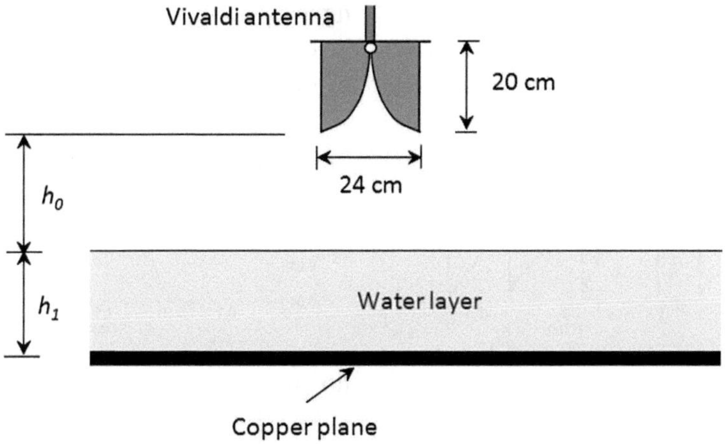

Fig. 7 Sketch of the laboratory experimental setup. The water layer thickness is $h_1 = 5$ cm, the antenna height above water ranges $h_0 = 0 \ldots 0.76$ m [modified from Lambot and Andre (2014)]. A similar setup also applied for the sand experiments

times, a first-order multiple between the antenna and the copper sheet can also be observed. The measured and modeled data agree remarkably well. The relative errors in terms of signal amplitude are lower than 8 % for the TD-GPR and 3 % for the FD-GPR and their correlation coefficients are 0.9983 and 0.9993, respectively.

For further validating the model, GPR measurements were performed with the antenna at different heights above a 5-cm water and 15-cm sand layer as illustrated in Fig. 7. The sand layer is subjected to 7 volumetric water content values ranging from 0.0 to 0.32 cm^3/cm^3. A copper plane was used as the bottom boundary condition. We use the Debye model (Debye 1929) for describing the frequency dependence of free water electrical properties with parameterizations as provided by Stogryn (1970) and Klein and Swift (1977) for the complex permittivity values and relaxation time as a function of temperature and salinity. The water temperature was 17.2 °C and its DC electrical conductivity was 0.0806 S/m. For sand, we combined the complex refractive index model (CRIM) with the Debye model to take into account the frequency dependence of soil electric properties. The modeled data were obtained by model inversion with the antenna height (h_0) and water thickness (h_1) as unknown variables for water, and sand thickness (h_1), water content (θ) and DC soil conductivity (σ_{DC}) for the sand.

Figure 8 shows, in frequency and time domains, the measured and modeled radar data for measurements collected with the antenna at a height of 5 cm above the water. For the FD-GPR, a good agreement can be observed with no visible differences between the model and the measurements. Results for the TD-GPR are slightly less satisfactory, which may be attributed to the lower dynamic range of the instrument, the inherent drift issues (amplitude and time drifts), and possibly the larger complexity of the antenna geometry which may require another

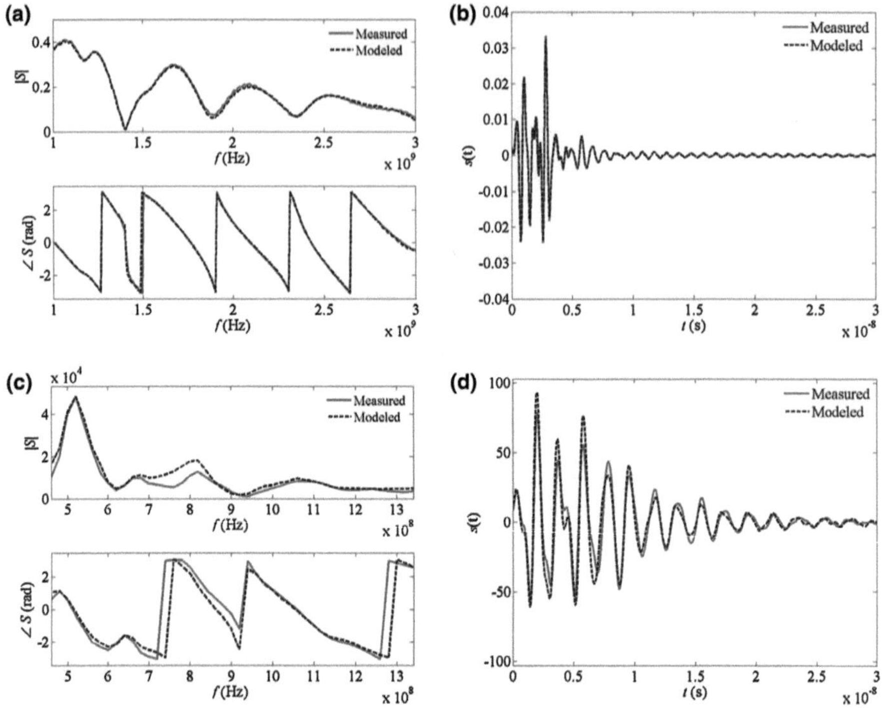

Fig. 8 Measured and modeled GPR data in the frequency (*left*) and time (*right*) domains with the antenna located at a height of 5 cm above a 5-cm water layer. The results were obtained by the frequency-domain Vivaldi (**a**, **b**) and time-domain GSSI GPR systems (**c**, **d**)

distribution of the source and field points over the antenna aperture. For both systems, the estimated heights were in very good agreement with direct height measurements (errors are less than 1/100 wave length).

Figure 9 presents an example of measured and modeled GPR data corresponding to sand in the frequency and time domains. The results were obtained by using the FD- and TD-GPR systems. The figure shows that the full-wave GPR inverse modeling well reproduces the measured radar signals in both time and frequency domains. The water content values obtained by both GPR systems are in close agreement with the reference sampling ones. The average and maximum errors of the GPR water content for the FD-GPR are 0.012 and 0.045 cm^3/cm^3. These errors are, respectively, 0.016 and 0.034 cm^3/cm^3 for the TD-GPR.

Comparing Figs. 8 and 9, we observe that the differences between modeled and measured GPR data corresponding to water are smaller than for the sand due to the fact that the water layer is much more homogeneous, thereby the homogeneous assumption of the near-field antenna model is better respected.

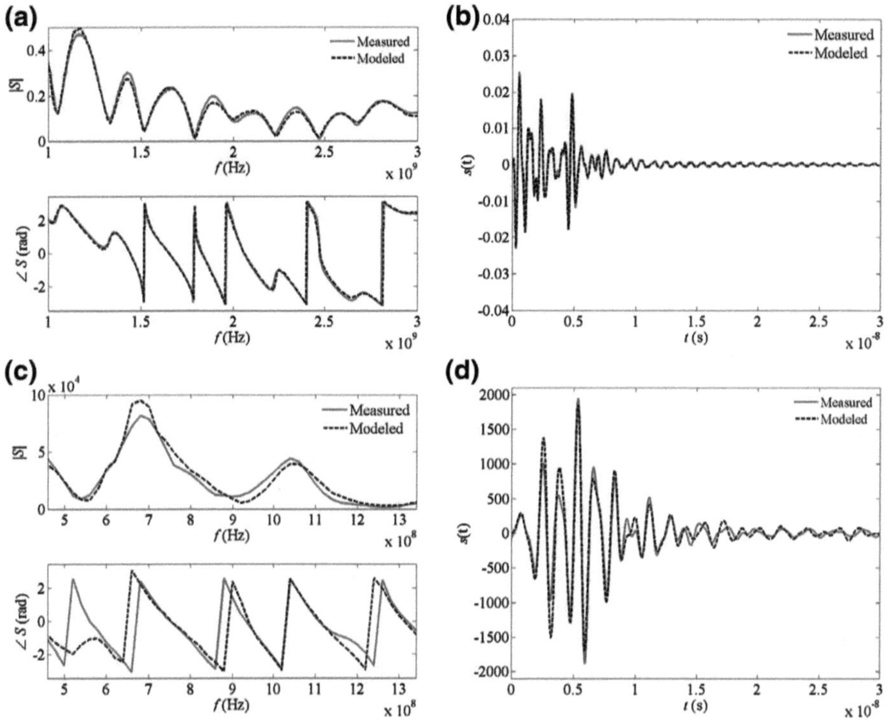

Fig. 9 Measured and modeled GPR data in both frequency (*left*) and time (*right*) domains. The antenna situated at a height of 5 cm above a 15-cm sand layer with $\theta = 0.11$ cm^3/cm^3. The results were obtained by the frequency-domain Vivaldi (**a**, **b**) and time-domain GSSI GPR systems (**c**, **d**)

3.3 Field Application

The study was carried out in the Belgian loam belt region along a cultivated transect of 308 m long (50.6669°N, 4.6331°W) (Fig. 10a). The site has a maritime temperate climate, with an average annual temperature of 9.7 °C and an average annual precipitation of 805 mm (Wiaux et al. 2014). Along the transect, the soil porosity at the 30 cm top is quite uniform, varying from 43 to 45 % due to the fact that the field is plowed every year. The elevation along the transect varies from 134 to 145 m. The higher topography is observed at the two extremities and the lower one around the middle of the transect.

The GPR field campaigns were conducted weekly during the period 23/03–08/06/2011 (11 datasets, one missing in May) to characterize the spatiotemporal variability of soil moisture. The GPR system consists of a VNA (ZVNA, Rohde & Schwarz, Munich, Germany) connected with a broadband horn antenna by a 50-Ω coaxial cable. The antenna has dimensions of 950 cm in length and 680×960 cm^2 in aperture area. For each measurement, the frequency-dependent

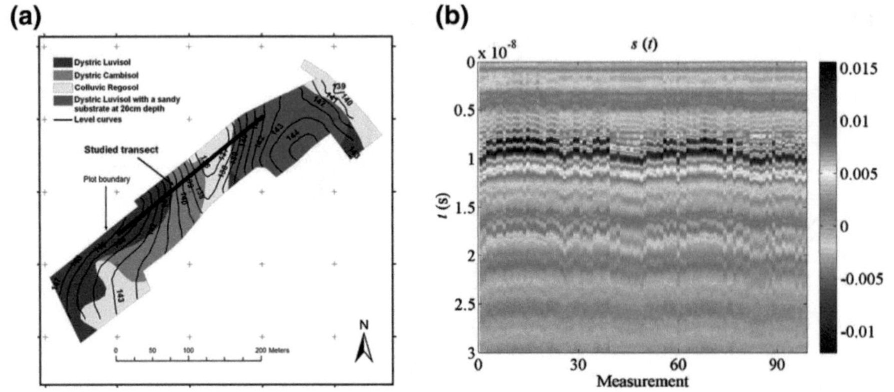

Fig. 10 **a** Topography and soil map of the site study [modified from Wiaux et al. (2014)]. **b** An example of B-scan GPR data along the transect [modified from Tran et al. (2015)]

complex backscattered signal S_{11} was measured at a frequency range 200–2000 MHz, with a frequency step of 6 MHz (301 measured frequencies). However, for soil moisture estimation, we selected GPR data within the frequency range 200–800 MHz to avoid noises in the signal occurring for larger frequencies due to the soil roughness and the presence of vegetation.

For rapid data acquisition, the GPR system was mounted on a 4-wheel motorcycle. A differential global positioning system device (DGPS) (Leica GPS1200, Leica Geosystems) was installed above the GPR radar to determine the measurement locations. Distance between the antenna aperture and soil surface was set at 0.3 m for the near-field soil moisture characterization. However, during the measurement process, this distance varied due to the effect of the local field topography. For each field occasion, there were from 85 to 120 GPR measurements collected in a period of around 1 h.

Figure 10b presents the time-domain B-scan GPR measurements collected on 08/06/2011 as an example. The figure indicates that the reflections from top soil layer, which appear around 7–12 ns, are the strongest ones. Above the soil surface reflections are the antenna internal reflections inside the antenna. These reflections are relatively constant both in amplitude and travel time. The reflections from layers below the topsoil are not visible, indicating that there is no abrupt change in the shallow soil moisture which might be explained by the plowing activities. As a result, we used the lower half-space to model the soil layer. Two unknowns for the optimization are the antenna height and soil moisture.

Together with the GPR measurements, we also used the portable frequency domain reflectometry sensor (FDR) ML2x (Delta-T, United Kingdom) to measure soil moisture at several positions along the transect. Volumetric soil moisture was measured at the topsoil layer (thickness of 6 cm) using a portable five-rod 6 cm long FDR. At each measurement location, we carried out 6 FDR measurements within a rectangle with a size of 5×1 m^2 next to the GPR measurement transect.

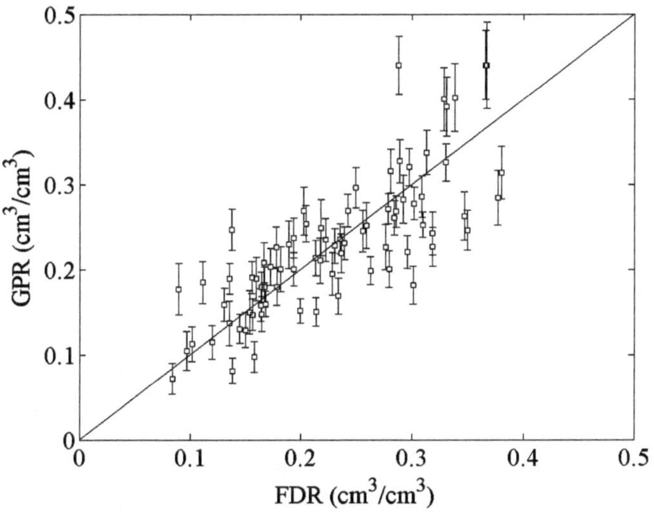

Fig. 11 Comparison of soil moisture derived from the GPR and FDR measurements during the field campaign period from 23/03 to 08/06/2011. The *vertical bars* represent the confidence intervals of the full-wave GPR inversion corresponding to a confidence level of 95 %

After that, these 6 measurements were averaged to represent FDR soil moisture. Totally, 78 FDR soil moisture values were performed during the field campaign period.

Figure 11 compares the soil moisture obtained from GPR and FDR measurements at the same locations over the field campaign period. Here, we compared the soil moisture at the nearest distance with the center of FDR rectangle. The vertical bars represent the uncertainty of the full-wave GPR inversion, i.e., the confidence intervals corresponding to a confidence level of 95 %. The figure shows that there is a relatively good agreement of soil moisture derived from the two techniques. The bias between them is relatively small. The correlation coefficient and root mean square error (*rmse*) are 0.79 and 0.05 cm^3/cm^3, respectively. This agreement proves that the near-field full-wave GPR model can provide reliable soil moisture data. However, when soil becomes wetter, the agreement between GPR and FDR soil moisture reduces with a maximum error up to 0.15 cm^3/cm^3. This might be, firstly, explained by the increasing heterogeneity of soil moisture under the antenna footprint, which causes inherent differences between GPR and FDR measurements. For GPR, the errors mainly come from the violation of the homogenous assumption of the GPR model and the decrease of sensitivity of the GPR reflection coefficient with respect to permittivity when permittivity increases. For FDR, the errors are caused by the lack of samples to cover the high spatial variability of soil moisture. Secondly, the wetter soil causes the difference of the vertical characterization scales between GPR and FDR to be larger, thereby increasing the discrepancy between soil moisture estimated by the two techniques (Tran et al. 2015).

As for the uncertainty of soil moisture estimated by the full-wave GPR inversion, Fig. 11 shows that the uncertainties are relatively small with the ratios between the standard deviation of soil moisture and its estimated values varying from 3 to 12 %. The large uncertainties usually appear at high soil moisture, implying that the GPR soil moisture estimation is less accurate and less reliable in the wet soil conditions.

4 Conclusion

A strong requirement is the accurate modeling of the radar data for quantitative reconstruction of medium properties using full-wave inversion as well as for improved radar imaging by removing artifacts that arise from the antenna effects, such as multiple reflections. Although available computing resources are ever increasing, the antenna modeling based on numerical methods is still impractical due to the fact that they require huge computational resources, especially when solving the inverse scattering problems. In that respect, a more effective solution to accurately describe the antenna is a necessity.

We introduced a closed-form equation for modeling antenna and wave propagation in the multilayered medium, which included in particular the antenna effects and its interactions with the medium. The model relies on a planar wave decomposition and global transmission and reflection coefficients that are characteristic to the antenna, thereby providing an intrinsic antenna representation and effective way that takes maximum benefit from the linearity of Maxwell's equations for fast computing. We summarized the results of some studies that successfully applied the model to quantify the electrical properties of medium under different conditions. These results showed the great promise for planar material property retrieval through full-wave inversion, e.g., for road inspection.

Yet, it is worth noting that this approach only applies to planar layered media (at least locally). Considering buried objects or more complex structures in the subsurface still requires further developments. This closed-form modeling approach can, however, be integrated with numerical formulations such as FDTD or other EFIE-based methods that efficiently model scattering by an embedded object.

Acknowledgments The author acknowledges the COST Action TU1208 "Civil Engineering Applications of Ground Penetrating Radar", supporting this work. This research was funded by the Fonds de la Recherche Scientifique (FNRS, Belgium) and the Université catholique de Louvain (UCL, Belgium).

References

Alvarez, Y., Las-Heras, F., Pino, M.R.: Reconstruction of equivalent currents distribution over arbitrary three-dimensional surfaces based on integral equation algorithms. IEEE Trans. Antennas Propag. **55**, 3460–3468 (2007)

Chen, Y., Yang, S., He, S., Nie, Z.: Fast analysis of microstrip antennas over a frequency band using an accurate mom matrix interpolation technique. Prog. Electromagnet. Res.-Pier **109**, 301–324 (2010)

Chew, W.C.: Waves and fields in inhomogeneous media. Van Nostrand Reinhold, New York (1990)

Craeye, C., Gonzalez-Ovejero, D.: A review on array mutual coupling analysis. Radio Sci. **46**, 25 (2011)

Craeye, C., Gilles, T., Dardenne, X.: Efficient full-wave characterization of arrays of antennas embedded in finite dielectric volumes. Radio Sci. **44**, RS1S90, pp. 25 (2009)

Crocco, L., D'Urso, M., Isernia, T.: Testing the contrast source extended Born inversion method against real data: the TM case. Inverse Prob. **21**, S33–S50 (2005)

Debye, P.: Polar molecules. Reinhold, New York (1929)

Diamanti, N., Giannopoulos, A.: Implementation of ADI-FDTD subgrids in ground penetrating radar FDTD models. J. Appl. Geophys. **67**, 309–317 (2009)

Diamanti, N., Giannopoulos, A.: Employing ADI-FDTD subgrids for GPR numerical modelling and their application to study ring separation in brick masonry arch bridges. Near Surf. Geophys. **9**, 245–256 (2011)

Fiaz, M.A., Frezza, F., Pajewski, L., Ponti, C., Schettini, G.: Asymptotic solution for a scattered field by cylindrical objects buried beneath a slightly rough surface. Near Surf. Geophys. **11**, 177–183 (2013)

Frezza, F., Pajewski, L., Ponti, C., Schettini, G., Tedeschi, N.: Electromagnetic scattering by a metallic cylinder buried in a lossy medium with the cylindrical-wave approach. IEEE Geosci. Remote Sens. Lett. **10**, 179–183 (2013a)

Frezza, F., Mangini, F., Pajewski, L., Schettini, G., Tedeschi, N.: Spectral domain method for the electromagnetic scattering by a buried sphere. J. Opt. Soc. America A-Opt. Image Sci. Vis. **30**, 783–790 (2013b)

Gentili, G.G., Spagnolini, U.: Electromagnetic inversion in monostatic ground penetrating radar: TEM horn calibration and application. IEEE Trans. Geosci. Remote Sens. **38**, 1936–1946 (2000)

Giannopoulos, A.: Modelling ground penetrating radar by GprMax. Constr. Build. Mater. **19**, 755–762 (2005)

Huisman, J.A., Hubbard, S.S., Redman, J.D., Annan, A.P.: Measuring soil water content with ground penetrating radar: a review. Vadose Zone J. **2**, 476–491 (2003)

Hyun, S.Y., Kim, S.Y., Kim, Y.S.: An equivalent feed model for the FDTD analysis of antennas driven through a ground plane by coaxial lines. IEEE Trans. Antennas Propag. **57**, 161–167 (2009)

Ilic, M.M., Djordjevic, M., Ilic, A.Z., Notaros, B.M.: Higher order hybrid FEM-MoM technique for analysis of antennas and scatterers. IEEE Trans. Antennas Propag. **57**, 1452–1460 (2009)

Klein, L.A., Swift, C.T.: Improved model for dielectric-constant of sea-water at microwave-frequencies. IEEE Trans. Antennas Propag. **25**, 104–111 (1977)

Lambot, S., Andre, F.: Full-wave modeling of near-field radar data for planar layered media reconstruction. Geosci. Remote Sens. IEEE Trans. **52**, 2295–2303 (2014)

Lambot, S., Slob, E.C., van den Bosch, I., Stockbroeckx, B., Vanclooster, M.: Modeling of ground-penetrating radar for accurate characterization of subsurface electric properties. IEEE Trans. Geosci. Remote Sens. **42**, 2555–2568 (2004)

Lambot, S., Weihermüller, L., Huisman, J.A., Vereecken, H., Vanclooster, M., Slob, E.C.: Analysis of air-launched ground-penetrating radar techniques to measure the soil surface water content. Water Resour. Res. **42**, W11403 (2006). doi:10.1029/2006WR005097

Lambot, S., Slob, E., Vereecken, H.: Fast evaluation of zero-offset green's function for layered media with application to ground-penetrating radar. Geophys. Res. Lett. **34**, L21405 (2007). doi:10.1029/2007GL031459

Lambot, S., Binley, A., Slob, E., Hubbard, S.: Ground penetrating radar in hydrogeophysics. Vadose Zone J. **7**, 137–139 (2008). doi:10.2136/vzj2007.0180

Meles, G., Greenhalgh, S., van der Kruk, J., Green, A., Maurer, H.: Taming the non-linearity problem in GPR full-waveform inversion for high contrast media. J. Appl. Geophys. **73**, 174–186 (2011)

Michalski, K.A., Mosig, J.R.: Multilayered media green's functions in integral equation formulations. IEEE Trans. Antennas Propag. **45**, 508–519 (1997)

Millington, T.M., Cassidy, N.J., Nuzzo, L., Crocco, L., Soldovieri, F., Pringle, J.K.: Interpreting complex, three-dimensional, near-surface GPR surveys: an integrated modelling and inversion approach. Near Surf. Geophys. **9**, 297–304 (2011)

Minet, J., Lambot, S., Slob, E.C., Vanclooster, M.: Soil surface water content estimation by full-waveform GPR signal inversion in the presence of thin layers. IEEE Trans. Geosci. Remote Sens. **48**, 1138–1150 (2010)

Moghadas, D., André, F., Vereecken, H., Lambot, S.: Efficient loop antenna modeling for zero-offset, off-ground electromagnetic induction in multilayered media. Geophysics, **75**, WA125–WA134 (2010)

Pantoja, M.F., Yarovoy, A.G., Bretones, A.R., Garcia, S.G.: Time domain analysis of thin-wire antennas over lossy ground using the reflection-coefficient approximation. Radio Sci. **44**, 14 (2009)

Patriarca, C., Lambot, S., Mahmoudzadeh, M.R., Minet, J., Slob, E.: Reconstruction of sub-wavelength fractures and physical properties of masonry media using full-waveform inversion of proximal penetrating radar. J. Appl. Geophys. **74**, 26–37 (2011)

Sarkar, T.K., Taaghol, A.: Near-field to near/far-field transformation for arbitrary near-field geometry utilizing an equivalent electric current and MoM. IEEE Trans. Antennas Propag. **47**, 566–573 (1999)

Sarkis, R., Craeye, C.: Circular array of wideband 3D Vivaldi antennas. In: 2010 URSI International Symposium on Electromagnetic Theory (EMTS), pp. 792–794

Serhir, M., Besnier, P., Drissi, M.: Antenna modeling based on a multiple spherical wave expansion method: application to an antenna array. IEEE Trans. Antennas Propag. **58**, 51–58 (2010)

Slob, E.C., Fokkema, J.: Coupling effects of two electric dipoles on an interface. Radio Sci. **37**, 1073 (2002). doi:10.1029/2001RS2529

Slob, E., Sato, M., Olhoeft, G.: Surface and borehole ground-penetrating-radar developments. Geophysics **75**, A103–A120 (2010)

Soldovieri, F., Lopera, O., Lambot, S.: Combination of advanced inversion techniques for an accurate target localization via GPR for demining applications. IEEE Trans. Geosci. Remote Sens. **49**, 451–461 (2011)

Stogryn, A.: Brightness temperature of a vertically structured medium. Radio Sci. **5**, 1397–1406 (1970)

Tran, A.P., Warren, C., André, F., Giannopoulos, A., Lambot, S.: Numerical evaluation of a full-wave antenna model for near-field applications. Near Surf. Geophys. (2013)

Tran, A.P., André, F., Lambot, S.: Validation of near-field ground-penetrating radar modeling using full-wave inversion for soil moisture estimation. IEEE Trans. Geosci. Remote Sens. **52**, 5483–5497 (2014)

Tran, A.P., Bogaert,P., Wiaux F., Vanclooster M., Lambot, S.: High-resolution space-time quantification of soil moisture along a hillslope using joint analysis of ground penetrating radar and frequency domain reflectometry data. J. Hydrol. **523**, 252–261 (2015)

Venkatarayalu, N.V., Gan, Y.B., Lee, R., Li, L.W.: Application of hybrid FETD-FDTD method in the modeling and analysis of antennas. IEEE Trans. Antennas Propag. **56**, 3068–3072 (2008)

Warren, C., Giannopoulos, A.: Creating finite-difference time-domain models of commecrical ground-penetrating radar antennas using Taguchi's optimization method. Geophysics **76**, G37–G47 (2011)

Wiaux, F., Cornelis, J.T., Cao, W., Vanclooster, M., Van Oost, K.: Combined effect of geomorphic and pedogenic processes on the distribution of soil organic carbon quality along an eroding hillslope on loess soil. Geoderma **216**, 36–47 (2014)

GPR Imaging Via Qualitative and Quantitative Approaches

Ilaria Catapano, Andrea Randazzo, Evert Slob and Raffaele Solimene

Abstract Ground Penetrating Radar (GPR) is a non-destructive imaging system able to provide high-resolution images of the subsurface. From a theoretical point of view, it requires to solve an inverse scattering problem, where a set of parameters describing the underground scenario must be retrieved starting from samples of the measured electromagnetic field. In this chapter, an overview of different methods/algorithms for quantitative and qualitative buried scatterer reconstruction widespread in literature is provided.

1 Introduction

The problem of imaging buried targets (subsurface imaging) finds application in a very large number of research fields. Civil engineering diagnostics (Pastorino 1998; Soldovieri et al. 2006; Hugenschmidt and Kalogeropoulos 2009; Randazzo and Estatico 2012), archaeological and geophysical prospecting (Conyers 1997;

I. Catapano (✉)
Institute for Electromagnetic Sensing of the Environment—National Research Council of Italy, Naples, Italy
e-mail: catapano.i@irea.cnr.it

A. Randazzo
Department of Electrical, Electronic, Telecommunications Engineering, and Naval Architecture, University of Genoa, Genoa, Italy
e-mail: andrea.randazzo@unige.it

E. Slob
Department of Geoscience and Engineering, Deft Univeristy of Technology, Delft, The Netherlands
e-mail: e.c.slob@tudelft.nl

R. Solimene
Department of Industrial and Information Engineering, Second University of Naples, Aversa, Italy
e-mail: raffaele.solimene@unina2.it

A. Benedetto and L. Pajewski (eds.), *Civil Engineering Applications of Ground Penetrating Radar*, Springer Transactions in Civil and Environmental Engineering,
DOI 10.1007/978-3-319-04813-0_10

Piro et al. 2003; Orlando and Soldovieri 2008; Catapano et al. 2012) and cultural heritage monitoring (Masini et al. 2010; Masini and Soldovieri 2011) are only a few of examples. Besides these "classical" fields of application, new fields such as through-wall imaging (Ahmad et al. 2005; Baranoski 2008; Solimene et al. 2009), contaminant detection (Bevan et al. 2003; Porsani et al. 2004), tunnel and underground facility detection (Lo Monte et al. 2010; Meschino et al. 2013), and planet exploration (Watters et al. 2006) are gaining increasing interest.

In order to succeed in subsurface imaging a number of requirements, which span from hardware to signal processing, must be complied with. For instance, it is often required to use imaging systems having relatively small dimension (for portability), ultra-wide-band behavior, and low levels of the direct coupling between TX/RX antennas. Moreover, as the targets are usually buried within extremely complicated (in-homogeneous) scene, with media that can be lossy and dispersive, before imaging a number of "preliminary duties" have to be accomplished. For example, clutter suppression procedures must be run in order to extract scattered field data (Lopera et al. 2007; Solimene et al. 2014b). In addition, medium estimation is mandatory in order to obtain properly focused reconstructions (Di Matteo et al. 2013; Gorriti and Slob 2005; Minet et al. 2010; Solimene et al. 2014c). Finally, antenna deconvolution is generally needed (Daniels 2004; Jadoon et al. 2011; Lambot et al. 2004; Patriarca et al. 2013), the latter being particularly difficult when antennas are deployed in close proximity of the air/soil interface (Lambot and André 2014).

In this contribution we focus only on the imaging step.

Basically, subsurface imaging entails solving an electromagnetic inverse scattering problem (Pastorino 2010). As is well known, such a problem is extremely difficult to solve due to its non-linearity (Chew 1999) and ill-posedness (Bertero and Boccacci 1998). A large number of research activities have been developed and documented in the literature to address non-linear inverse scattering problems (Caorsi et al. 1994, 1995; Franchois and Pichot 1997; Lobel et al. 1997; Dourthe et al. 2000; Liu et al. 2002; Benedetti et al. 2006; Pastorino 2010; Ali and Moghaddam 2010; Meles et al. 2010; Patriarca et al. 2011; Crocco and Soldovieri 2011). In this framework, the reconstruction/imaging is usually recast as an optimization problem where the minimum of a suitable cost function accounting for the misfit between the scattered field data and the model data must be found. Newton-like iterative methods (Cui et al. 2001; Estatico et al. 2005, 2012; Ostadrahimi et al. 2013; Roger 1981; Rubek et al. 2007; De Zaeytijd et al. 2007) and gradient based solution procedures (Kleinman and van den Berg 1992; Harada et al. 1995; Lobel et al. 1996; Isernia et al. 1997; van den Berg and Abubakar 2001) are generally the approaches upon which deterministic non-linear inversions are based. It is well known that, such a kind of inversion methods can suffer from convergence problems and can experience limited reliability due to the occurrence of false solutions (Isernia et al. 2001). The latter problem is due to the non-quadratic structure of the function to be minimized. These problems are further exacerbated by the ill-posedness, which makes the solution strongly sensitive to the uncertainties and the noise that "always" corrupt data. Nonetheless, if "quantitative" reconstruction (i.e., in terms of the shape and material composition of scatterers) are required, non-linear inversions is the only way to follow. This stimulated researchers in finding out strategies to overcome/mitigate the drawbacks mentioned above. For example, in order to reduce non-linearity, approximate (quadratic) and novel non-linear scattering models have been developed (Leone et al. 2003; Isernia et al. 2004; Estatico et al. 2005; Catapano et al. 2006a; Crocco et al. 2012). Again, in order to exploit in a clever way the ratio between independent data/number of unknowns (which affects the curvature of the functional range and hence the occurrence of local minima) enlarging the number of unknowns as minimization procedure evolves (Baussard et al. 2004b) or multi-resolution unknown representation have been employed (Catapano et al. 2004; Donelli et al. 2006; Oliveri et al. 2012a). Also stochastic minimization schemes have been largely used (Rahmat-Samii and Michielssen 1999; Pastorino et al. 2000; Robinson and Rahmat-Samii 2004; Pastorino 2007; Rekanos 2008; Randazzo 2012). These methods are in principle able to overcome the problem of local minima, but their numerical complexity is often much higher than the one of deterministic approaches. Finally, despite advancement in numerical modeling and in computer science, effective use of non-linear inversion methods is still limited to relatively small spatial region.

The problem is drastically simplified if linearized scattering models are exploited. Linearization can be achieved by invoking the Born or Rytov approximations (Chew 1999; Pastorino 2010) in the case of penetrable scatterers and the Kirchhoff one for metallic objects (Pierri et al. 2006). The resulting imaging methods can be coarsely grouped in "migration" methods and "inverse-filtering" procedures (Soldovieri and Solimene 2010).

Migration algorithms (Gazdag 1978; Stolt1978; Bleistein2001) were first developed within the framework of seismic prospecting. For a lossless medium they exploit the time "reversibility" to define the conjugate of the wavefield extrapolation operator as a suitable inverse. Related imaging algorithms are the back-propagation (Hansen and Johansen 2000), the time-reversal (Fink 1992), and the omega-kappa algorithm, the latter usual in SAR imaging (Gazdag 1978; Cafforio et al.1991). When the scenario becomes lossy, FFT cannot be any longer used and migration implementation can be conveniently cast by approximating the inverse operator in terms of the adjoint of the scattering operator (Bloemenkamp and Slob 2003; Soldovieri and Solimene 2010). Inverse filtering cast the imaging as a linear inverse problem where reconstructions are obtained by inverting the linearized scattering operator for some object function that accounts for the support and the electromagnetic features of the scatterers (Bloemenkamp and Slob 2003; Soldovieri and Solimene 2010).

Linear inversion imaging methods are computationally effective as compared to non-linear methods and hence allow to deal with scattering scenarios which can be very large (in terms of wavelength) and do not suffer from reliability problems. However, ill-posedness still remains an issue. To this end, linear inversions must always be paired with some regularization scheme in order to establish a trade-off between accuracy and stability against noise (Bertero and Boccacci 1998). In particular, while migration "implicitly" regularizes the problem, inverse filtering requires explicit use of a regularization scheme. This latter is obtained by enforcing some constraint on the solution, dictated by a priori information, or by adopting iterative solving procedures where regularization is achieved by controlling the number of iterations. However, it must be remarked that, since the class of scatterers for which linear models work is extremely limited, such imaging procedures generally do not allow to obtain quantitative reconstructions. Therefore, detection, localization and rough information about the scatterers' shape is what one can expect to retrieve through a linear inversion methods (Slaney et al. 1984; Pike and Sabatier 2002; Pastorino 2010). Fortunately, in many applicative contexts this is exactly what one is interested in. In many other frameworks this is not enough. However, linear reconstructions can be adopted to gain information about the setting of the starting point of non-linear inversions, the latter can strongly affect the outcome of a quantitative reconstruction (Belkebir et al. 2000; Pike and Sabatier 2002; Pastorino 2010).

The computational advantages of linear model based inversion schemes are also shared by the so-called "qualitative" methods (Cakoni 2006). Such methods directly aim at retrieving the scatterers' supports and do not require simplified models of the scattering phenomenon. In particular, they identify the spatial support of the scatterers by adopting certain indicator functions, which assume very different values in norm depending whether they are evaluated within or outside the scatterer supports. The linear sampling (Colton et al. 2000) and the factorization (Kirsch 1999) methods as well as MUSIC based approaches (Devaney 2000; Marengo et al. 2007) belong to this class of inversion algorithms. They require a multistatic/multiview/single-frequency configuration and it has been noted that their performances are reduced when aspect-limited configurations are adopted (Brandfass et al. 2001; Liseno and Pierri 2004; Catapano et al. 2011).

In this chapter, an overview of different methods for subsurface imaging within the framework of the electromagnetic scattering equations is provided. Of course, material presentation will be unavoidable biased from authors expertise and is not intended to provide an exhaustive account of subsurface imaging literature.

First, different measurement configurations used to gather data for subsurface imaging are recalled (Sect. 2). The very popular migration algorithms are described in Sect. 3. In Sect. 4, the scattering equations are briefly introduced. It is shown that the subsurface imaging is a special case of inverse scattering problem and, as such, it is non-linear and ill-posed. Despite of that, we start by presenting imaging algorithms based on simplified linearized scattering models and establish a connection with the migration ones (Sect. 5). Non-linear inversion methods are then presented in Sect. 6, whereas Sect. 7 is devoted to the description of a more recent class of imaging algorithms, the so called "qualitative" methods. Finally, in Sect. 8 hybrid schemes are discussed, while new prospective are addressed in Sect. 9. Concluding remarks and open issues end the chapter.

2 Measurement Configurations

The basic/standard principle of GPR is the transmission of a short electromagnetic pulse into the ground and the measurement of the reflected/scattered field as a function of time or frequency for different spatial positions over which

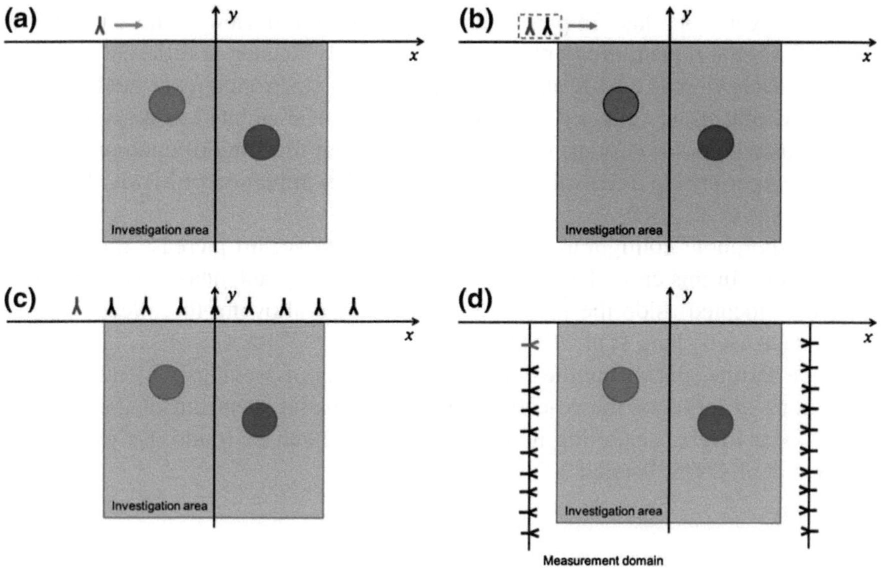

Fig. 1 GPR measurement configurations. **a** Monostatic; **b** bistatic; **c** multistatic; **d** cross borehole

transmitting (TX) and receiving (RX) antennas can be arranged. Depending on the way TX and RX antennas are spatially deployed, different configurations can be adopted in GPR surveys (Fig. 1). The choice of the configuration to be adopted is dictated by different applicative motivations (e.g., the type of investigation to be made, the kind of targets to be detected, the extent of the investigated region, the hardware system at ones disposal, etc.).

The simplest setup, and by far the more common, consists of a single antenna moving along a line (or a surface) at the air-ground interface (monostatic configuration) (Soldovieri et al. 2011a). In this case, the TX and RX antennas are co-located or even a single antenna acting as TX/RX can be used. A variant of this acquisition scheme is the bistatic setting. In this case, the scanning can be achieved by synthesizing the measurement aperture by moving simultaneously the TX and RX antennas while keeping fixed the spatial offset between them (this scheme is also addressed as Common Offset (CO)), or by moving the RX antenna while keeping fixed the TX one (Common Source), or vice versa (Common Receiver) (Huisman et al. 2003; Neal 2004). The Common Source and Common Receiver methods for GPR survey are also addressed in literature as Wide-Angle-Reflection-Refraction (WARR) technique and generally used to delineate the penetration speed (velocity) of the radar waves in a rugged terrain (Neal 2004). Again, the offset between the TX and the RX antennas can be progressively increased by moving the TX antenna and the RX one in opposite directions so as to achieve the so-called Common Mid Point (CMP) configuration (Bristow and Jol 2003).

The main advantage of the above setups reside in their simplicity. However, in some cases it is useful to employ more complicated measurement configurations in order to get more data. In particular, multi-static/multi-view setups can be used. In this case, the transmitting antenna is successively located at different positions, and, for each view (i.e., TX antenna position) the electromagnetic field scattered by buried objects is collected in several measurement points (an array of antennas can also be used) (Counts et al. 2007). Note that this kind of configuration has recently experiencing a revival under the trendy denomination of MIMO GPR systems (Zhuge et al. 2010).

Cross-borehole configurations can also be employed to increase the available information. In this case, the transmitting and receiving antennas are buried inside boreholes located aside the investigation area, thus allowing to collect transmission data as well (Jang et al. 2011).

Some results concerning the comparison of many of previously mentioned configurations in terms of the so-called retrievable spatial spectrum of the unknown when linear inverse scattering imaging is exploited can be found, for example, in (Persico et al. 2005; Persico 2006).

3 Migration Algorithms

This section is devoted to recall migration subsurface imaging methods. For the sake of simplicity, discussion is going to be developed for a two-dimensional and scalar configuration. Moreover, only the case of multimonostatic configuration will be considered with Tx and RX antennas being just located at the air/soil interface. This simplifies discussion as scattering can be roughly assumed to occur within a homogeneous medium (the one coinciding with the lower half-space representing the soil). Of course, generalization to stand-off configurations or to the case of layered soils do not put formal problems and only requires accounting for refractions law in algorithm development.

Consider a point-like scatterer located in the object space at $\mathbf{r}_{sc} = (x_{sc}, y_{sc})$ and denote as $\mathbf{r}_O = (x_O, y_O)$ the positions where the scattered field is recorded. If $s_T(t)$ is the transmitted signal, the corresponding B-scan measurement (the so-called radargram) appears (in the $x_O - t$ data space) as a diffraction hyperbola whose apex is in $(x_{sc}, 2y_{sc}/v)$, v being the soil propagation speed.

The basic idea of migration is to collapse such hyperbola on its apex in the object space. Migration was first conceived as a graphical method (Hagedoorn 1954) based on high frequency assumption. Basically, it is noted that for each A scan (each time-trace), the scatterer's position should lie on a semicircle centered on the source-receiver position and of radius equal to the distance obtained by multiplying the travel-time by half the wave-speed in the soil. Accordingly, each $x_O - t$ data point is equally distributed over a semicircle in the image space so that all the semicircle intersect at $\mathbf{r}_{sc}$. This procedure is schematically depicted in Fig. 2. Since the pioneering graphical method worked out by Hagendoorn, many migration algorithms that implement such a procedure have been developed (Bleistein 2001). In particular, algorithms implementing the Hagendoorn

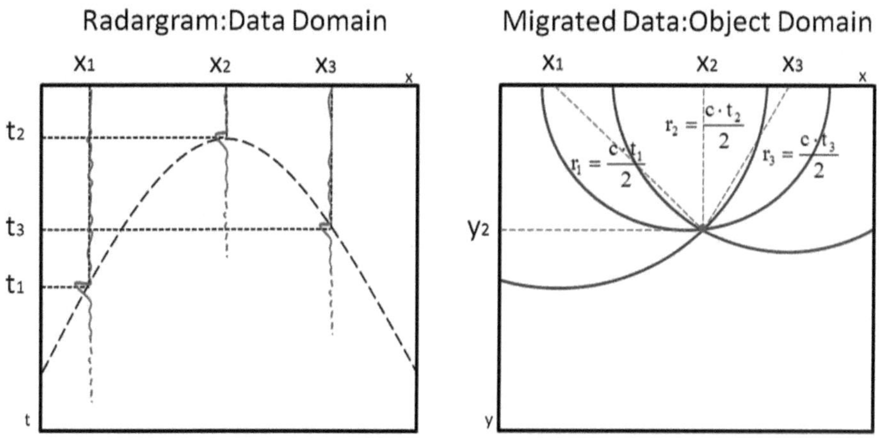

Fig. 2 Illustrating the migration procedure. *Left panel* reports radargram whereas the *right* one shown how hyperbola is collapsed over scatterer's position by A-scan driven approach

procedure are known as Wave Interference Migration (Gazdag and Sguazzero 1984), also addressed in the literature as A-scan-driven approach (Marklein et al. 2002).

A counterpart of Wave Interference Migration is the so-called Diffraction Summation (Gazdag and Sguazzero 1984) [also known as pixel-driven approach (Marklein et al. 2002)]. In this method the object space is divided in pixels and for each of them a diffraction hyperbola is constructed in the data space. The reconstruction at each pixel is then obtained by summing up all the traces (A-scans) that the synthetic hyperbola intersects. This procedure can be implemented automatically and requires the evaluation of the following summation integral for each pixel (x, y)

$$R(x, y) = \int_{\Sigma} \int_{T} s_R(x_O, t)\delta\left(t - \frac{2}{v}\sqrt{(x - x_O)^2 + y^2}\right) dx_O dt \qquad (1)$$

where Σ and T are the measurement aperture and the interval of time during which data are being collected, respectively.

If we denote as $S_R(x_O, k)$ the Fourier transform of $s_R(x_O, t)$ then Eq. (1) can be recast as

$$R(x, y) = \int_{\Sigma} \int_{\Omega_k} S_R(x_O, k) \exp(j2k\sqrt{(x - x_O)^2 + y^2}) dx_O dk \qquad (2)$$

where Ω_k denotes the frequency band in the k domain. Equation (2) points out the equivalence between the Diffraction Summation and the Range Migration Technique presented in (Lopez-Sanchez and Fortuny-Guasch 2000) and also to the Synthetic Aperture Focusing Technique (SAFT) (Marklein et al. 2002).

After computing the spatial convolution in the spatial Fourier domain, the following equation is obtained

$$R(x,y) = \int_{\Omega_{kx}} \int_{\Omega_k} f(k_x, k) \frac{k_z}{\sqrt{k_x^2 + k_y^2}} S_R(k_x, k_y) \exp(-jk_x x) \exp(jk_y y) dk_x dk_y \tag{3}$$

where Ω_{kx} is the selected frequency band in the spatial spectral k_x,$f(k_x, k)$ is an amplitude factor and $k_y = \sqrt{4k^2 - k_x^2}$. Equation (3) has the computational advantage of being computed by FFT algorithms but also requires data to be interpolated and re-sampled according to a rectangular grid in the $k_x - k_y$ spatial spectral domain (Stolt 1978). It is also noted that Eq. (3) is identical to the Synthetic Aperture Radar (SAR) imaging algorithm presented in (Soumekh 1999) and also very similar to the F-K Migration (Gilmore et al. 2006).

Another way to derive migration starts from the wave equation and the so-called "exploding source model". Accordingly, the field spatial spectrum at quota y is related to the one at quota $y'(y < y')$ as

$$S_R(k_x, \omega, y) = S_R(k_x, \omega, y') \exp[jk_z(y - y')] \tag{4}$$

Therefore, the field in the object space can be determined by back-propagating the field measurements. Finally, a superimposition along the frequency ω returns

$$R(x,y) = \int_{\Omega_{kx}} \int_{\Omega} S_R(k_x, \omega) \exp(-jk_x x) \exp(jk_y y) dk_x d\omega \tag{5}$$

This is the so-called F-K Migration [also known as Phase Shift Migration (Gazdag 1978)] which has been shown to be a generalization of the Doppler compression technique typical in SAR imaging (Cafforio et al. 1991).

As can been seen, Phase Shift Migration is very similar to the Diffraction Summation after it is recast as a double Fourier integral in the spatial/spectral domain. However, it requires less approximations on the underlying model. Moreover, in the case of layered medium, it offers the great advantage of avoiding to solve transcendent equations necessary for the Diffraction Summation in order to take into account refraction while computing round-trip delay.

By inverting the Fourier transformation with respect to k_x, Eq. (5) can be rewritten as

$$R(x,y) = \int_{\Sigma} \int_{\Omega} s_R(x_O, y = 0, \omega) \frac{\partial}{\partial z} G^*(x - x_O, y, \omega) dx_O d\omega \tag{6}$$

where G^* is the complex conjugated Green's function.

Migration of Eq. (6) is known as the Rayleigh-Sommerfeld holography which is a particular case (when data are collected over an infinite line) of the so-called Generalized Holography (Langenberg 1987) which in turn is founded on the Porter-Bojarski integral equation. This connection is import because it establishes, in rigorous way, the relationship between the migrated field and the secondary sources, that is the ones arising from the interaction between the incident field and the scatterers, and hence between the migrated field and the scatterers. Finally, we observe that the time domain version of Eq. (6) is nothing else that the well-known Kirchhoff Migration (Berkhout 1986).

4 The Scattering Problem

In the previous section, different migration algorithms have been presented. It is shown that all those imaging schemes are very similar and one can pass from the different migration implementations by Fourier (spatial and/or temporal) transform operators. However, the link between the migrated field and the scatterers to be reconstructed has not been clearly shown and remained only supported by intuitive arguments. To cope with this question the equations describing accurately the scattering phenomenon are needed and are herein introduced. For the sake of brevity, only the case of penetrable scatterers is considered. Similar arguments follow for strong and metallic scatterers. What is more, migration schemes actually arise from linearized scattering model pertaining to the case of penetrable scatterers, even when they are exploited to image metallic objects (Solimene et al. 2014a).

Let us consider the scattering scenario shown in Fig. 3. The buried scatterers consists of in-homogeneities characterized by relative permittivity $\varepsilon_x(\mathbf{r})$ and conductivity $\sigma_x(\mathbf{r})$ supported over a portion of the investigation area (termed as D), which is embedded in a background medium having relative permittivity ε_b and conductivity σ_b. The scatterers and the background medium are supposed to be isotropic and non-magnetic, so that their permeability μ is everywhere equal to that of free space μ_0.

When the region D is probed by an incident electric field E_{inc}, an electric current is induced into the scatterers, or on their surfaces in the case of metallic

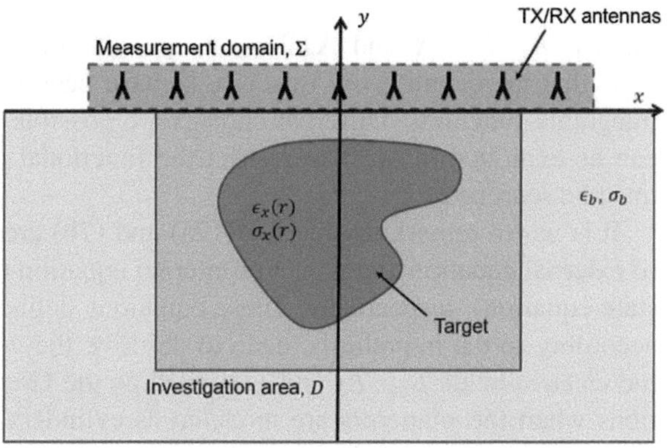

Fig. 3 Schematic representation of the scattering scenario

objects. Such a current radiates an electric scattered field, E_S, that is measured by probes located on the measurement surface Σ outside the investigated region D. Whatever the measurement configuration one may/want/can adopt, the scattering phenomenon can be described by means of two coupled integral equations, which are directly derived by the Maxwell's equations by taking into account the constitutive equations and whose expression depends on the considered reference scenario (Pastorino 2010).

In the frequency domain, as far as penetrable scatterers are concerned, for the TX antenna located at $\mathbf{r}_t$ and the angular frequency ω, the "total" electric field E inside the investigation area and the scattered electric field E_s at the generic observation point $\mathbf{r}_o$ are linked by the following couple of Electric Field Integral Equations (EFIE):

$$E_s(\mathbf{r}_O,\mathbf{r}_\mathbf{t},\omega) = k_b^2 \int\limits_D G_e(\mathbf{r}_O,\mathbf{r},\omega)\chi(\mathbf{r},\omega)E(\mathbf{r},\mathbf{r}_\mathbf{t},\omega)d\mathbf{r} = \mathbf{A}_e[\chi E] \quad \mathbf{r}_O \in \Sigma,\ \mathbf{r} \in D \tag{7a}$$

$$\begin{aligned} E(\mathbf{r},\mathbf{r}_\mathbf{t},\omega) - E_{inc}(\mathbf{r},\mathbf{r}_\mathbf{t},\omega) &= k_b^2 \int\limits_D G_i\left(\mathbf{r},\mathbf{r}',\omega\right)\chi\left(\mathbf{r}',\omega\right)E\left(\mathbf{r}',\mathbf{r}_\mathbf{t},\omega\right)d\mathbf{r}' \\ &= \mathbf{A}_i[\chi E] \quad \mathbf{r},\ \mathbf{r}' \in D \end{aligned} \tag{7b}$$

In Eqs. (7a) and (7b), k_b is the background (complex) wave-number, whereas G_i and G_e denote the so-called internal and external Green's functions which correspond to the field radiated by an elementary electric source inside and outside D, respectively, whose expressions depends on the considered reference scenario (Chew 1999). Moreover, χ is the contrast function accounting for the discontinuity between the (complex) permittivity of the scatterers, ε_x^{eq}, and of the background medium, ε_b^{eq}, whose expression is given by:

$$\chi(\mathbf{r}) = \frac{\varepsilon_x^{eq}(\mathbf{r})}{\varepsilon_b^{eq}} - 1 \tag{8}$$

Finally, $\mathbf{A}_i: X \to X$ and $\mathbf{A}_e: X \to E$ are integral, linear and compact operators. Note that the quantities in Eqs. (7a, b) have been implicitly assumed to be square integrable functions. Different choices are possible as well if a priori information can be exploited in order to restrict the functional set within which the unknown must be searched for.

It is worth remarking that Eqs. (7a) and (7b) are commonly known as the data or external equation and object or internal equation (sometimes also referred as the state equation), respectively. These equations define scalar or vector relationships according to the hypothesis made to describe the wave propagation. In particular, the electric fields E_{inc}, E_S and E as well as the Green's functions are scalar functions when the scatterers are modeled as cylinders of arbitrary cross section and the primary source of the scattering experiment is modeled as a filamentary line source directed along the invariance axis (TM polarization). Conversely, the electric field are vectorial quantities and the Green's functions are dyads when a full 3D scattering model is taken into account.

An equivalent scattering model is obtained by rewriting Eqs. (7a) and (7b) in terms of the "contrast source" current J_{cs}, which is related to the electrical field by the constitutive relation $J_{cs}(\mathbf{r}, \mathbf{r}_t, \omega) = \chi(\mathbf{r}, \omega)E(\mathbf{r}, \mathbf{r}_t, \omega)$. It is worth noting that J_{cs} is not formally a current being the multiplicative factor $j\omega\varepsilon_b^{eq}$ missing. According to the Contrast Source (CS) formulation, the scattering phenomenon is described by the following two equations (van den Berg and Kleinman 1997)

$$E_s(\mathbf{r_O}, \mathbf{r_t}, \omega) = -k_b^2 \int\limits_D G_e(\mathbf{r_O}, \mathbf{r}, \omega)J_{cs}(\mathbf{r}, \mathbf{r_t}, \omega)d\mathbf{r} = \mathbf{A}_e[J_{cs}] \quad \mathbf{r}_O \in \Sigma, \mathbf{r} \in D \tag{9a}$$

$$\begin{aligned} J_{cs}(\mathbf{r}, \mathbf{r_t}, \omega) - \chi(\mathbf{r}, \omega)E_{inc}(\mathbf{r}, \mathbf{r_t}, \omega) &= -k_b^2\chi(\mathbf{r}, \omega) \int\limits_D G_i\left(\mathbf{r}, \mathbf{r}',\omega\right)J_{cs}\left(\mathbf{r}',, \mathbf{r_t}, \omega\right)d\mathbf{r}' \\ &= \mathbf{A}_i[J_{cs}] \quad \mathbf{r}, \mathbf{r}' \in D \end{aligned} \tag{9b}$$

A further alternative source-type formulation is the so-called Contrast Source—Extended Born Model and it is obtained by simply rewriting Eq. (9b), while Eq. (9a) does not change. In particular, by adding and subtracting J_{cs} into the integral operator in Eq. (9b), simple passages lead to (Isernia et al. 2004; D'Urso et al. 2007)

$$\begin{aligned} &J_{cs}(\mathbf{r}, \mathbf{r_t}, \omega) - p(\mathbf{r}, \omega)E_{inc}(\mathbf{r}, \mathbf{r_t}, \omega) = \\ &p(\mathbf{r})\left[k_b^2 \int\limits_D G_i\left(\mathbf{r}, \mathbf{r}',\omega\right)J_{cs}\left(\mathbf{r}'\mathbf{r_t}, \omega\right)d\mathbf{r}' - J_{cs}(\mathbf{r}, \mathbf{r_t}, \omega)f_\Omega(\mathbf{r},\omega)\right] = \mathbf{A}_{i\mathrm{MOD}}[J_{cs}] \quad \mathbf{r}, \mathbf{r}' \in D \end{aligned} \tag{10}$$

where

$$p(\mathbf{r}, \omega) = \frac{\chi(\mathbf{r}, \omega)}{1 - f_D(\mathbf{r}, \omega)} \text{ and}$$

$$f_D(\mathbf{r}, \omega) = \begin{cases} k_b^2 \int\limits_D G_i\left(\mathbf{r}, \mathbf{r}', \omega\right)d\mathbf{r} & \text{2D scalar case} \\ \mathrm{mean}\left\{\mathrm{diag}\left(k_b^2 \int\limits_D G_i\left(\mathbf{r}, \mathbf{r}', \omega\right)d\mathbf{r}'\right)\right\} & \text{3D vectorial case} \end{cases} \tag{11}$$

In order to stress that the three considered models are alternative and equivalent way to describe the wave propagation, let us observe that whatever is the adopted integral formulation (EFIE, CS, CS-EB) we take into account the integral equation can be synthetically expressed as

$$\gamma(r) = \eta E_{inc}(r) + \eta\mathbf{B}\left[\gamma(r)\right] \tag{12}$$

wherein the function γ, the scalar quantity η and the operator **B** are specified as:

$$\gamma = \begin{cases} E & \text{EFIE} \\ J & \text{CS/CS-EB-s} \end{cases}, \quad \eta = \begin{cases} 1 & \text{EFIE} \\ \chi & \text{CS} \\ p & \text{CS-EB-s} \end{cases}, \\ \mathbf{B}[\cdot] = \begin{cases} \mathbf{A}_{\mathrm{i}}[\chi \cdot] & \text{EFIE} \\ \mathbf{A}_{\mathrm{i}}[\cdot] & \text{CS} \\ \mathbf{A}_{\mathrm{iMOD}}[\cdot] & \text{CS-EB} \end{cases} \tag{13}$$

Finally, it is worth remarking that several other forward models can also be considered. In particular, efficient analytical and semi-analytical solutions have been devised for objects with canonical shapes, e.g., multilayer circular and elliptical cylinders (Chew 1999; Caorsi et al. 1997; Di Vico et al. 2005; Frezza et al. 2007).

5 Linear Inverse Scattering Schemes

Linear inverse scattering methods rely on linearized scattering equations. As mentioned above, different approximations can be invoked in order to linearize the scattering problem. However, here, the analysis is restricted to the case of penetrable scatterers. Accordingly, linearization can be achieved by arresting at the linear term the Neumann series expansion of the state equation (i.e., Eq. 7a) (Krasnov et al. 1976). Basically, this consists in assuming $E \cong E_{inc}$ within the scatterer region and the so-called Born linear model is obtained (Chew 1999). Accordingly, the scattering model is now described by

$$E_s(\mathbf{r}_O, \mathbf{r_t}, \omega) = -k_b^2 \int_D G_e(\mathbf{r}_O, \mathbf{r}, \omega)\chi(\mathbf{r})E_{inc}(\mathbf{r}, \mathbf{r_t}, \omega)d\mathbf{r} \\ = \mathbf{A}_e[\chi E_{inc}] = \mathbf{L}[\chi] \quad \mathbf{r}_O \in \Sigma, \mathbf{r} \in D \tag{14}$$

wherein the contrast function frequency dependence has been also ignored. The imaging problem is then cast as the inversion of the linear operator

$$\mathbf{L}\colon \chi \in \mathbf{X} \to \mathbf{E}_S \in \mathbf{E} \tag{15}$$

where χ and E_S are the unknown and the data of the problem, respectively. Moreover, **X** and **E** represent the functional sets within we search for the contrast function and the one we assume the scattered field data belong to. Usually, they are assumed to be Hilbert spaces of square integrable functions. The first one is made by complex valued functions defined on the investigation domain D, whereas the second one is the set of those functions supported over $\Lambda = \Sigma \times \Omega$. In general, the data space depends on the adopted configuration, that is on the choice of the measurement domain as well as the adopted strategy of illumination and observation.

It is worth remarking that the choice of **X** and **E** as Hilbert spaces of square integrable functions accommodates the circumstance of a priori information available on the unknown except that one on the finiteness of its "energy" dictated by physical consideration. On the other side, it assures that **E** is "broad" enough to include the effect of uncertainties and noise on data. If a priori information about the unknown is available, **X** can be chosen in order to accommodate those prioris. For example, if it is known that contrast function has some degree of regularity, i.e., it is differentiable up to a certain order, then **X** can be chosen to be a suitable Sobolev space and so on. Thus, the problem amounts to inverting Eq. (15) to determine the contrast function. Since the kernel of the operator in (15) is a continuous function on $\mathbf{X} \times \mathbf{E}$, then the linear operator is compact (Taylor and Lay 1980). As stated above this means that the inverse problem is ill-posed (Bertero 1989). For compact and non-symmetric operator (as the one at hand) the singular value decomposition is a powerful tool to analyze and solve the problem.

Let us denote as $\{\sigma_n, u_n, v_n\}_{n=0}^{\infty}$ the singular system of operator **L**. In particular, $\{\sigma_n\}_{n=0}^{\infty}$ is the sequence of the singular values ordered in non-increasing way, $\{u_n\}_{n=0}^{\infty}$ and $\{v_n\}_{n=0}^{\infty}$ are orthonormal set of functions that are solution of the following shifted eigenvalue problems

$$\begin{aligned} \mathbf{L}u_n &= \sigma_n v_n \\ \mathbf{L}^+ u_n &= \sigma_n v_n \end{aligned} \tag{16}$$

where $\mathbf{L}^+$ is the adjoint operator of **L** (Taylor and Lay 1980). **L+** and **L** span the orthogonal complement of the null of **L**, $N(\mathbf{L})^{\perp}$, and the closure of the range of **L**, $\overline{R(\mathbf{L})}$, respectively.

A formal solution of Eq. (15) has the following representation (Bertero 1989)

$$\chi = \sum_{n=0}^{\infty} \frac{\langle E_S, v_n \rangle_{\mathrm{E}}}{\sigma_n} u_n \tag{17}$$

where $\langle \cdot, \cdot \rangle_{\mathrm{E}}$ denotes the scalar product in the data space **E**.

By virtue of the compactness of **L**, $R(\mathbf{L})$ is not a closed set (Krasnov et al. 1976). This implies that the Picard's conditions is not fulfilled for any data functions (i.e., the ones having non null component orthogonal to $R(\mathbf{L})$). Hence, the solution may not exist and does not depend continuously on data (Bertero 1989). This is just a mathematical re-statement of ill-posedness. From another point of view, we have to take into account that the actual data of the problem are corrupted by uncertainties and noise n, hence

$$\tilde{\chi} = \sum_{n=0}^{\infty} \frac{\langle E_S, v_n \rangle_{\mathrm{E}}}{\sigma_n} u_n + \sum_{n=0}^{\infty} \frac{\langle n, v_n \rangle_{\mathrm{E}}}{\sigma_n} u_n \tag{18}$$

Now, because of the compactness, the singular values tend to zero as their index increases. This implies that, the second series term (noise-related) in (18) does not converge generally. This leads to an unstable solution since even small error on data are amplified by the singular values close to zero.

The lack of existence and stability of solution can be remedied by regularizing the addressed ill-posed inverse problem (De Mol 1992). For example, this can be achieved by discarding, in the inversion procedure, the "projections" of data on the singular functions corresponding to the "less significant" singular values. This means to filter out the singular functions corresponding to the singular values which are below to a prescribed noise dependent threshold. This regularizating scheme is known as Numerical Filtering or Truncated Singular Value Decomposition (TSVD) and is the simplest one within the large class of windowing based regularizing algorithms (Bertero 1989). More in general, the basic idea of regularization theory is to replace an ill-posed problem by a parameter dependent family of well-posed neighboring problems

$$\chi = \mathbf{L}_{\alpha}^{n} E_S \tag{19}$$

In Eq.(19) α denotes the so-called regularization parameter (in the TSVD this corresponds to the truncation index N_T) and n is the noise level, so that to establish a compromise between accuracy and stability (Groetsch 1993). In particular, as $n \to 0$ also $\alpha \to 0$, and the regularized reconstruction must tends to the generalized inverse whose outcome is just shown in Eq. (17).

Another, is a widespread adopted regularization scheme is the Tikhonov one, which takes advantages from exploiting a priori information about the unknown (Tikhonov and Arsenine 1977). In this case the inversion problem is cast as a constrained optimization problem

$$\tilde{\chi} = \min\{\|\mathbf{L}\chi - E_S\|^2_{L^2(\Sigma)} + \alpha\|\mathbf{C}E_S\|^2_{L^2(\Sigma)}\} \tag{20}$$

Here the constraint arises from the available a priori information expressed by the operator **C**. Note that for energy constraint only, **C** coincides with the identity operator. Problem of convergence in this case has been elegantly studied in (De Micheli et al. 1998).

Finally, the Landweber regularization scheme recasts the first kind integral equation to be inverted as an integral equation of second kind so that a well-posed problem is obtained (Groetsch 1993). Accordingly, Eq. (20) is recast as

$$\tilde{X} = \mathbf{L}^{+}E_S + (I - \mathbf{L}^{+}\mathbf{L}E_S)\chi \tag{21}$$

and a solution is obtained by means of an iterative procedure. In this case the regularization parameter is the number of iteration N_I.

Tikhonov and Landweber regularization schemes can be compared if all are analyzed in terms of the operator properties. This can be done by expressing the different regularized reconstruction in terms of the singular system. By doing so, one obtains

$$\tilde{\chi} = \sum_{n=0}^{\infty} \frac{\sigma_n}{\sigma_n^2 + \alpha} \langle E_S, v_n \rangle_{\mathrm{E}} u_n \tag{22}$$

for the Tikhonov regularization with $\mathbf{C} = \mathbf{I}$, and

$$\tilde{\chi} = \sum_{n=0}^{\infty} \frac{1-(1-\sigma_n^2)^{N_I}}{\sigma_n} \langle E_S, v_n \rangle_{\mathrm{E}} u_n \tag{23}$$

for the Landweber method. Other regularizzation methods are based on metric or statistic information criteria (De Micheli and Viano 2002). All these regularization methods result (in a different) filtering of the unknown spectral expansion.

It is clear that, apart from the computational convenience that can dictate the regularization algorithm to adopt, the key question is the choice of the regularization parameter. This choice must compromises the noise level, the mathematical features of the operator to be inverted and available a priori information about the unknown. Different methods exist to select the regularization parameter. Such methods can explicitly exploit the knowledge of the noise level (such as the Morozov discrepancy principle) or not (such as the generalized cross validation) (Hansen et al. 2006).

It is worth noting that the scattered field has finite number of degrees of freedom (Bucci and Franceschetti 1989; Bucci and Isernia 1997; Bucci et al. 1998; Solimene and Pierri 2007; Solimene et al. 2013b, c) and the amount of available independent data depends on the parameter of the adopted measurement configuration. Such information can be exploited to select the regularization parameter.

The singular system formalism can be also employed to compare migration and inverse filtering. It can be easily recognized that migration (see for example Range Migration) substantially, corresponds to achieve the inversion by means of the adjoint operator, that is

$$\tilde{\chi} = \mathbf{L}^{+} E_S \tag{24}$$

whose expression in terms of the singular system is

$$\tilde{\chi} = \sum_{n=0}^{\infty} \sigma_n \langle E_S, v_n \rangle_{\mathrm{E}} u_n \tag{25}$$

It is readily noted that, migration allows to obtain a stable reconstruction because the singular values now appear at the numerator. However, it is not a regularization scheme in the sense of Tikhonov (Tikhonov and Arsenine 1977) because even in absence of noise the actual unknown is not retrieved. From a practical point of view, this entails an intrinsic limit on the achievable resolution regardless of the noise level (Soldovieri and Solimene 2010).

Finally, in the framework of inverse filtering, it is worth mentioning the class of inversion algorithms that have being developed on the basis of a new paradigm subtended by compressive sampling (Donoho 2006).

The approach is of relevant interest when the scattering scene is sparse with respect to a suitable representation dictionary (roughly saying, if the target occupies a small fraction of the area to be imaged). In this case, the imaging problem is formulated as in Eq. (20), but the regularization term is based on the L^1norm (Soldovieri et al. 2011b), that is

$$\tilde{\chi} = \min\{\|\mathbf{L}\chi - E_S\|^2_{L^2(\Sigma)} + \alpha\|\mathbf{C}E_S\|_{L^1(\Sigma)}\} \tag{26}$$

Details on why such a kind of inverse filtering allows to obtain highly resolved reconstructions with sparse set of data is left to the pertinent literature. Here, we just mention the paper (Baraniuk 2007) where an intuitive geometrical interpretation is provided in terms of the "shape" of the iper-spheres in L^1. Of course, also for this scheme, the choice of the regularization parameter α is crucial, since it dictates the degree of sparsity at which the scene is reconstructed.

6 Non-linear Inversion Schemes

Linear inversion schemes are computationally effective and flexible. However, they generally allow to retrieve only approximate information about targets' geometric features. In order to obtain "quantitative" reconstructions, the full non-linearity of the problem must be tackled. A first step towards this direction is to consider high order Born models. For example, several works considered the second-order Born approximation as direct model for relating the measured scattered-field data to the contrast function (Chew 1999; Pastorino 2010). In this case, the scattering equations reduces to

$$E_S = \mathbf{A}_e\chi(E_{inc} + \mathbf{A}_i\chi E_{inc}). \tag{27}$$

The main advantage of this approach is that the dependence upon the internal electric field is removed. Moreover, it has been found that such an approximation is able to provide better results than its first-order counterpart (Caorsi et al. 2001a; Leone et al. 2003; Estatico et al. 2005). Higher-order approximations have successfully been used, too (Azaro et al. 2006).

However, in general the full-nonlinearity must be addressed. To this end, two main classes of approaches have been developed in the scientific literature: deterministic and stochastic methods.

6.1 Deterministic Inversion Procedures

Deterministic approaches usually start from an initial guess (e.g., an empty investigation area if no a priori information is available) and iteratively modify the solution according to some deterministic rule. It is worth noting that most of this approaches aim at obtaining a solution, at each iteration, that minimizes the error between the measured data and the model ones, i.e., the scattered fields computed by using the current estimate of the unknown contrast profile. In this frame, two main classes of approaches can be identified.

In the first one, the formal solution of the object equation, i.e.,

$$E(\mathbf{r}_O) = (I - \mathbf{A}_i\chi)^{-1}E_{inc}(\mathbf{r}_O), \quad \mathbf{r}_O \in D, \tag{28}$$

is substituted in the data equation in order to obtain the following non-linear model

$$E_S(\mathbf{r}) = \mathbf{A}_e\chi(I - \mathbf{A}_i\chi)^{-1}E_{inc}(\mathbf{r}), \quad \mathbf{r} \in \Sigma \tag{29}$$

Then the imaging problem is solved by looking for the contrast function that minimizes the L^2 nom of the difference between the left and right hand side terms of Eq. (29). In particular, such a kind of approaches minimize the quantity:

$$f(\mathbf{x}) = w_d \left\| E_S - \mathbf{A}_i\chi(I - \mathbf{A}_i\chi)^{-1}E_{inc} \right\|^2_{L^2(\Sigma)} = w_c f_c(\mathbf{x}), \tag{30}$$

where $\mathbf{x} = (\chi)$. The normalization constant is usually set equal to $w_d = \|E_S\|^{-2}_{L^2(\Sigma)}$. In order to face the ill posedness of the inverse problem, a proper Tikhonov term is often added in Eq. (30) as a way to regularize the problem and obtain a stable solution against noise and measurement uncertainties affecting the data.

An example of this type of approaches is the Born or the distorted-Born iterative method (which are equivalent to the Newton-Kantorovich method) (Chew and Wang 1990; Remis and van den Berg 2000). In such solving strategy, Eq. (28) is solved by iteratively constructing a linearized problem by means of the distorted-Born approximation. Given a reference contrast χ_k (e.g., the solution at the kth iteration), such approximation is defined as

$$E_S \cong E_S^k + \mathbf{A}_e^k \delta\chi E^k \tag{31}$$

where E_S^k and E^k denote the external scattered and internal total fields in the inhomogeneous medium χ_k and $\mathbf{A}_e^k(\mathbf{r}) = \int_D (\cdot) g_{db}^k(\mathbf{r}, \mathbf{r}')d\mathbf{r}'$, being g_{db}^k the Green's function for an inhomogeneous background characterized by the contrast function χ_k. The obtained linear problem is then solved, for example, by using a conjugate gradient method and the solution is then updated with $\chi_{k+1} = \chi_k + \delta\chi$. Clearly, at every step it is necessary to solve a forward problem in order to update the Green's function needed in the operator $\mathbf{A}_e^k$ and the scattering model (28).

It is worth noting that, although substituting the formal solution of the object equation into the data equation allows to remove the explicit dependence upon the internal electric field, it has some disadvantages. In fact, the strong nonlinear dependence (especially when dealing with very large contrasts) could make the inversion algorithms converge to false solutions corresponding to local minima. In order to avoid such a problem, it is necessary the use of a priori knowledge for properly initializing the iterative procedure.

In the second class of approaches, the two equations describing the scattering phenomenon are kept separate and solved simultaneously, i.e., the contrast function is the solution of the following system of equations

$$\begin{cases} E_S(\mathbf{r}) = \mathbf{A}_e\chi E(\mathbf{r}) & \mathbf{r} \in \Sigma \\ E(\mathbf{r}) = E_{inc}(\mathbf{r}) + \mathbf{A}_i\chi E(\mathbf{r}) & \mathbf{r} \in D \end{cases}. \tag{32}$$

Clearly, both the contrast function χ and the electric field E inside the investigation area are unknown. Consequently, the inversion algorithms must retrieve the unknown array $\mathbf{x} = (\chi, E)$.

In this case, the necessity of solving a forward problem at each iteration of the inversion procedure (e.g., for computing the Frechét derivatives or for updating the internal electric field) is avoided. In addition a moderate degree of nonlinearity (Bucci et al. 2001a) is achieved, since the imaging problem is linear with respect to the product χE (Isernia et al. 1997). Conversely, this kind of approaches could lead to account for search spaces of very high dimensions. In fact, the number of unknowns is as twice as that used by methods that try to solve Eq. (30) in the single-view case, and increases linearly with the number of views in the multi-view case, since the total field E changes with the direction of the probing wave (i.e., the location of the primary source of the scattering experiment). Moreover, since χ and E are physical quantities, which vary in different ranges, the overall number of iterations needed by the inversion algorithm for converging to the final solution could increase significantly.

In this framework, several methods have been devised. One of the first developed approaches is the modified conjugate gradient method, in which at each iteration the field and the contrast function are alternatively updated by using a conjugate gradient scheme (Kleinman and van den Berg 1992; Isernia et al. 1997). By this approach, the solution to the inverse scattering problem is looked for by minimizing the following cost function

$$\begin{aligned} f(\mathbf{x}) &= w_d \| E_S - \mathbf{A}_e \chi E \|^2_{L^2(\Sigma)} + w_s \| E - E_{inc} - \mathbf{A}_i \chi E \|^2_{L^2(D)} \\ &= w_d f_d^{EFIE}(\mathbf{x}) + w_s f_s^{EFIE}(\mathbf{x}). \end{aligned} \tag{33}$$

The weights w_d and w_s should balance the two terms of the cost function. A common choice is $w_d = \| E_S \|^{-2}_{L^2(\Sigma)}$ and $w_s = \| E_{inc} \|^{-2}_{L^2(D)}$, for which $w_d f_d = w_s f_s = 1$ when $\chi = E = 0$. Moreover, such a choice also ensures that the approach is insensitive to the magnitude of the incident field. More recently, inversion methods based on new inner-outer schemes have also been proposed (Bozza et al. 2006, 2007; Pastorino and Randazzo 2012). In such inversion approaches, an outer Gauss-Newton linearization loop is used to linearize the system of Eq. (32). In the inner loop, the obtained linearized problem is solved in a regularized sense by means of a truncated Landweber method. Such approach has been found to exhibit very good regularization capabilities.

In the previous class of approaches, the unknowns were the contrast function and the internal electric field. However, it is also possible to consider inversion approaches based on the CS form of the scattering equations. Accordingly, it is possible to recast the imaging problem as the solution of the following system of equations:

$$\begin{cases} E_S(\mathbf{r}) = \mathbf{A}_e J(\mathbf{r}) & \mathbf{r} \in \Sigma \\ J(\mathbf{r}) = \chi E_{inc}(\mathbf{r}) + \mathbf{A}_i J(\mathbf{r}) & \mathbf{r} \in D \end{cases}. \tag{34}$$

In this case, the data equation is now linear with respect to the contrast source J. However, the inversion of such an equation is still ill-posed. Moreover, it has been showed that it admits nontrivial solutions and that solving directly such equation (e.g., by finding the minimum norm solution) could lead to false solutions (see (Pike and Sabatier 2002) and the reference therein). Consequently, it is necessary to jointly solve the two equations.

In this framework, one of the main successful approaches is the contrast source inversion method proposed in (van den Berg and Abubakar 2001). In this solving procedure, the inverse problem is recast as an optimization problem, which is solved by using a conjugate-gradient strategy for updating, at each iteration, the contrast source and the contrast function in an alternate way. In this case, the cost function to be minimized is:

$$\begin{aligned} f(\mathbf{x}) &= w_d \|E_S - \mathbf{A}_e J\|^2_{L^2(\Sigma)} + w_s \|J - \chi E_{inc} - \mathbf{A}_i J\|^2_{L^2(D)} \\ &= w_d f_d^{CS}(\mathbf{x}) + w_s f_s^{CS}(\mathbf{x}). \end{aligned} \tag{35}$$

where, the weights of the cost functional are $w_d = \|E_S\|^{-2}_{L^2(\Sigma)}$ and $w_s = \|\chi E_{inc}\|^{-2}_{L^2(D)}$. It must be remarked, however, that as the cost function is not defined for $\chi = 0$, it is not possible to start the inversion with an empty domain. A common choice for the initial value is to use the estimate obtained by means of backpropagation (van den Berg and Abubakar 2001) or other linear inversion schemes.

Other approaches based on the contrast source equations have also been proposed in the literature. As an example, the outer/inner inexact-Newton approach proposed in (Bozza and Pastorino 2009) has been found to be very effective in reconstructing dielectric structures. In fact, such approach combine the advantages of the CS formulation with the good regularization capabilities of the outer/inner inexact-Newton method. Moreover it is worth noting that differently from (van den Berg and Abubakar 2001), where the two unknowns are updated in an alternate way, the updates of all the problem unknowns are contemporarily found by solving a single linear problem.

Finally, it is worth remarking that the minimization problem defined in Eq. (35) can be rewritten by using the Contract Sources–Extended Born model as an alternative form of the scattering equations and solved by looking for the auxiliary function p and the contrast source J, simultaneously. By doing so, improved reconstruction accuracy, as compared to those obtained by using the traditional EFIE and CS forms, can be achieved (Isernia et al. 2004; Crocco et al. 2005; Catapano et al. 2006a; D'Urso et al. 2010). This is due to the fact that the rearrangement of the scattering equations, on which the CSEB model is based, leads to a reduction of the "degree of nonlinearity" (Bucci et al. 2001a), in many cases of practical interest and in particular when lossy media are surveyed, (Isernia et al. 2004; D'Urso et al. 2007, 2010). Accordingly, an increased robustness against false solutions is obtained.

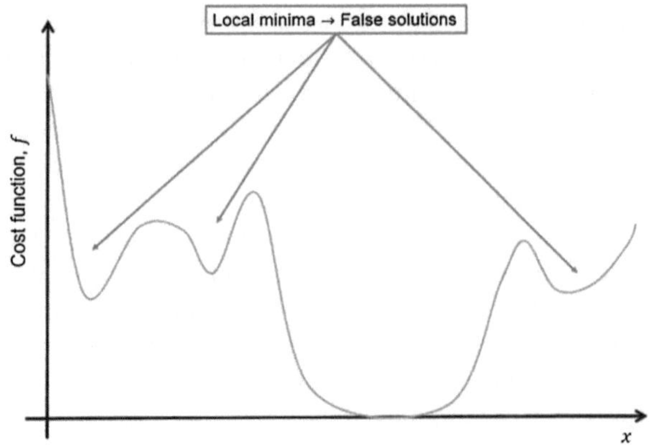

Fig. 4 Representation of multimodality of the minimized functional. Local minima corresponds to false solutions of the inverse problem

Whatever deterministic inversion scheme is adopted, crucial issues are to avoid local minima, i.e. that the reconstruction procedure can be trapped into a false solution (Fig. 4), and to properly account for the ill-posed nature of the inverse problem. Of course this means that the inversion algorithms have to be carefully initialized, but also that the use of a regularization strategy is mandatory to prevent unreliable results. A possible and widely adopted regularization scheme is to consider an unknown function belonging to a finite dimensional space whose dimension does not exceed the essential dimension of the data space (Bucci and Isernia 1997). In most cases, the unknown quantities, i.e. the contrast function and the electric field (or, correspondingly, the contrast source) are discretized by using pulse basis functions. By doing so, a simple relationship for computing the matrices representing the scattering operators is obtained but also a large number of unknowns has to be retrieved, i.e., the unknown array has very large dimensions.

A convenient strategy to reduce the number of unknowns without affecting the accuracy of the contrast representation is the regularization by projection. In this case, different basis functions, such as sinusoidal, Fourier, wavelets and spline functions have been considered. It is worth noting that the use of wavelets and spline functions allows a quite straightforward implementation of multi-resolution or sparse representation of the unknowns (Catapano et al. 2004; Baussard et al. 2004a, b; Winters et al. 2010, Scapaticci et al. 2012). Another effective possibility is the use of a multi-resolution strategy (Caorsi et al. 2003a). In this approach, the inversion is performed at different scales, and at each resolution step the investigation is focused only on the estimated region containing the target. By doing so, a better exploitation of the information available from collected data is obtained, thus yielding both accurate reconstructions and high computational efficiency. Such strategies can also be combined with different inversion schemes in order to exploit their regularization properties (Randazzo et al. 2011) and the advantages of the contrast source formulation (Oliveri et al. 2012b).

A further relevant aspect is the choice of the measurement configuration. Since the goal is to collect as much independent data as possible, the use of a multiview, multistatic and multi-frequency measurement setup is commonly preferable (Bucci et al.

2001b). On the other hand, while the contrast function does not depend on the direction of the probing waves, due to the dispersive nature of the materials, in general it changes with the frequency. As a consequence the exploitation of frequency diversity is not obvious. In this framework, the most common strategy is the frequency hopping procedure, wherein the available multiple frequency data are sequentially processed (Chew and Lin 1995). In particular, the optimization problem is solved at each single frequency starting from the lowest considered one. Then, one moves up in frequency by using the obtained solution as starting point to perform the minimization of the cost functional at the higher frequency. It is worth noting that usually in the frequency hopping scheme the number of unknown coefficients adopted to represent the contrast function is commonly progressively enlarged. This is consistent to the fact that the amount of independent data increases with frequency (Bucci et al. 2001b).

Further regularization strategies are those exploiting a penalty term that is added to the cost functional to be minimized.

In this frame, the exploitation of the total variation of the image has been found to provide significant enhancements in the reconstructions. To this end, the minimized cost function is modified by adding a total variation term, which in its basic form is defined as

$$f_{TV}(\mathbf{x}) = \int\limits_{\Omega} \left(\sqrt{|\nabla \chi(\mathbf{r})|^2 + \delta^2} \right)^p d\mathbf{r}, \quad 1 \leq p \leq 2. \tag{36}$$

Such term can be added (van den Berg and Kleinman 1995) or multiplied (Abubaker and Van Den Berg 2001; Abubakar et al. 2012) to the original residual error term. The multiplicative approach has however been found to be more effective. Moreover, it does not require to define a proper weight for balancing the various cost terms, as needed in the additive case.

An additive penalty term enforcing a priori knowledge on the homogeneous nature of the scatterers has also been proposed in (Catapano et al. 2004), to face the problem of imaging objects embedded in a homogeneous or layered medium from aspect limited data. In this frame, regularization by projection and multi-frequency data have been also exploited. In particular, to make it possible to accommodate the unknowns in a non-uniform way within the investigated domain, Haar wavelet basis functions have been used to represent the contrast function. Moreover, multi-frequency data have been processed, without neglecting the dispersive behavior of the materials. The reconstruction capabilities of this approach have been assessed against experimental data gathered in laboratory controlled conditions (Catapano et al. 2006b).

Finally, it is worth mentioning that, more recently, the use of different regularization strategies has also been introduced in the field of electromagnetic imaging. In particular, the regularization schemes working in Banach spaces has been found to be very effective, especially in reconstructing targets with small dimensions and with sharp edges (Estatico et al. 2012, 2013). In fact, they are able to provide a reconstruction characterized by less over smoothing and background ripples than standard formulation based on square integrable Hilbert spaces.

6.2 Stochastic Inversion Procedures

In stochastic approaches, the inversion of the scattering equations is recast as an optimization problem (Rahmat-Samii and Michielssen 1999; Robinson and Rahmat-Samii 2004; Pastorino 2007; Qing and Lee 2010; Randazzo 2012), which is solved by using stochastic minimization schemes such as evolutionary or swarm algorithms. The cost function is usually defined as the difference between the measured data and the field computed by means of the assumed propagation model. As an example, when considering the EFIE formulation, the cost function in Eq. (33) is usually used. Several terms can also be added, e.g., regularization terms, a priori information, total variation, etc. Clearly, in order to avoid the need of considering the internal electric field as an unknown of the optimization problem, the two scattering equations can also be combined as in Eq. (29). However, such choice has the drawback of requiring the solution of a forward problem every time the cost function is evaluated, thus leading to significant computational requirements.

Stochastic approaches have several advantages with respect to the deterministic one (Rahmat-Samii and Michielssen 1999; Pastorino 2007). First of all, they are in principle able to reach the global minimum of the cost function, even starting from solutions very far from the actual one. On the contrary, deterministic approaches, as previously stated, are strongly dependent on the initialization. Moreover, a great advantage of stochastic approaches is the simplicity of including a priori information in the inversion procedure. In fact, it is possible to introduce arbitrary penalty terms in the cost function, even if they do not allow the computation of the derivatives (which are needed in most deterministic approaches). Moreover, it is possible to introduce a priori information on the search ranges and on the physical/geometrical structure under test by simply forcing some of the unknowns. The main drawback of stochastic approaches is however the high amount of computational resources needed to perform the inversion, especially when considering a pixel-based representation of the contrast function since good spatial resolutions produce very high dimensions of the discrete representation of the unknowns. However, it should be noted that, in several cases, only a limited subset of parameters can be used to describe the underground scenario [e.g., the position, size and dielectric properties of buried targets (Li et al. 2008) or the coefficients of a parametric curve describing the shape of the object (Sun et al. 2010)]. Consequently, the use of simplified scattering models allows to efficiently reduce the unknowns' number, allowing for faster reconstructions. Moreover, stochastic algorithms can be easily implemented on parallel computers for speeding up the computation (Massa et al. 2005).

In the scientific literature, several different stochastic optimization schemes have been proposed. In almost all cases a set of trial solutions

$$\wp_k = \left\{ \mathbf{x}_p^{(k)}, p = 1, \ldots, P \right\} \tag{37}$$

is defined (where k denotes the kth iteration of the algorithm). Usually, the initial population $\wp_0$ is randomly generated, e.g., by creating random distributions of the contrast function. Clearly, a priori information on some of the characteristics of the investigated structures can be also added in this phase. Such population is then iteratively modified by applying stochastic operators. How such operators act on the population depends on the particular optimization scheme and impact on the convergence properties of the algorithms. The iterations are stopped when some predefined stopping criteria is fulfilled. In particular, the stopping criteria can be composed by several conditions. Some of the most commonly used are the following:

- Maximum number of iterations—The method is stopped when a given number of iteration k_{max} is reached;
- Cost function threshold—The method is stopped when the value of cost function of the best trial solution falls below a given threshold f_{th};
- Cost function improvement threshold—The method is stopped if the improvement of the cost function of the best individual after k_{th} iterations is below a fixed threshold Δf_{th}.

Concerning the population update mechanism, several algorithms have been proposed and used in electromagnetic imaging inversion schemes applied to GPR. One of the first algorithm employed in this field is the genetic algorithm (GA) (Haupt 1995; Chiu and Liu 1996; Weile and Michielssen 1997; Rahmat-Samii and Michielssen 1999; Caorsi et al. 2000, 2001a, b; Caorsi and Pastorino 2000; Chen and Chiu 2000; Qing et al. 2001; Pastorino et al. 2004). In GAs, the trial solutions are combined together with stochastic rules that mimic the crossover and mutation of biological genes. In this framework, several different implementations have been developed. In particular, both binary (i.e., the unknowns are encoded in binary strings) and real (i.e., the unknowns are encoded in arrays of real values) representations have been successfully used. However, in electromagnetic prospecting applications real-coded GAs are usually preferred since they avoid the need of decoding the unknowns when computing the cost function (which could increase the computational time needed to perform the inversion). Moreover, different crossover and mutation operators can be used to modify the population (Pastorino 2007).

Although GAs have been widely applied and very good results have been obtained, they usually require several iterations to converge. In order to reduce the time needed to perform the inversion, other optimization algorithms have been considered. An approach that have been found to be quite effective is the Differential Evolution (DE) algorithm. In fact, in the DE algorithm the crossover and mutation operators ensures a better exploration of the search space by also limiting the "destructive" effects of the mutation strategies usually employed in GAs (Qing and Lee 2010). The DE algorithm has been applied to subsurface imaging by considering simplified models of the scatterers, such as circular and elliptical cylinders and spheres (Michalski 2000, 2001; Breard et al. 2008). Moreover, it has been successfully used for the identification of the locations and shapes of PEC cylinders (Anyong 2003, 2006). Finally, DE-based strategies has been developed also considering a pixel-driven discretization of the unknown contrast function (Massa et al. 2004; Semnani et al. 2010).

More recently, swarm optimization algorithms, which are inspired by the collective behavior of real entities such as particles, birds, ants, etc., have been found to be very effective for solving optimization problems (Clerc 2006; Blum and Merkle 2008; Randazzo 2012). In particular, the particle swarm optimization (PSO), which is inspired by the behavior of flocks of birds and shoals of fish, has been extensively used in electromagnetic problems (Robinson and Rahmat-Samii 2004), and in particular for electromagnetic imaging (Caorsi et al. 2004; Rekanos and Kanaki 2006; Huang and Mohan 2007; Huang et al. 2008; Rekanos 2008; Semnani et al. 2009). The Ant Colony Optimization (ACO) has also been successfully used for the reconstruction of 2D cylindrical structures and 3D objects (Pastorino and Randazzo 2013). In (Pastorino 2007), the ACO-based inversion approach is compared with GA and DE, showing that ACO "reaches a more accurate reconstruction than those obtained by the other two methods" and "requires a lower number of function evaluations". Finally, the Artificial Bee Colony (ABC) has also been used in imaging applications, both considering 2D problems (Randazzo 2012) and 3D (Donelli et al. 2011) configurations. In the latter case, it has also been shown that ABC outperforms GA, PSO, and DE approaches.

7 Qualitative Inversion Schemes

The general title of qualitative inverse scattering methods group those imaging procedures avoiding approximate scattering models, on which the linear approaches are based, but, differently from the nonlinear procedures, they only look for limited information about the scattering objects.

In this Section the attention is restricted on those qualitative approaches that provide information on the geometrical features (i.e. location and shape) of unknown scatterers by taking into account the characteristic behavior of an indicator function, whose construction depends on the adopted approach.

Generally speaking, this kind of qualitative approaches are attractive because, they face the imaging problem without introducing any approximation on the scattering model, as opposed to linear inversion procedures. At the same time, non-linearity of inverse scattering problem is avoided as they do not require to explicitly solve the scattering equation for the scatterer function. Indeed, imaging is cast as the inversion of a suitable auxiliary linear problem. Moreover, they are claimed to be able to achieve a characterization of the support of the scatterers (dielectric or metallic) even if such a support is not convex nor connected (i.e., in the case of multiple targets). Finally, at the expenses of some conceptual complication, they are effective in terms of the required processing time and computational resources.

Amongst the others, the so-called "sampling methods" are the most famous ones. They originated from the Simple Method (SM) introduced by Colton and Kirsch in (Colton and Kirsch 1996) and then developed into the Linear Sampling Method (LSM) (Colton et al. 1997) and the Factorization Method (FM) (Kirsch 1998), which are among the most famous ones.

According to the mathematical theory on which these methods are based, imaging through LSM and FM consists in building up an indicator function I which assumes low values for points located within the targets' support and it is large elsewhere (Colton et al. 1997; Kirsch 1998). Therefore, location and shape of the objects are retrieved by sampling the spatial domain under investigation into an arbitrary grid of points, by computing in each of them the value of I and by observing its spatial map.

Let $E_s(\theta,\varphi)$ be the scattered field measured in the angular direction φ when the investigated region is probed by an incident field impinging from the angular direction θ. By introducing the integral operator

$$F\left[x(\cdot,\mathbf{r_p})\right]: x(\cdot,\mathbf{r_p}) \in \Sigma \rightarrow \int_{\Sigma} E_s(\theta,\varphi)x(\theta,\mathbf{r_p}) \quad d\vartheta \in \Sigma \tag{38}$$

the LSM and the FM consist in solving, for each point $\mathbf{r_p}$ in D, the linear equation:

$$L\left[x(\cdot,\mathbf{r_p})\right] = g(\mathbf{r_p},\varphi) \tag{39}$$

In the case of the LSM, L is equal to the operator F as defined in Eq. (38), whereas it is defined as $(F^{+}F)^{1/4}$ when the FM is applied. In Eq. (38), $x(\cdot,\mathbf{r_p})$ is the unknown quantity whose energy defines the indicator function I, whereas the right hand side term $g(\mathbf{r_p},\varphi)$ is the field radiated along the angular direction φ by an elementary source located in $\mathbf{r_p}$, i.e. the Green's function, whose expression depends on the considered reference scenario (Chew 1999).

By using the Tikhonov regularization in order to obtain a stable solution of Eq. (39), for each sampling point $\mathbf{r_p}$, the indicator function is given by:

$$I(\mathbf{r_p}) = \left\|\mathbf{x}(\cdot,\mathbf{r_p})\right\|^2 = \sum_{n=1}^{N}\left(\frac{\lambda_n^p}{\lambda_n^q + \alpha^2}\right)^2 \left|\left\langle \mathbf{g}(\mathbf{r_p}),\mathbf{u}_n\right\rangle\right|^2$$
$$p = \begin{cases} 1 & \text{LSM} \\ 1/2 & \text{FM} \end{cases}, \quad q = \begin{cases} 2 & \text{LSM} \\ 1 & \text{FM} \end{cases} \tag{40}$$

In Eq. (39), λ_n and $\mathbf{u}_n$ denote the n-th singular value and right singular vector of the MxV multistatic/multiview data matrix, whose i-th row is the scattered field collected at a fixed working frequency at M measurement points when the i-th of the V transmitters is radiating; $N = \min(M,V)$; $\|\cdot\|^2$ denotes the L^2-norm on Γ, $\langle\cdot,\cdot\rangle$ is the scalar product on Σ; α is the Tikhonov regularization parameter; $\mathbf{g}$ is the M dimensional vector containing the values of the electric field radiated at the M measurement points by an elementary source located in $\mathbf{r_p}$.

In the original implementation of the LSM and FM, the Tikhonov regularization parameter is determined for each sampling point by means of the Morozov's generalized discrepancy principle. However, in (Aramini et al. 2006; Catapano et al. 2007; Brignone et al. 2009) it has been shown that the same regularization parameter can be used for all the sampling points and an empirical practical rule to fix α has been proposed in (Catapano and Crocco 2010).

It is worth remarking that, in the full 3D vectorial case, **g** depends on the spatial orientation d of the elementary source (i.e., the electric dipole) and such an orientation can be considered as a degree of freedom. Hence, Eq. (39) represents a family of equations, having d as parameter. By taking this feature into account, the LSM and FM image is actually built by combining the results from at least two orthogonal orientations of the elementary source. For instance, one could use the indicator obtained as:

$$\mathbf{I} = \frac{1}{3}\left[I_x + I_y + I_z\right] \tag{41}$$

where $I_{x,}$ I_y and I_z are the indicator functions defined as in Eq. (39) and corresponding to elementary sources oriented along the three Cartesian coordinate axis. Notably, such a combined indicator does not require to perform additional measurements, since the different orientations only regards the quantity **g**, which is computed and not measured.

From discussion above, it is evident that the main computational effort involved in the LSM and FM implementation is the computation of the SVD of the data matrix. However, since the dimension of such a matrix is dictated by the number of probing and measurement points, such an effort would be almost negligible even for a full 3-D reconstruction.

Summarizing, the main advantages of the LSM and FM are:

- the flexibility, as they can be applied regardless of the peculiar (i.e., metallic or dielectric) nature of the scatterers without requiring a priori information about the nature and number of the scatterers;
- a very simple implementation and a remarkable computational speed, since they mainly involve several repetitions (for each 'sampling' point) of a matrix-times-vector product.

Despite these advantages, the widespread use of the LSM and FM has been significantly limited up to now due to two main issues.

The first one is the complexity of the mathematical theory explaining why these methods work. The theoretical proof of the applicability of the LSM and FM is, indeed, based on not trivial mathematical concepts (Colton et al. 1997, 2003; Kirsch 1998, 1999, 2004; Collino et al. 2003; Arens 2004; Liseno and Pierri 2004; Arens and Lechleiter 2009) and currently it has not yet provided for all the possible configurations of scatterers and reference scenarios, in which it has been empirically shown that LSM and FM can provide accurate results. To overcome this drawback, several efforts have been done to provide a physical interpretation of these imaging approaches. In this framework, an interpretation of the LSM restricted to the case of perfect electric conductor targets and not valid for the complementary case of dielectric (i.e., penetrable) objects has been proposed in (Shelton and Warnick 2003). Conversely, a physical meaning dealing with penetrable scatterers and assimilating the LSM to the problem of focusing into a point the field radiated by a collection of sources has been given in (Catapano et al. 2007). An alternative physical interpretation based on the energy conservation principle, which does not depend on the penetrable or impenetrable nature of the scatterers, has been discussed in (Aramini et al. 2010, 2011).

The second issue limiting the use of both LSM and FM in applicative contests is that these imaging strategies require multiview multistatic data collected all around the investigated domain, otherwise their performance dramatically get worse. In this framework, modified indicator functions compensating for the decreasing energy value of the right had side of Eq. (39), occurring when the distance between the sampling point and the measurement surface increases, as well as indicator functions exploiting frequency diversity have been proposed (Coyle 2000; Gebauer et al. 2005; Catapano et al. 2008; Guzina et al. 2010; Sun 2012).

On the other hand, even if a modified indicator is used, a deterioration of the achievable reconstruction capabilities has been observed as long as the number of antennas, i.e. the number of source and measurement points, or the aperture of the (synthetic) array are reduced (Colton et al. 2000; Brandfass et al. 2001; Liseno and Pierri 2004; Fischer et al. 2007). An analytic tool to foresee the achievable LSM performance for a fixed array size and number of antennas has been proposed in (Catapano et al. 2011). Such a tool together with the recent development of array based GPR systems, which allows to collect multiview and multistatic data, support the adoption of LSM and FM as GPR data processing tools when the data are collected by means of short arrays moved above the investigated domain.

Further qualitative methods, belonging to the considered class and inspired by the LSM and FM or in some way related to them, are the Time Reversal-Multiple Signal Classification (TR-MUSIC) (Marengo et al. 2007), the probe method (Ikehata 1998), the method of singular sources (Potthast 2001), the no response test (Luke and Potthast 2003) and the multipole-expansion based LSM (Agarwal et al. 2010).

Among these approaches, in the following we limit ourselves to briefly recall the basic concepts of the MUSIC and the multipole-expansion based LSM. Such a choice is dictated by the fact that the former approach has been widely considered in the inverse scattering literature, while the second one provides the basis to understand, from a physical point of view, some mathematical details of the LSM and to improve its performances.

The MUSIC is a well-known procedure in the framework of signal processing and can be considered as a precursor of the FM as far as point-like scatterers are considered (Cheney 2001). In particular, as it has been assessed in (Devaney 2000), in the case of exact data and provided that the number of point like scatterers is lower than that of measurement points, the rank of data matrix coincides with the number of targets. Accordingly, one or more unknown point like objects can be retrieved by determining the essential range and the noise subspace of the multiview/multistatic data matrix.

More in detail, let $\mathbf{g}$ be the vector whose elements are the values of the field radiated by an elementary source located in a generic point $\mathrm{r_p}$ of the investigated domain D as measured at the M measurement points on Σ. The point $\mathrm{r_p}$ coincides with one of the scatterers if and only if $\mathbf{g}$ belongs to the essential range of the multiview, multistatic data matrix. Therefore, let $\boldsymbol{P}_{\text{noise}}$ the operator projecting the vector **g** onto the noise subspace of the data matrix, the MUSIC implementation requires to verify, for each one of the points $\mathbf{r_p}$ in D, if the quantity $\boldsymbol{P}_{\text{noise}}\,\mathbf{g}$ is negligible or not. Then, the image of the unknown objects is provided by plotting the quantity:

$$\mathbf{W} = \frac{1}{\|\mathbf{P}_{noise}\mathbf{g}\|} \tag{42}$$

The points wherein **W** exhibits large values identify the scatterers locations. The adoption of the MUSIC algorithm as an effective tool for locating small inclusions buried in a half-space has been considered in (Ammari et al. 2005) with reference to the applicative field of antipersonnel mines localization. In such a paper, a numerical analysis assessing the achievable reconstruction capabilities has been provided. A two-step MUSIC procedure has been instead presented in (Solimene et al. 2013a) in for a cylindrical layered background medium in the framework of rebar detection.

MUSIC method is extremely appealing as it shares the computational simplicity of other qualitative methods. What is more, it allows the achievement of a resolution which can be much below diffraction limits that generally characterize linear inversion schemes. However, this is true for isolated small scatterers. When the scene becomes crowed this does not hold anymore, even for a little amount of noise (Solimene and Dell'Aversano 2014). Finally, it has been adopted to retrieve geometrical features of extended objects in (Marengo et al. 2007).

The multipole-expansion based LSM is a variant of the LSM exploiting a multipole expansion of the scattered field and has been formulated for the canonical 2D scalar case and homogeneous medium. In particular, first of all the field $E_s(\theta,\varphi)$ is rewritten as:

$$E_s(\theta,\varphi) = \sum_{q=-\infty}^{\infty} a_q(\mathbf{r_p},\theta)\Phi_q(\varphi,\mathbf{r_p}) \tag{43}$$

where

$$\begin{aligned} a_q(\mathbf{r_p},\theta) &= -\frac{\omega\mu}{4}\int_S J(r,\theta)\mathrm{J_q}(k_b|\mathbf{r_p}-r|)e^{-jq\arg(\mathbf{r_p}-r)}dr \\ \Phi_q(\varphi,\mathbf{r_p}) &= -\frac{i}{4}\mathrm{H_q^2}(k_b|r_\varphi-\mathbf{r_p}|)e^{jq\arg(r_\varphi-\mathbf{r_p})} \end{aligned} \tag{44}$$

$J(r,\theta)$ being the current induced in the generic point r belonging to the support of the scatterers S, $\mathrm{J_q}$ and $\mathrm{H_q^2}$ being the Bessel and second kind Hankel function of order q. Then, Eq. (38) is rewritten by adopting an approximate expression of the scattered field wherein only monopole and dipole terms are taken into, while all the other terms of the summation in Eq. (43) are neglected. For a detailed description of the approach the reader is referred to (Agarwal et al. 2010).

8 Hybrid Inversion Schemes

Inversion approaches, where standard optimization algorithms are hybridized with other methods, have been proposed to overcome limits of standard stochastic and deterministic approaches.

In the frame of deterministic approaches, strategies facing the imaging according to a stepwise scheme have been proposed to reduce the overall complexity of the inverse problem (Fig. 5) and tested against experimental data (Crocco et al. 2005; Catapano et al. 2009). In (Crocco et al. 2005) the limited bandwidth properties of the scattered field (Bucci et al. 1998), which are determined by accounting for the amplitude spectrum of the data, are taken into account for achieving a preliminary estimation of the region wherein the scatterers are located. Differently, in (Catapano et al. 2009) the imaging process is split into two steps. In the first step, the LSM is used to estimate the support of the unknown targets. Then, such an information is exploited in the second step, where a CSEB inversion scheme is adopted to retrieve the (complex) permittivity of the targets. At this stage, morphologic information is also exploited, together with the bandwidth properties of the scattered field, to set a simple procedure aimed at removing (to some extent) noise possibly present on the collected measures. Recently, a different way to combine different imaging procedures has been proposed (Crocco et al. 2012). In this approach, the LSM is exploited to devise an effective approximation of the scattering phenomenon, that leads to a noniterative linear inversion method for the estimation of the target's electric contrast.

As far as stochastic approaches, whose main limit is the high amount of needed computational resources, two main strategies have been considered. In the first one, the hybrid schemes aim at reducing the number of unknowns. To this end, similarly to the approaches discussed above, it is possible to firstly localize the targets and then reconstructing only the distributions of dielectric parameters

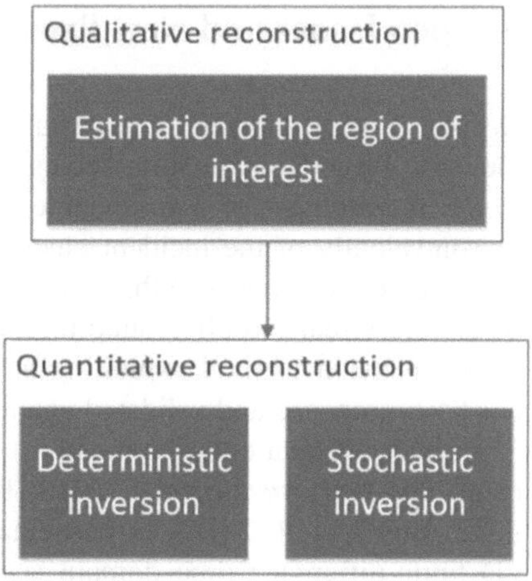

Fig. 5 Example of hybridization scheme

inside the limited area previously identified. As an example of such strategy, the linear sampling method can be successfully used to find the support of the scatterers and successively only the dielectric properties of the identified regions are reconstructed (Brignone et al. 2008). Another approach is to use iterative multiscale strategies (Donelli et al. 2006). In such techniques, the investigation area is iteratively reconstructed at different scales. At every scale, the stochastic inversion methods are used to obtain quantitative reconstructions of the distribution of the dielectric properties. A clustering techniques is then used to identify the scatterers and then the investigation area is refined in order to focus only on the objects. In this way, the "limited amount of the information content of inverse scattering data" is efficiently used, guaranteeing a sufficient resolution level in the retrieved image of the investigation domain.

In the second strategy, stochastic algorithms are hybridized with deterministic (quantitative and qualitative) approaches. In particular, different schemes can be followed. The first one is to use the stochastic algorithm to find a starting solution for the subsequent deterministic inversion (Caorsi et al. 2000). Clearly, since it is not required to obtain a high-resolution reconstruction in the first step, it is possible to use a limited discretization, allowing for a faster execution of the stochastic algorithms. In the second scheme, quantitative deterministic procedures are inserted inside the population update mechanism, in order to refine the trial solutions in the population. The so-called memetic algorithm (MA) belongs to this class of hybrid approaches. In this case, the genetic operators are used and all the solutions are locally optimized by using a local deterministic optimization procedure. All the elements of the population are then local minima and the best solution evolves by "jumping" from a local minima to another until the globally optimal solution is found (Caorsi et al. 2003b, c). Consequently, the dimensionality of the search space is greatly reduced and a significant improvement is obtained.

9 Open Issues and New Prospective

Despite a large number of quantitative imaging approaches have been proposed, their use in GPR application seems to be still an open issue. This is mainly due to the requirement of a reasonable knowledge of the antenna radiation pattern, or equivalently of the incident electric field into the investigated domain. A first advancement to overcome this gap has been given in (Lambot et al. 2004). Herein, a laboratory realized off-ground monostatic GPR system has been used and a full-wave inversion method exploiting a sophisticated model of the probing wavelet has been proposed and validated against experimental data.

Further relevant contributions have been presented in (Ernst et al. 2007; Meles et al. 2010; Klotzsche et al. 2010; Patriarca et al. 2011; van der Wielen et al. 2012; Busch et al. 2012). In particular, a full-wave inversion scheme based on a 2D finite-difference time domain scattering model has been proposed in (Ernst et al. 2007) and it has been extended to the 3D vectorial case (Meles et al. 2010) and applied to process experimental cross-hall data (Klotzsche et al. 2010). With respect to the problem of detecting and characterizing a thin delamination layer in a concrete slab, a strategy based on an antenna calibration procedure has been considered in (Patriarca et al. 2011), while an inversion procedure accounting for a model of the probing antenna effects has been considered in (van der Wielen et al. 2012). Finally, a full wave inversion procedure, whose peculiarity is that amplitude and phase of the probing wavelet is optimized simultaneously with dielectric permittivity and electric conductivity of a single layered subsurface has been proposed in (Busch et al. 2012). Such a procedure processes frequency domain data gathered under a common-midpoint measurement configuration.

Finally, it is worth noting that learning-based approaches have also been recently applied to GPR prospection. The first developed algorithms were based on neural networks (Gamba and Lossani 2000; Mydur and Michalski 2001). In this case, the network can be directly trained from measurements, thus avoiding the need of a reliable model of the scattering phenomena. Moreover, real-time detection can be achieved once the network is trained. Support Vector Machines (SVM) have also been used in buried target detection (Bermani et al. 2003). Finally, subarrays processing and triangularization technique have also been found to quite effective in the detection of tunnels and pipes (Sahin and Miller 2001; Meschino et al. 2012, 2013).

Acknowledgments This work is a contribution to COST Action TU1208 "Civil Engineering Applications of Ground Penetrating Radar".

References

Abubaker, A., Van Den Berg, P.M.: Total variation as a multiplicative constraint for solving inverse problems. IEEE Trans. Image Process. **10**, 1384–1392 (2001). doi:10.1109/83.941862

Abubakar, A., Habashy, T.M., Pan, G., Li, M.-K.: Application of the multiplicative regularized gauss-newton algorithm for three-dimensional microwave imaging. IEEE Trans. Antennas Propag. **60**, 2431–2441 (2012). doi:10.1109/TAP.2012.2189712

Agarwal, K., Chen, X., Zhong, Y.: A multipole-expansion based linear sampling method for solving inverse scattering problems. Opt. Express **18**, 6366 (2010). doi:10.1364/OE.18.006366

Ahmad, F., Amin, M.G., Kassam, S.A.: Synthetic aperture beamformer for imaging through a dielectric wall. IEEE Trans. Aerosp. Electron. Syst. **41**, 271–283 (2005). doi:10.1109/TAES.2005.1413761

Ali, M.A., Moghaddam, M.: 3D nonlinear super-resolution microwave inversion technique using time-domain data. IEEE Trans. Antennas Propag. **58**, 2327–2336 (2010). doi:10.1109/TAP.2010.2048848

Ammari, H., Iakovleva, E., Lesselier, D.: A MUSIC algorithm for locating small inclusions buried in a half-space from the scattering amplitude at a fixed frequency. Multiscale Model. Simul. **3**, 597–628 (2005). doi:10.1137/040610854

Anyong, Q.: Electromagnetic inverse scattering of multiple two-dimensional perfectly conducting objects by the differential evolution strategy. IEEE Trans. Antennas Propag. **51**, 1251–1262 (2003). doi:10.1109/TAP.2003.811492

Aramini, R., Brignone, M., Piana, M.: The linear sampling method without sampling. Inverse Probl. **22**, 2237–2254 (2006). doi:10.1088/0266-5611/22/6/020

Aramini, R., Caviglia, G., Massa, A., Piana, M.: The linear sampling method and energy conservation. Inverse Probl. **26**, 055004 (2010). doi:10.1088/0266-5611/26/5/055004

Aramini, R., Brignone, M., Caviglia, G., et al.: The linear sampling method in a lossy background: an energy perspective. Inverse Probl Sci. Eng. **19**, 963–984 (2011). doi:10.1080/17415977.2011.565875

Arens, T.: Why linear sampling works. Inverse Probl. **20**, 163–173 (2004). doi:10.1088/0266-5611/20/1/010

Arens, T., Lechleiter, A.: The linear sampling method revisited. J. Integral Equ. Appl. **21**, 179–202 (2009). doi:10.1216/JIE-2009-21-2-179

Azaro, R., Bozza, G., Estatico, C., et al.: New results on electromagnetic imaging based on the inversion of synthetic and measured scattered-field data. IEEE Trans. Instrum. Meas. **55**, 1085–1093 (2006). doi:10.1109/TIM.2006.876576

Baraniuk, R.: Compressive sensing. IEEE Signal Process. Mag. **24**, 118–121 (2007). doi:10.1109/MSP.2007.4286571

Baranoski, E.J.: Through-wall imaging: historical perspective and future directions. J. Frankl Inst. **345**, 556–569 (2008). doi:10.1016/j.jfranklin.2008.01.005

Baussard, A., Miller, E.L., Lesselier, D.: Adaptive multiscale reconstruction of buried objects. Inverse Probl. **20**, S1–S15 (2004a). doi:10.1088/0266-5611/20/6/S01

Baussard, A., Miller, E.L., Prémel, D.: Adaptive B-spline scheme for solving an inverse scattering problem. Inverse Probl. **20**, 347–365 (2004b). doi:10.1088/0266-5611/20/2/003

Belkebir, K., Bonnard, S., Pezin, F., et al.: Validation of 2D inverse scattering algorithms from multi-frequency experimental data. J. Electromagn. Waves Appl. **14**, 1637–1667 (2000). doi:10.1163/156939300X00437

Benedetti, M., Donelli, M., Martini, A., et al.: An innovative microwave-imaging technique for nondestructive evaluation: applications to civil structures monitoring and biological bodies inspection. IEEE Trans. Instrum. Meas. **55**, 1878–1884 (2006). doi:10.1109/TIM.2006.884287

Berkhout, A.J.: Seismic inversion in terms of pre-stack migration and multiple elimination. Proc. IEEE **74**, 415–427 (1986). doi:10.1109/PROC.1986.13483

Bermani, E., Boni, A., Caorsi, S., Massa, A.: An innovative real-time technique for buried object detection. IEEE Trans. Geosci. Remote Sens. **41**, 927–931 (2003). doi:10.1109/TGRS.2003.810928

Bertero, M.: Linear inverse and ill-posed problems. In: Kazan, B. (ed.) Advances in Electronics and Electron Physics, vol. 75, pp. 1–120 (1989)

Bertero, M., Boccacci, P.: Introduction to Inverse Problems in Imaging. IOP Publishing, Bristol (1998)

Bevan, M.J., Endres, A.L., Rudolph, D.L., Parkin, G.: The non-invasive characterization of pumping-induced dewatering using ground penetrating radar. J. Hydrol. **281**, 55–69 (2003). doi:10.1016/S0022-1694(03)00200-2

Bleistein, N.: Mathematics of Multidimensional Seismic Imaging, Migration, and Inversion. Springer, New York (2001)

Bloemenkamp, R.F., Slob, E.C.: Imaging of high-frequency fullvectorial GPR data using measured footprints: 23rd IEEE Geoscience and Remote Sensing Symposium IGARS, 1362–1364 (2003)

Blum, C., Merkle, D.: Swarm Intelligence Introduction and Applications. Springer, Berlin (2008)

Bozza, G., Pastorino, M.: An inexact newton-based approach to microwave imaging within the contrast source formulation. IEEE Trans. Antennas Propag. **57**, 1122–1132 (2009). doi:10.1109/TAP.2009.2015820

Bozza, G., Estatico, C., Pastorino, M., Randazzo, A.: An inexact newton method for microwave reconstruction of strong scatterers. IEEE Antennas Wirel. Propag. Lett. **5**, 61–64 (2006). doi:10.1109/LAWP.2006.870360

Bozza, G., Estatico, C., Massa, A., et al.: Short-range image-based method for the inspection of strong scatterers using microwaves. IEEE Trans. Instrum. Meas. **56**, 1181–1188 (2007). doi:10.1109/TIM.2007.900127

Brandfass, M., Lanterman, A.D., Warnick, K.F.: A comparison of the Colton-Kirsch inverse scattering methods with linearized tomographic inverse scattering. Inverse Probl. **17**, 1797–1816 (2001). doi:10.1088/0266-5611/17/6/316

Breard, A., Perrusson, G., Lesselier, D.: Hybrid differential evolution and retrieval of buried spheres in subsoil. IEEE Geosci. Remote Sens. Lett. **5**, 788–792 (2008). doi:10.1109/LGRS.2008.2005790

Brignone, M., Bozza, G., Randazzo, A., et al.: A hybrid approach to 3D microwave imaging by using linear sampling and ACO. IEEE Trans. Antennas Propag. **56**, 3224–3232 (2008). doi:10.1109/TAP.2008.929504

Brignone, M., Bozza, G., Aramini, R., et al.: A fully no-sampling formulation of the linear sampling method for three-dimensional inverse electromagnetic scattering problems. Inverse Probl. (2009). doi:10.1088/0266-5611/25/1/015014

Bristow, C.S., Jol, H.M.: An introduction to ground penetrating radar (GPR) in sediments. Geol. Soc. Lond. Spec. Publ. **211**, 1–7 (2003). doi:10.1144/GSL.SP.2001.211.01.01

Bucci, O.M., Franceschetti, G.: On the degrees of freedom of scattered fields. IEEE Trans. Antennas Propag. **37**, 918–926 (1989). doi:10.1109/8.29386

Bucci, O.M., Isernia, T.: Electromagnetic inverse scattering: retrievable information and measurement strategies. Radio Sci. **32**, 2123–2137 (1997). doi:10.1029/97RS01826

Bucci, O.M., Gennarelli, C., Savarese, C.: Representation of electromagnetic fields over arbitrary surfaces by a finite and nonredundant number of samples. IEEE Trans. Antennas Propag. **46**, 351–359 (1998). doi:10.1109/8.662654

Bucci, O.M., Cardace, N., Crocco, L., Isernia, T.: Degree of nonlinearity and a new solution procedure in scalar two-dimensional inverse scattering problems. J. Opt. Soc. Am. A: **18**, 1832 (2001a). doi:10.1364/JOSAA.18.001832

Bucci, O.M., Crocco, L., Isernia, T., Pascazio, V.: Subsurface inverse scattering problems: quantifying, qualifying, and achieving the available information. IEEE Trans. Geosci. Remote Sens. **39**, 2527–2538 (2001b). doi:10.1109/36.964991

Busch, S., van der Kruk, J., Bikowski, J., Vereecken, H.: Quantitative conductivity and permittivity estimation using full-waveform inversion of on-ground GPR data. Geophysics **77**, H79–H91 (2012). doi:10.1190/geo2012-0045.1

Cafforio, C., Prati, C., Rocca, F.: SAR data focusing using seismic migration techniques. IEEE Trans. Aerosp. Electron. Syst. **27**, 194–207 (1991). doi:10.1109/7.78293

Cakoni, F.: Qualitative Methods in Inverse Scattering Theory: An Introduction. Springer, Berlin (2006)

Caorsi, S., Pastorino, M.: Two-dimensional microwave imaging approach based on a genetic algorithm. IEEE Trans. Antennas Propag. **48**, 370–373 (2000). doi:10.1109/8.841897

Caorsi, S., Gragnani, G.L., Medicina, S., et al.: Microwave imaging based on a Markov random field model. IEEE Trans. Antennas Propag. **42**, 293–303 (1994). doi:10.1109/8.280714

Caorsi, S., Ciaramella, S., Gragnani, G.L., Pastorino, M.: On the use of regularization techniques in numerical inverse-scattering solutions for microwave imaging applications. IEEE Trans. Microw. Theory Tech. **43**, 632–640 (1995). doi:10.1109/22.372110

Caorsi, S., Pastorino, M., Raffetto, M.: Electromagnetic scattering by a multilayer elliptic cylinder under transverse-magnetic illumination: series solution in terms of Mathieu functions. IEEE Trans. Antennas Propag. **45**, 926–935 (1997). doi:10.1109/8.585739

Caorsi, S., Massa, A., Pastorino, M.: A computational technique based on a real-coded genetic algorithm for microwave imaging purposes. IEEE Trans. Geosci. Remote Sens. **38**, 1697–1708 (2000). doi:10.1109/36.851968

Caorsi, S., Costa, A., Pastorino, M.: Microwave imaging within the second-order Born approximation: stochastic optimization by a genetic algorithm. IEEE Trans. Antennas Propag. **49**, 22–31 (2001a). doi:10.1109/8.910525

Caorsi, S., Massa, A., Pastorino, M.: A crack identification microwave procedure based on a genetic algorithm for nondestructive testing. IEEE Trans. Antennas Propag. **49**, 1812–1820 (2001b). doi:10.1109/8.982464

Caorsi, S., Donelli, M., Franceschini, D., Massa, A.: A new methodology based on an iterative multiscaling for microwave imaging. IEEE Trans. Microw. Theory Tech. **51**, 1162–1173 (2003a). doi:10.1109/TMTT.2003.809677

Caorsi, S., Massa, A., Pastorino, M., et al.: Detection of buried inhomogeneous elliptic cylinders by a memetic algorithm. IEEE Trans. Antennas Propag. **51**, 2878–2884 (2003b). doi:10.1109/TAP.2003.817984

Caorsi, S., Massa, A., Pastorino, M., Randazzo, A.: Electromagnetic detection of dielectric scatterers using phaseless synthetic and real data and the memetic algorithm. IEEE Trans. Geosci. Remote Sens. **41**, 2745–2753 (2003c). doi:10.1109/TGRS.2003.815676

Caorsi, S., Donelli, M., Lommi, A., Massa, A.: Location and imaging of two-dimensional scatterers by using a particle swarm algorithm. J. Electromagn. Waves Appl. **18**, 481–494 (2004). doi:10.1163/156939304774113089

Catapano, I., Crocco, L., Isernia,T.: A simple two-dimensional inversion technique for imaging homogeneous targets in stratified media: inverse scattering in stratified media. Radio Sci. **39**, RS1012 (2004). doi: 10.1029/2003RS002917

Catapano, I., Crocco, L., D'Urso, M., Isernia, T.: A novel effective model for solving 3-d nonlinear inverse scattering problems in lossy scenarios. IEEE Geosci. Remote Sens. Lett. **3**, 302–306 (2006a). doi:10.1109/LGRS.2006.869976

Catapano, I., Crocco, L., Persico, R., et al.: Linear and nonlinear microwave tomography approaches for subsurface prospecting: validation on real data. Antennas Wirel Propag. Lett. **5**, 49–53 (2006b). doi:10.1109/LAWP.2006.870363

Catapano, I., Crocco, L., Isernia, T.: On simple methods for shape reconstruction of unknown scatterers. IEEE Trans. Antennas Propag. **55**, 1431–1436 (2007). doi:10.1109/TAP.2007.895563

Catapano, I., Crocco, L., Isernia, T.: improved sampling methods for shape reconstruction of 3-D buried targets. IEEE Trans. Geosci. Remote Sens. **46**, 3265–3273 (2008). doi:10.1109/TGRS.2008.921745

Catapano, I., Crocco, L., Urso, M.D., Isernia, T.: 3D microwave imaging via preliminary support reconstruction: testing on the Fresnel 2008 database. Inverse Probl. **25**, 024002 (2009). doi:10.1088/0266-5611/25/2/024002

Catapano, I., Crocco, L.: A qualitative inverse scattering method for through-the-wall imaging. IEEE Geosci. Remote Sens.Lett. **7**, 685–689 (2010). doi:10.1109/LGRS.2010.2045473

Catapano, I., Soldovieri, F., Crocco, L.: On the feasibility of the linear sampling method for 3d Gpr surveys. Prog. Electromagn. Res. **118**, 185–203 (2011). doi:10.2528/PIER11042704

Catapano, I., Crocco, L., Napoli, R.D., et al.: Microwave tomography enhanced GPR surveys in Centaur's Domus, Regio VI of Pompeii, Italy. J. Geophys. Eng. **9**, S92–S99 (2012). doi:10.1088/1742-2132/9/4/S92

Chen, W.-T., Chiu, C.-C.: Electromagnetic imaging for an imperfectly conducting cylinder by the genetic algorithm. IEEE Trans. Microw. Theory Tech. **48**, 1901–1905 (2000). doi:10.1109/22.883869

Cheney, M.: The linear sampling method and the MUSIC algorithm. Inverse Probl. **17**, 591–595 (2001). doi:10.1088/0266-5611/17/4/301

Chew, W.C.: Waves and Fields in Inhomogeneous Media. IEEE Press, London (1999)

Chew, W.C., Wang, Y.M.: Reconstruction of two-dimensional permittivity distribution using the distorted Born iterative method. IEEE Trans. Med. Imaging **9**, 218–225 (1990). doi:10.1109/42.56334

Chew, W.C., Lin, J.H.: A frequency-hopping approach for microwave imaging of large inhomogeneous bodies. IEEE Microw. Guid. Wave Lett. **5**, 439–441 (1995). doi:10.1109/75.481854

Chiu, C.-C., Liu, P.-T.: Image reconstruction of a perfectly conducting cylinder by the genetic algorithm. IEE Proc.—Microw. Antennas Propag. **143**, 249 (1996). doi:10.1049/ip-map:19960363

Clerc, M.: Particle Swarm Optimization. ISTE, New Delhi (2006)

Collino, F., Fares, M.B., Haddar, H.: Numerical and analytical studies of the linear sampling method in electromagnetic inverse scattering problems. Inverse Probl. **19**, 1279–1298 (2003). doi:10.1088/0266-5611/19/6/004

Colton, D., Kirsch, A.: A simple method for solving inverse scattering problems in the resonance region. Inverse Probl. **12**, 383–393 (1996). doi:10.1088/0266-5611/12/4/003

Colton, D., Piana, M., Potthast, R.: A simple method using Morozov's discrepancy principle for solving inverse scattering problems. Inverse Probl. **13**, 1477–1493 (1997). doi:10.1088/0266-5611/13/6/005

Colton, D., Giebermann, K., Monk, P.: A regularized sampling method for solving three-dimensional inverse scattering problems. SIAM J. Sci. Comput. **21**, 2316–2330 (2000). doi:10.1137/S1064827598340159

Colton, D., Haddar, H., Piana, M.: The linear sampling method in inverse electromagnetic scattering theory. Inverse Probl. **19**, S105–S137 (2003). doi:10.1088/0266-5611/19/6/057

Conyers, L.B.: Ground-Penetrating Radar: An Introduction for Archaeologists. AltaMira Press, Walnut Creek (1997)

Counts, T., Gurbuz, A.C., Scott, W.R., et al.: Multistatic ground-penetrating radar experiments. IEEE Trans. Geosci. Remote Sens. **45**, 2544–2553 (2007). doi:10.1109/TGRS.2007.900677

Coyle, J.: Locating the support of objects contained in a two-layered background medium in two dimensions. Inverse Probl. **16**, 275–292 (2000). doi:10.1088/0266-5611/16/2/301

Crocco, L., D'Urso, M., Isernia, T.: Testing the contrast source extended Born inversion method against real data: the TM case. Inverse Probl. **21**, S33–S50 (2005). doi:10.1088/0266-5611/21/6/S04

Crocco L., Soldovieri, F.: Nonlinear Inversion Algorithms. Subsurf. Sens. **1**, 365–376 (2011)

Crocco, L., Catapano, I., Di Donato, L., Isernia, T.: The linear sampling method as a way to quantitative inverse scattering. IEEE Trans. Antennas Propag. **60**, 1844–1853 (2012). doi:10.1109/TAP.2012.2186250

Cui, T.J., Chew, W.C., Aydiner, A.A., Chen, S.: Inverse scattering of two-dimensional dielectric objects buried in a lossy earth using the distorted Born iterative method. IEEE Trans. Geosci. Remote Sens. **39**, 339–346 (2001). doi:10.1109/36.905242

D'Urso, M., Catapano, I., Crocco, L., Isernia, T.: Effective solution of 3-D scattering problems via series expansions: applicability and a new hybrid scheme. IEEE Trans. Geosci. Remote Sens. **45**, 639–648 (2007). doi:10.1109/TGRS.2006.888144

D'Urso, M., Isernia, T., Morabito, A.F.: On the solution of 2-D inverse scattering problems via source-type integral equations. IEEE Trans. Geosci. Remote Sens. **48**, 1186–1198 (2010). doi:10.1109/TGRS.2009.2032175

Daniels, D.J.: Ground Penetrating Radar, 2nd edn. Institution of Electrical Engineers, London (2004)

Devaney, A.J.: Super-resolution processing of multi-static data using time reversal And MUSIC. Northeastern University (2000)

De Micheli, E., Magnoli, N., Viano, G.A.: On the regularization of fredholm integral equations of the first kind. SIAM J. Math. Anal. **29**, 855–877 (1998). doi:10.1137/S0036141096301749

De Micheli, E., Viano, G.A.: Metric and probabilistic information associated with fredholm integral equations of the first kind. J. Integral Equ. Appl. **14**, 283–310 (2002). doi:10.1216/jiea/1181074917

De Zaeytijd, J., Franchois, A., Eyraud, C., Geffrin, J.-M.: Full-wave three-dimensional microwave imaging with a regularized gauss-newton method—theory and experiment. IEEE Trans. Antennas Propag. **55**, 3279–3292 (2007). doi:10.1109/TAP.2007.908824

Di Matteo, A., Pettinelli, E., Slob, E.: Early-time GPR signal attributes to estimate soil dielectric permittivity: a theoretical study. IEEE Trans. Geosci. Remote Sens. **51**, 1643–1654 (2013). doi:10.1109/TGRS.2012.2206817

Di Vico, M., Frezza, F., Pajewski, L., Schettini, G.: Scattering by a finite set of perfectly conducting cylinders buried in a dielectric half-space: a spectral-domain solution. IEEE Trans. Antennas Propag. **53**, 719–727 (2005). doi:10.1109/TAP.2004.841315

Donelli, M., Franceschini, G., Martini, A., Massa, A.: An integrated multiscaling strategy based on a particle swarm algorithm for inverse scattering problems. IEEE Trans. Geosci. Remote Sens. **44**, 298–312 (2006). doi:10.1109/TGRS.2005.861412

Donelli, M., Craddock, I., Gibbins, D., Sarafianou, M.: A three-dimensional time domain microwave imaging method for breast cancer detection based on an evolutionary algorithm. Prog. Electromagn. Res. **M 18**, 179–195 (2011)

Donoho, D.L.: Compressed sensing. IEEE Trans. Inf. Theory **52**, 1289–1306 (2006). doi:10.1109/TIT.2006.871582

Dourthe, C., Pichot, C., Dauvignac, J.Y., Cariou, J.: Inversion algorithm and measurement system for microwave tomography of buried object. Radio Sci. **35**, 1097–1108 (2000). doi:10.1029/1999RS002244

Ernst, J.R., Maurer, H., Green, A.G., Holliger, K.: Full-waveform inversion of crosshole radar data based on 2-D finite-difference time-domain solutions of Maxwell's equations. IEEE Trans. Geosci. Remote Sens. **45**, 2807–2828 (2007). doi:10.1109/TGRS.2007.901048

Estatico, C., Pastorino, M., Randazzo, A.: An inexact-Newton method for short-range microwave imaging within the second-order Born approximation. IEEE Trans. Geosci. Remote Sens. **43**, 2593–2605 (2005). doi:10.1109/TGRS.2005.856631

Estatico, C., Pastorino, M., Randazzo, A.: A novel microwave imaging approach based on regularization in Lp Banach spaces. IEEE Trans. Antennas Propag. **60**, 3373–3381 (2012). doi:10.1109/TAP.2012.2196925

Estatico, C., Fedeli, A., Pastorino, M., Randazzo, A.: Microwave imaging of elliptically shaped dielectric cylinders by means of an Lp Banach-space inversion algorithm. Meas. Sci. Technol. **24**, 074017 (2013). doi:10.1088/0957-0233/24/7/074017

Fink, M.: Time reversal of ultrasonic fields. I. Basic principles. IEEE Trans. Ultrason. Ferroelectr. Freq. Control **39**, 555–566 (1992). doi:10.1109/58.156174

Fischer, C., Herschlein, A., Younis, M., Wiesbeck, W.: Detection of antipersonnel mines by using the factorization method on multistatic ground-penetrating radar measurements. IEEE Trans. Geosci. Remote Sens. **45**, 85–92 (2007). doi:10.1109/TGRS.2006.883464

Franchois, A., Pichot, C.: Microwave imaging-complex permittivity reconstruction with a Levenberg-Marquardt method. IEEE Trans. Antennas Propag. **45**, 203–215 (1997). doi:10.1109/8.560338

Frezza, F., Martinelli, P., Pajewski, L., Schettini, G.: Short-pulse electromagnetic scattering by buried perfectly conducting cylinders. IEEE Geosci. Remote Sens. Lett. **4**, 611–615 (2007). doi:10.1109/LGRS.2007.903078

Gamba, P., Lossani, S.: Neural detection of pipe signatures in ground penetrating radar images. IEEE Trans. Geosci. Remote Sens. **38**, 790–797 (2000). doi:10.1109/36.842008

Gazdag, J.: Wave equation migration with the phase shift method: Geophysics, **43**, 1342–1351 (1978)

Gazdag, J., Sguazzero, P.: Migration of seismic data. Proc. IEEE **72**, 1302–1315 (1984). doi:10.1109/PROC.1984.13019

Gebauer, B., Hanke, M., Kirsch, A., et al.: A sampling method for detecting buried objects using electromagnetic scattering. Inverse Probl. **21**, 2035–2050 (2005). doi:10.1088/0266-5611/21/6/015

Gilmore, C., Jeffrey, I., LoVetri, J.: Derivation and comparison of SAR and frequency-wavenumber migration within a common inverse scalar wave problem formulation. IEEE Trans. Geosci. Remote Sens. **44**, 1454–1461 (2006). doi:10.1109/TGRS.2006.870402

Gorriti, A.G., Slob, E.C.: A new tool for accurate S-parameters measurements and permittivity reconstruction. IEEE Trans. Geosci. Remote Sens. **43**, 1727–1735 (2005). doi:10.1109/TGRS.2005.851163

Groetsch, C.W.: Inverse Problems in the Mathematical Sciences. Vieweg, Braunschweig (1993)

Guzina, B.B., Cakoni, F., Bellis, C.: On the multi-frequency obstacle reconstruction via the linear sampling method. Inverse Probl. **26**, 125005 (2010). doi:10.1088/0266-5611/26/12/125005

Hagedoorn, J.G.: A process of seismic reflection interpretation. Geophys. Prospect. **2**, 85–127 (1954). doi:10.1111/j.1365-2478.1954.tb01281.x

Hansen, P.C., Nagy, J.G., O'Leary, D.P.: Deblurring Images: Mmatrices, Spectra, and Filtering. SIAM, Philadelphia (2006)

Hansen, T.B., Johansen, P.M.: Inversion scheme for ground penetrating radar that takes into account the planar air-soil interface. IEEE Trans. Geosci. Remote Sens. **38**, 496–506 (2000). doi:10.1109/36.823944

Harada, H., Wall, D.J.N., Takenaka, T., Tanaka, M.: Conjugate Gradient method applied to inverse scattering problem. IEEE Trans. Antennas Propag. **43**, 784–792 (1995). doi:10.1109/8.402197

Haupt, R.L.: An introduction to genetic algorithms for electromagnetics. IEEE Antennas Propag. Mag. **37**, 7–15 (1995). doi:10.1109/74.382334

Huang, C.-H., Chiu, C.-C., Li, C.-L., Chen, K.-C.: Time domain inverse scattering of a two-dimensional homogenous dielectric object with arbitrary shape by particle swarm optimization. Prog Electromagn Res. **82**, 381–400 (2008). doi:10.2528/PIER08031904

Huang, T., Mohan, A.S.: A Microparticle swarm optimizer for the reconstruction of microwave images. IEEE Trans. Antennas Propag. **55**, 568–576 (2007). doi:10.1109/TAP.2007.891545

Hugenschmidt, J., Kalogeropoulos, A.: The inspection of retaining walls using GPR. J. Appl. Geophys. **67**, 335–344 (2009). doi:10.1016/j.jappgeo.2008.09.001

Huisman, J.A., Hubbard, S.S., Redman, J.D., Annan, A.P.: Measuring soil water content with ground penetrating radar: a review. Vadose Zone J. **2**, 476–491 (2003). doi:10.2113/2.4.476

Ikehata, M.: Reconstruction of an obstacle from the scattering amplitude at a fixed frequency. Inverse Probl. **14**, 949–954 (1998). doi:10.1088/0266-5611/14/4/012

Isernia, T., Pascazio, V., Pierri, R.: A nonlinear estimation method in tomographic imaging. IEEE Trans. Geosci. Remote Sens. **35**, 910–923 (1997). doi:10.1109/36.602533

Isernia, T., Pascazio, V., Pierri, R.: On the local minima in a tomographic imaging technique. IEEE Trans. Geosci. Remote Sens. **39**, 1596–1607 (2001). doi:10.1109/36.934091

Isernia, T., Crocco, L., D'Urso, M.: New tools and series for forward and inverse scattering problems in Lossy media. IEEE Geosci. Remote Sens. Lett. **1**, 327–331 (2004). doi:10.1109/LGRS.2004.837008

Jadoon, K.Z., Lambot, S., Slob, E.C., Vereecken, H.: Analysis of Horn antenna transfer functions and phase-center position for modeling off-ground GPR. IEEE Trans. Geosci. Remote Sens. **49**, 1649–1662 (2011). doi:10.1109/TGRS.2010.2089691

Jang, H., Kuroda, S., Kim, H.J.: SVD inversion of zero-offset profiling data obtained in the Vadose Zone using cross-borehole radar. IEEE Trans. Geosci. Remote Sens. **49**, 3849–3855 (2011). doi:10.1109/TGRS.2011.2134855

Kirsch, A.: Characterization of the shape of a scattering obstacle using the spectral data of the far field operator. Inverse Probl. **14**, 1489–1512 (1998). doi:10.1088/0266-5611/14/6/009

Kirsch, A.: Factorization of the far-field operator for the inhomogeneous medium case and an application in inverse scattering theory. Inverse Probl. **15**, 413–429 (1999). doi:10.1088/0266-5611/15/2/005

Kirsch, A.: The factorization method for Maxwell's equations. Inverse Probl. **20**, S117–S134 (2004). doi:10.1088/0266-5611/20/6/S08

Kleinman, R.E., van den Berg, P.M.: A modified gradient method for two- dimensional problems in tomography. J. Comput. Appl. Math. **42**, 17–35 (1992). doi:10.1016/0377-0427(92)90160-Y

Klotzsche, A., Van der Kruk, J., Meles, G.A., et al.: Full-waveform inversion of cross-hole ground-penetrating radar data to characterize a gravel aquifer close to the Thur River, Switzerland. Surf. Geophys. **8**, 631–646 (2010). doi:10.3997/1873-0604.2010054

Krasnov, M.L., Kiselev, A.I., Makarenko, G.I.: Integral Equations. MIR, Moscow (1976)

Lambot, S., Slob, E.C., van den Bosch, I., et al.: Modeling of ground-penetrating Radar for accurate characterization of subsurface electric properties. IEEE Trans. Geosci. Remote Sens. **42**, 2555–2568 (2004). doi:10.1109/TGRS.2004.834800

Lambot, S., André, F.: Full-wave modeling of near-field radar data for planar layered media reconstruction. IEEE Trans. Geosci. Remote Sens. **52**, 2295–2303 (2014). doi:10.1109/TGRS.2013.2259243

Langenberg, K.J.: Applied inverse problems for acoustic, electromagnetic and elastic wave scattering. In: Sabatier P.C. (ed.) Basic Methods of Tomography and Inverse Problems, pp. 127–467. Hilger, Bristol (1987)

Leone, G., Persico, R., Solimene, R.: A quadratic model for electromagnetic subsurface prospecting. AEU—Int. J. Electron Commun. **57**, 33–46 (2003). doi:10.1078/1434-8411-54100138

Li, F., Chen, X., Huang, K.: Microwave imaging a buried object by the GA and using the S11 parameter. Prog. Electromagn. Res. **85**, 289–302 (2008). doi:10.2528/PIER08081401

Liseno, A., Pierri, R.: Shape reconstruction by the spectral data of the far-field operator: analysis and performances. IEEE Trans. Antennas Propag. **52**, 899–903 (2004). doi:10.1109/TAP.2004.824674

Liu, Q.H., Zhang, Z.Q., Wang, T.T., et al.: Active microwave imaging. I. 2-D forward and inverse scattering methods. IEEE Trans. Microw. Theory Tech. **50**, 123–133 (2002). doi:10.1109/22.981256

Lobel, P., Kleinman, R.E., Pichot, C., et al.: Conjugate-gradient method for solving inverse scattering with experimental data. IEEE Antennas Propag. Mag. **38**, 48 (1996). doi:10.1109/MAP.1996.511954

Lobel, P., Blanc-Féraud, L., Pichot, C., Barlaud, M.: A new regularization scheme for inverse scattering. Inverse Probl. **13**, 403 (1997). doi:10.1088/0266-5611/13/2/013

Lopera, O., Milisavljević, N., Lambot, S.: Clutter reduction in GPR measurements for detecting shallow buried landmines: a Colombian case study. Surf. Geophys. **5**, 57–64 (2007). doi:10.3997/1873-0604.2006018

Lopez-Sahcnez, J.M., Fortuny-Guasch, J.: 3-D radar imaging using range migration techniques. IEEE Trans. Antennas Propag. **48**, 728–737 (2000). doi:10.1109/8.855491

Luke, D.R., Potthast, R.: The no response test—a sampling method for inverse scattering problems. SIAM J Appl Math **63**, 1292–1312 (2003). doi:10.1137/S0036139902406887

Marengo, E.A., Gruber, F.K., Simonetti, F.: Time-reversal MUSIC imaging of extended targets. IEEE Trans. Image Process. **16**, 1967–1984 (2007). doi:10.1109/TIP.2007.899193

Marklein, R., Mayer, K., Hannemann, R., et al.: Linear and nonlinear inversion algorithms applied in nondestructive evaluation. Inverse Probl. **18**, 1733–1759 (2002). doi:10.1088/0266-5611/18/6/319

Masini, N., Persico, R., Rizzo, E.: Some examples of GPR prospecting for monitoring of the monumental heritage. J. Geophys. Eng. **7**, 190–199 (2010). doi:10.1088/1742-2132/7/2/S05

Masini, N., Soldovieri, F.: Integrated non-invasive sensing techniques and geophysical methods for the study and conservation of architectural, archaeological and artistic heritage. J. Geophys. Eng. (2011). doi:10.1088/1742-2140/8/3/E01

Massa, A., Pastorino, M., Randazzo, A.: Reconstruction of two-dimensional buried objects by a differential evolution method. Inverse Probl. **20**, S135–S150 (2004). doi:10.1088/0266-5611/20/6/S09

Massa, A., Franceschini, D., Franceschini, G., et al.: Parallel GA-based approach for microwave imaging applications. IEEE Trans. Antennas Propag. **53**, 3118–3127 (2005). doi:10.1109/TAP.2005.856311

Meles, G.A., Van der Kruk, J., Greenhalgh, S.A., et al.: A new vector waveform inversion algorithm for simultaneous updating of conductivity and permittivity parameters from combination crosshole/borehole-to-surface GPR Data. IEEE Trans. Geosci. Remote Sens. **48**, 3391–3407 (2010). doi:10.1109/TGRS.2010.2046670

Meschino, S., Pajewski, L., Schettini, G.: A direction-of-arrival approach for the subsurface localization of a dielectric object. J. Appl. Geophys. **85**, 68–79 (2012). doi:10.1016/j.jappgeo.2012.07.002

Meschino, S., Pajewski, L., Pastorino, M., et al.: Detection of subsurface metallic utilities by means of a sap technique: comparing MUSIC- and SVM-based approaches. J. Appl. Geophys. **97**, 60–68 (2013). doi:10.1016/j.jappgeo.2013.01.011

Michalski, K.A.: Electromagnetic imaging of circular-cylindrical conductors and tunnels using a differential evolution algorithm. Microw. Opt. Technol. Lett. **27**, 330–334 (2000). doi:10.1002/1098-2760(20001205)27:5<330:AID-MOP13>3.0.CO;2-H

Michalski, K.A.: Electromagnetic imaging of elliptical-cylindrical conductors and tunnels using a differential evolution algorithm. Microw. Opt. Technol. Lett. **28**, 164–169 (2001). doi:10.1002/1098-2760(20010205)28:3<164:AID-MOP5>3.0.CO;2-D

Minet, J., Lambot, S., Slob, E.C., Vanclooster, M.: Soil surface water content estimation by full-waveform GPR signal inversion in the presence of thin layers. IEEE Trans. Geosci. Remote Sens. **48**, 1138–1150 (2010). doi:10.1109/TGRS.2009.2031907

De Mol, C.: A critical survey of regularized inversion method. In: Bertero, M., Pike, E.R. (eds.) Inverse Problems in Scatterring Imaging, pp. 345–370. Hilger, Bristol (1992)

Lo Monte, L., Erricolo, D., Soldovieri, F., Wicks, M.C.: Radio frequency tomography for tunnel detection. IEEE Trans. Geosci. Remote Sens. **48**, 1128–1137 (2010). doi:10.1109/TGRS.2009.2029341

Mydur, R., Michalski, K.A.: A neural-network approach to the electromagnetic imaging of elliptic conducting cylinders. Microw Opt. Technol. Lett. **28**, 303–306 (2001). doi:10.1002/1098-2760(20010305)28:5<303:AID-MOP1024>3.0.CO;2-C

Neal, A.: Ground-penetrating radar and its use in sedimentology: principles, problems and progress. Earth-Sci. Rev. **66**, 261–330 (2004). doi:10.1016/j.earscirev.2004.01.004

Oliveri, G., Lizzi, L., Pastorino, M., Massa, A.: A nested multi-scaling inexact-newton iterative approach for microwave imaging. IEEE Trans. Antennas Propag. **60**, 971–983 (2012a). doi:10.1109/TAP.2011.2173131

Oliveri, G., Randazzo, A., Pastorino, M., Massa, A.: Electromagnetic imaging within the contrast-source formulation by means of the multiscaling inexact newton method. J. Opt. Soc. Am. A: **29**, 945–958 (2012b). doi:10.1364/JOSAA.29.000945

Orlando, L., Soldovieri, F.: Two different approaches for georadar data processing: a case study in archaeological prospecting. J. Appl. Geophys. **64**, 1–13 (2008). doi:10.1016/j.jappgeo.2007.10.002

Ostadrahimi, M., Mojabi, P., Zakaria, A., et al.: Enhancement of Gauss-Newton inversion method for biological tissue imaging. IEEE Trans. Microw. Theory Tech. **61**, 3424–3434 (2013). doi:10.1109/TMTT.2013.2273758

Pastorino, M.: Short-range microwave inverse scattering techniques for image reconstruction and applications. IEEE Trans. Instrum. Meas. **47**, 1419–1427 (1998). doi:10.1109/19.746706

Pastorino, M., Massa, A., Caorsi, S.: A microwave inverse scattering technique for image reconstruction based on a genetic algorithm. IEEE Trans. Instrum. Meas. **49**, 573–578 (2000). doi:10.1109/19.850397

Pastorino, M., Caorsi, S., Massa, A., Randazzo, A.: Reconstruction algorithms for electromagnetic imaging. IEEE Trans. Instrum. Meas. **53**, 692–699 (2004). doi:10.1109/TIM.2004.827093

Pastorino, M.: Stochastic optimization methods applied to microwave imaging: a review. IEEE Trans. Antennas Propag. **55**, 538–548 (2007). doi:10.1109/TAP.2007.891568

Pastorino, M.: Microwave Imaging. Wiley, Hoboken (2010)

Pastorino, M., Randazzo, A.: Buried object detection by an inexact-newton method applied to nonlinear inverse scattering. Int J. Microw. Sci. Technol. **2012**, 1–7 (637301) (2012)

Pastorino, M., Randazzo, A.: Nondestructive analysis of dielectric bodies by means of an ant colony optimization method. In: Fornarelli, G., Mescia, L. (eds.) Swarm Intelligence for Electric and Electronic Engineering, pp 308–325. Engineering Science Reference, Hershey, PA (2013). doi:10.4018/978-1-4666-2666-9

Patriarca, C., Lambot, S., Mahmoudzadeh, M.R., et al.: Reconstruction of sub-wavelength fractures and physical properties of masonry media using full-waveform inversion of proximal penetrating radar. J. Appl. Geophys. **74**, 26–37 (2011). doi:10.1016/j.jappgeo.2011.03.001

Patriarca, C., Miorali, M., Slob, E., Lambot, S.: Uncertainty quantification in off-ground monostatic ground penetrating radar. IEEE Trans. Antennas Propag. **61**, 3334–3344 (2013). doi:10.1109/TAP.2013.2251597

Persico, R., Bernini, R., Soldovieri, F.: The role of the measurement configuration in inverse scattering from buried objects under the Born approximation. IEEE Trans. Antennas Propag. **53**, 1875–1887 (2005). doi:10.1109/TAP.2005.848468

Persico, R.: On the role of measurement configuration in contactless GPR data processing by means of linear inverse scattering. IEEE Trans. Antennas Propag. **54**, 2062–2071 (2006). doi:10.1109/TAP.2006.877170

Pierri, R., Liseno, A., Solimene, R., Soldovieri, F.: Beyond physical optics SVD shape reconstruction of metallic cylinders. IEEE Trans. Antennas Propag. **54**, 655–665 (2006). doi:10.1109/TAP.2005.863121

Pike, E.R., Sabatier, P.C.: Scattering: Scattering and Inverse Scattering in Pure and Applied Science. Academic Press, San Diego (2002)

Piro, S., Goodman, D., Nishimura, Y.: The study and characterization of Emperor Traiano's Villa (Altopiani di Arcinazzo, Roma) using high-resolution integrated geophysical surveys. Archaeol. Prospect. **10**, 1–25 (2003). doi:10.1002/arp.203

Porsani, J.L., Filho, W.M., Elis, V.R., et al.: The use of GPR and VES in delineating a contamination plume in a landfill site: a case study in SE Brazil. J. Appl. Geophys. **55**, 199–209 (2004). doi:10.1016/j.jappgeo.2003.11.001

Potthast, R.: Point Sources and Multipoles in Inverse Scattering Theory. Chapman & Hall/CRC, Boca Raton (2001)

Qing, A., Lee, C.K., Jen, L.: Electromagnetic inverse scattering of two-dimensional perfectly conducting objects by real-coded genetic algorithm. IEEE Trans. Geosci. Remote Sens. **39**, 665–676 (2001). doi:10.1109/36.911123

Qing, A.: Dynamic differential evolution strategy and applications in electromagnetic inverse scattering problems. IEEE Trans. Geosci. Remote Sens. **44**, 116–125 (2006). doi:10.1109/TGRS.2005.859347

Qing, A., Lee, C.K.: Differential Evolution in Electromagnetics. Springer, Berlin (2010)

Rahmat-Samii, Y., Michielssen, E.: Electromagnetic Optimization by Genetic Algorithms. J. Wiley, New York (1999)

Randazzo, A., Oliveri, G., Massa, A., Pastorino, M.: Electromagnetic inversion with the multiscaling inexact newton method-experimental validation. Microw. Opt. Technol. Lett. **53**, 2834–2838 (2011). doi:10.1002/mop.26435

Randazzo, A.: Swarm optimization methods in microwave imaging. Int. J. Microw. Sci. Technol. **2012**, 1–12 (491713) (2012). doi: 10.1155/2012/491713

Randazzo, A., Estatico, C.: A regularisation scheme for electromagnetic inverse problems: application to crack detection in civil structures. Nondestruct. Test Eval. **27**, 189–197 (2012). doi :10.1080/10589759.2012.665920

Rekanos, I.T., Kanaki, M.: Microwave imaging of two-dimensional conducting scatterers using particle swarm optimization. In: Krawczyk, A., Wiak, S., Fernandez, L.M. (eds.) Electromagnetic Fields in Mechatronics, Electrical and Electronics Engineering, pp. 84–89. IOS Press, Amsterdam (2006)

Rekanos, I.T.: Shape reconstruction of a perfectly conducting scatterer using differential evolution and particle swarm optimization. IEEE Trans. Geosci. Remote Sens. **46**, 1967–1974 (2008). doi:10.1109/TGRS.2008.916635

Remis, R.F., van den Berg, P.M.: On the equivalence of the Newton-Kantorovich and distorted Born methods. Inverse Probl. **16**, L1–L4 (2000). doi:10.1088/0266-5611/16/1/101

Robinson, J., Rahmat-Samii, Y.: Particle swarm optimization in electromagnetics. IEEE Trans. Antennas Propag. **52**, 397–407 (2004). doi:10.1109/TAP.2004.823969

Roger, A.: Newton-Kantorovitch algorithm applied to an electromagnetic inverse problem. IEEE Trans. Antennas Propag. **29**, 232–238 (1981). doi:10.1109/TAP.1981.1142588

Rubek, T., Meaney, P.M., Meincke, P., Paulsen, K.D.: Nonlinear microwave imaging for breast-cancer screening using gauss-newton's method and the CGLS inversion algorithm. IEEE Trans. Antennas Propag. **55**, 2320–2331 (2007). doi:10.1109/TAP.2007.901993

Sahin, A., Miller, E.L.: Object detection using high resolution near-field array processing. IEEE Trans. Geosci. Remote Sens. **39**, 136–141 (2001). doi:10.1109/36.898675

Scapaticci, R., Catapano, I., Crocco, L.: Wavelet-based adaptive multiresolution inversion for quantitative microwave imaging of breast tissues. IEEE Trans. Antennas Propag. **60**(8), 3717–3726 (2012)

Semnani, A., Kamyab, M., Rekanos, I.T.: Reconstruction of one-dimensional dielectric scatterers using differential evolution and particle swarm optimization. IEEE Geosci. Remote Sens. Lett. **6**, 671–675 (2009). doi:10.1109/LGRS.2009.2023246

Semnani, A., Rekanos, I.T., Kamyab, M., Papadopoulos, T.G.: Two-dimensional microwave imaging based on hybrid scatterer representation and differential evolution. IEEE Trans. Antennas Propag. **58**, 3289–3298 (2010). doi:10.1109/TAP.2010.2055793

Shelton, N., Warnick, K.F.: Behavior of the regularized sampling inverse scattering method at internal resonance frequencies. J. Electromagn. Waves Appl. **17**, 487–488 (2003). doi:10.1163/156939303767868991

Slaney, M., Kak, A.C., Larsen, L.E.: Limitations of imaging with first-order diffraction tomography. IEEE Trans. Microw. Theory Tech. **32**, 860–874 (1984). doi:10.1109/TMTT.1984.1132783

Soldovieri, F., Persico, R., Utsi, E., Utsi, V.: The application of inverse scattering techniques with ground penetrating radar to the problem of rebar location in concrete. NDT E Int. **39**, 602–607 (2006). doi:10.1016/j.ndteint.2005.12.005

Soldovieri, F., Solimene, R.: Ground penetrating radar subsurface imaging of buried objects. Radar Technol. 105–126 (2010)

Soldovieri, F., Lopera, O., Lambot, S.: Combination of advanced inversion techniques for an accurate target localization via GPR for demining applications. IEEE Trans. Geosci. Remote Sens. **49**, 451–461 (2011a). doi:10.1109/TGRS.2010.2051675

Soldovieri, F., Solimene, R., Lo Monte, L., et al.: Sparse reconstruction from GPR data with applications to rebar detection. IEEE Trans. Instrum. Meas. **60**, 1070–1079 (2011b). doi:10.1109/TIM.2010.2078310

Solimene, R., Pierri, R.: Number of degrees of freedom of the radiated field over multiple bounded domains. Opt. Lett. **32**, 3113 (2007). doi:10.1364/OL.32.003113

Solimene, R., Soldovieri, F., Prisco, G., Pierri, R.: Three-dimensional through-wall imaging under ambiguous wall parameters. IEEE Trans. Geosci. Remote Sens. **47**, 1310–1317 (2009). doi:10.1109/TGRS.2009.2012698

Solimene, R., Leone, G., Dell'Aversano, A.: MUSIC algorithms for rebar detection. J. Geophys. Eng. **10**, 064006 (2013a). doi:10.1088/1742-2132/10/6/064006

Solimene, R., Maisto, M.A., Pierri, R.: Role of diversity on the singular values of linear scattering operators: the case of strip objects. J. Opt. Soc. Am. A **30**, 2266 (2013b). doi:10.1364/JOSAA.30.002266

Solimene, R., Maisto, M.A., Romeo, G., Pierri, R.: On the singular spectrum of the radiation operator for multiple and extended observation domains. Int. J. Antennas Propag. **2013**, 1–10 (2013c). doi:10.1155/2013/585238

Solimene, R., Catapano, I., Gennarelli, G., et al.: A unified mathematical overview of SAR imaging algorithms and some unconventional applications. IEEE Signal Process (2014a) (In print)

Solimene, R., Cuccaro, A., Dell'Aversano, A., et al.: Ground clutter removal in GPR surveys. IEEE J. Sel. Top Appl. Earth Obs. Remote Sens. **7**, 792–798 (2014a). doi:10.1109/JSTARS.2013.2287016

Solimene, R., D'Alterio, A., Gennarelli, G., Soldovieri, F.: Estimation of soil permittivity in presence of antenna-soil interactions. IEEE J. Sel Top. Appl. Earth Obs. Remote Sens. **7**, 805–812 (2014b). doi:10.1109/JSTARS.2013.2268576

Solimene, R., Dell'Aversano, A.: Some remarks on time-reversal MUSIC for two-dimensional thin PEC scatterers. IEEE Geosci. Remote Sens. Lett. **11**, 1163–1167 (2014). doi:10.1109/LGRS.2013.2288516

Soumekh, M.: Synthetic Aperture Radar Signal Processing with MATLAB Algorithms. Wiley, New York (1999)

Stolt, R.H.: Migration by fourier transform. Geophysics **43**, 23–48 (1978). doi:10.1190/1.1440826

Sun, C.-H., Chiu, C.-C., Li, C.-L.: Time-domain inverse scattering of a two-dimensional metallic cylinder in slab medium using asynchronous particle swarm optimization. Prog. Electromagn. Res. M **14**, 85–100 (2010). doi:10.2528/PIERM10051101

Sun, J.: An eigenvalue method using multiple frequency data for inverse scattering problems. Inverse Probl. **28**, 025012 (2012). doi:10.1088/0266-5611/28/2/025012
Taylor, A.E., Lay, D.C.: Introduction to Functional Analysis, 2d edn. Wiley, New York (1980)
Tikhonov, A.N., Arsenine, V.I.: Solution to Ill-posed Problems. Halsted, York (1977)
Watters, T.R., Leuschen, C.J., Plaut, J.J., et al.: MARSIS radar sounder evidence of buried basins in the northern lowlands of Mars. Nature **444**, 905–908 (2006). doi:10.1038/nature05356
Weile, D.S., Michielssen, E.: Genetic algorithm optimization applied to electromagnetics: a review. IEEE Trans. Antennas Propag. **45**, 343–353 (1997). doi:10.1109/8.558650
Van den Berg, P.M., Kleinman, R.E.: A total variation enhanced modified gradient algorithm for profile reconstruction. Inverse Probl. **11**, L5–L10 (1995). doi:10.1088/0266-5611/11/3/002
Van den Berg, P.M., Kleinman, R.E.: A contrast source inversion method. Inverse Probl. **13**, 1607–1620 (1997). doi:10.1088/0266-5611/13/6/013
Van den Berg, P.M., Abubakar, A.: Contrast source inversion method: state of art. Prog. Electromagn. Res. **34**, 189–218 (2001). doi:10.2528/PIER01061103
Van der Wielen, A., Courard, L, Nguyen, F.: Detection of thin layers into concrete with static and CMP measurements. In: Proceedings of 14th International Conference on Ground Penetrating Radar GPR, pp 530–535. IEEE, Shanghai, China (2012)
Winters, D.W., Van Veen, B.D., Hagness, S.C.: A sparsity regularization approach to the electromagnetic inverse scattering problem. IEEE Trans. Antennas Propag. **58**, 145–154 (2010). doi:10.1109/TAP.2009.2035997
Zhuge, X., Yarovoy, A.G., Savelyev, T., Ligthart, L.: Modified Kirchhoff migration for UWB MIMO array-based radar imaging. IEEE Trans. Geosci. Remote Sens. **48**, 2692–2703 (2010). doi:10.1109/TGRS.2010.2040747

GPR Data Processing Techniques

Nikos Economou, Antonis Vafidis, Francesco Benedetto and Amir M. Alani

Abstract Ground penetrating radar (GPR) is a non-destructive geophysical method that uses radar pulses to image the subsurface. Notwithstanding that it is particularly promising for soil studies, GPR is characterised by notoriously difficult automated data analysis. Hence, the focus of this chapter is to provide the reader with a deep understanding of the state of the art and open issues in the field of GPR data processing techniques as well as of the interesting application of GPR in the field of civil engineering. In particular, we present an overview on noise suppression, deconvolution, migration, attribute analysis and classification techniques for GPR data.

1 Introduction

Ground penetrating radar (GPR) consists primarily of a radar device which transmits and receives electromagnetic pulses into the subsurface. Although the principle of the GPR method is identical to the seismic reflection method, many

N. Economou (✉) · A. Vafidis
Applied Geophysics Lab, School of Mineral Resources Engineering, Technical University of Crete, Polytechnioupolis-Chania, 731 00 Chania, Crete, Greece
e-mail: neconom@mred.tuc.gr

A. Vafidis
e-mail: vafidis@mred.tuc.gr

F. Benedetto
Signal Processing for Telecommunications and Economics Lab, Department of Economics, Roma Tre University, Via Vito Volterra 62, 00146 Rome, Italy
e-mail: francesco.benedetto@uniroma3.it

A.M. Alani
School of Computing and Technology, University of West London, 8th floor, Villiers House, Ealing Broadway, London, W5 2PA, UK
e-mail: Amir.Alani@uwl.ac.uk

A. Benedetto and L. Pajewski (eds.), *Civil Engineering Applications of Ground Penetrating Radar*, Springer Transactions in Civil and Environmental Engineering,
DOI 10.1007/978-3-319-04813-0_11

differences exist in terms of both data acquisition and processing. The GPR method is characterised by the rapidly decaying amplitude of the electromagnetic waves together with the loss of the relevant higher frequency harmonics. Thus, GPR data are highly non-stationary.

GPR technology has seen tremendous progress in its range of applications, as well in its analysis and processing algorithms, over the past 20 years. GPR applications include sedimentology, ground water contamination, glaciology, archaeology and geotechnical engineering (Jol 2009). Notwithstanding the undoubted promise for the technology in most of the above-mentioned applications, the most successful results have come from glaciology and geotechnical engineering.

In particular, GPR has proved to be very useful in road monitoring applications (Diamanti and Redman 2012; Le Bastard et al. 2007; Benedetto et al. 2005; Saarenketo and Scullion 2000), in the delineation of buried objects such as pipes, cables and tunnels (Zheng et al. 2011), the condition monitoring of railway ballast (Shao et al. 2011; Benedetto and Benedetto 2011), the inspection of concrete structures and bridge decks (Alani et al. 2013), the mapping of buried archaeological ruins (Li et al. 2012) and many other relevant applications which have already been applied or are projected for the future.

The remainder of this chapter is organised as follows. First, we will present the most basic and commonly used signal processing techniques, before concentrating our focus on the enhancement of the GPR signal by the use of advanced signal processing methods such as denoising, deconvolution, velocity analysis, migration and attribute analysis and classification. Finally, we will review some interesting applications of GPR data processing techniques and conclude the chapter by summarising what has been undertaken, what further work is necessary and what further expectations the field holds.

2 GPR Data Processing

The interpretation of GPR data is affected by many factors related to the complexity of the subsurface. The data interpretation requires dense grids of measurements, optimum choice of transmitting frequency and effective signal processing techniques. The processing of GPR measurements is enhanced in the present day by powerful hardware. The choice of transmitting frequency antennas depends mainly on the target properties (size and depth) and the radio frequency properties of the target and the surrounding media. Accepting the fact that GPR measurements are dense enough for the delineation of the target and that the optimum frequency is selected (Annan 2005), processing should enhance the GPR signal, providing sufficient data for interpretation. The GPR data processing system consists of time-zero corrections, dewow, band-pass filtering, spatial filters and gain.

Drift of the zero time along the profile can occur due to temperature differences between the instrument electronics and the air temperature, or as a result of damaged cables. This drift causes a misalignment of the reflections and the zero time has to be reset for all traces along the profile (Jol 2009). Dewow removes the low frequency harmonics caused by electromagnetic induction. Many automatic algorithms exist for dewowing but, even though they correct for the low frequency undesired harmonics, they sometimes cause reverberations. Time-varying low cut filtering can solve this problem (Battista et al. 2009). Spatial filters are mainly used for muting specific forms of reflections (horizontal or dipping). The most commonly used gain functions are exponential gain (SEC) and windowed gain functions (AGC). Even though SEC retains information that is lost by AGC, it also slightly alters the shape of the wavelet, making it time varying (Annan 2005). Conversely, AGC enhances low amplitudes, but can be misleading because it can also enhance unwanted information and lose the initial spatial relationship of amplitudes.

GPR data processing can also include the following techniques: time varying band-pass filtering (Schimmel and Gallart 2007; Bradford and Wu 2007; Economou and Vafidis 2010), time and frequency analysis (Bradford and Wu 2007; Economou and Vafidis 2010; Bano 2004), deconvolution (Xia et al. 2004; Chahine et al. 2011; Economou and Vafidis 2011, 2012; Economou et al. 2012; Schmelzbach et al. 2011), velocity analysis (Pipan et al. 1999; Bradford 2008; Brown et al. 2009) and migration (Sena et al. 2006; Brooke and Maillol 2007) as well as attribute analysis and classification (Massa et al. 2005; Spanoudakis 2007; Economou et al. 2007; Pipan et al. 2001). The main purpose of all the above is to enhance the signal in order to improve the GPR images.

Time varying band-pass filtering mainly suppresses noise outside the selected band. Deconvolution acquires the reflectivity series by decreasing the time duration of the dominant wavelet. Velocity analysis in combination with migration is necessary for placing the reflections at the correct recording time, for collapsing diffractions to their apex and, further, for giving depth information. Attribute analysis aids in the interpretation of the GPR sections. Velocity analysis, aside from being a processing method together with the migration process, can also be treated as an attribute analysis method. In the following sections, we will review in detail each of the afore-mentioned techniques.

In summary, GPR data processing and subsequent display/presentation of the results are of paramount importance to GPR operators and end users and also should be considered as an integrated part as well as a prerequisite for successful surveying and data processing.

From a practical and applications point of view, GPR data processing depends upon a number of parameters and considerations which predominantly relate to planning, data collection and site operation as well as choice of antenna. Basically, data processing goes hand-in-hand with quality of the raw data collected. In real terms, regardless of choice of antenna system and skill of the GPR operator, data processing is achieved by means of suitable software. GPR manufacturers use and recommend specific software, which may or may not be a product of their own. No doubt, the knowledge and experience of individuals (GPR users) play a significant role in achieving the optimum outcome within the context of data processing and presentation. However, as mentioned above, the success of any data

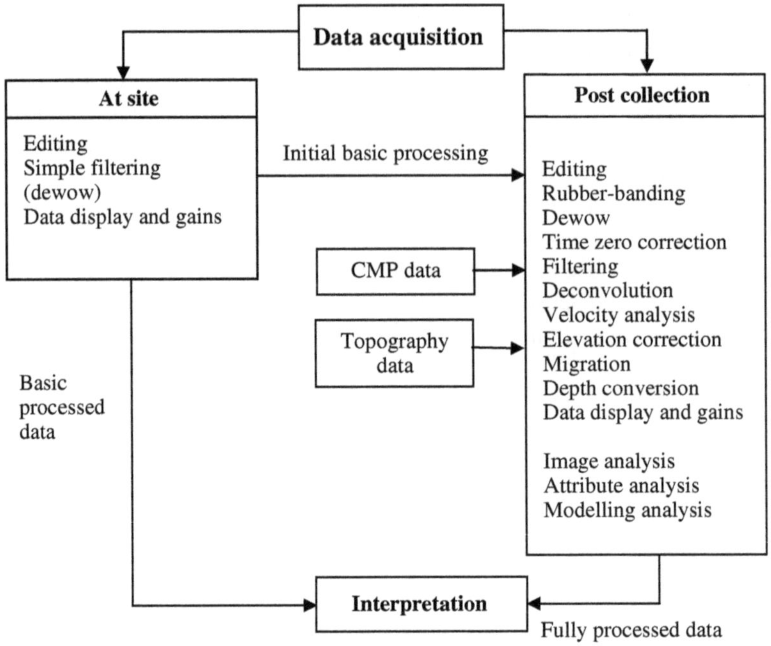

Fig. 1 Typical GPR data processing and analysis steps, after Cassidy (2009)

processing is directly related to parameters that one should be fully aware and have extended knowledge of. It must be emphasised that the nature and type of projects, in turn, will have an effect on the quality of data collection and the processing of any GPR survey too.

The diagram depicted below by Cassidy (2009) can still be considered to thoroughly capture the GPR data processing steps, which in turn summarises what has been presented above (Fig. 1).

3 Deconvolution

Deconvolution is the process needed to improve the temporal resolution, in order to clearly visualise the reflections of interest (Economou and Vafidis 2012). The vertical resolution of reflection horizons in a GPR profile depends on the bandwidth used and, thus, on the dominant frequency transmitted by the antenna which determines pulse duration. Relatively low frequency antennas may record strong signals from waves returning from depths greater than several metres, but the resolution of the GPR section is reduced. A section additionally encounters a gradual reduction of resolution as the recording time increases because the attenuation of the electromagnetic (EM) waves is proportional to the frequency and the travel distance.

When measuring thin layers or closely spaced targets using relatively low frequency antennas, the reflections from the above mentioned structures may overlap. A deconvolution algorithm extracts the reflectivity series and thus improves the temporal resolution (Chahine et al. 2011). It is well understood that increased resolution is necessary when dealing with geotechnical GPR data. If deconvolution satisfactorily improves the resolution for the relatively low frequency antennas, which penetrate deeper and have stronger responses (Diamanti and Redman 2012), the higher frequency antennas may prove to be unnecessary and the number of involved antennas can be reduced. Additionally, it is necessary to distinguish overlapping reflections in the GPR sections for any antenna's dominant frequency.

Despite the popularity of GPR reflection, successful deconvolution applications of GPR data are very rare (e.g. Xia et al. 2004; Turner 1994; Todoeschuck et al. 1992). The authors in Chahine et al. (2011) and Schmelzbach et al. (2011) performed blind deconvolution with GPR data. GPR data time varying deconvolution related to geotechnical engineering was also applied in Economou et al. (2012), in order to successfully increase the bandwidth and hence the temporal resolution of the time series. Deconvolution of GPR data has been notoriously unsuccessful for many years mainly due to the non-stationarity of the GPR traces and the mixed phase of EM wavelets (Economou and Vafidis 2011). Even though Bano (2004), Turner (1994) and Irving and Knight (2003) presented methods for inverse Q filtering, in most cases this inverse Q filter suffers in the estimation of the absorption value (Bradford and Wu 2007). The authors in Economou and Vafidis (2010) developed a narrow time-window methodology applied in the t-f domain (Stockwell et al. 1996) for spectral balancing (Neto and Medeiros 2006; Belina et al. 2008). This approach is adequate for enhancing the amplitudes and restoring the dominant frequency of the GPR traces. This technique is a precondition rather than an alternative technique for deconvolution as presented in Belina et al. (2008). Furthermore, wavelet deconvolution is successful when applied after spectral balancing and followed by the application of the maximum kurtosis method (Economou and Vafidis 2011, 2012; Van der Baan 2008; White 1988; Longbottom et al. 1998; Levy and Oldenburg 1987; Economou 2010). In particular, the authors in Economou and Vafidis (2011) used a reference wavelet, obtained experimentally, in the time varying deconvolution scheme. This reference wavelet is antenna dependent and is adequate for datasets measured with the same antenna.

Further developments of the methods for automatic enhancement of the temporal resolution of GPR data must be achieved. For instance, spectral balancing can be achieved by automatically selecting the frequency range where spectral balancing must be applied, in order to restore the GPR signal suffering attenuation.

In general, there is a lack of successful applications of deconvolution methods on GPR data. Further research should focus on the application of deconvolution operators such as predictive deconvolution on GPR data, which is a standard tool for seismic reflection data, in order to suppress multiples which mask the reflections of interest.

4 Velocity Analysis of GPR Data

This section of the chapter discusses velocity analysis techniques for GPR data. Velocity estimation requires common-midpoint (CMP) data. The estimated velocities are useful for the migration of GPR data as well as for the time-to-depth conversion, providing depth sections using migration techniques.

The travel time curve of reflected waves is approximated by a hyperbola. The time difference of this travel time at a given and at zero offsets is called normal moveout (NMO). The hyperbolic travel time equation is linear in t^2-x^2. The RMS velocity for a given reflector can be estimated from the time that best fits the travel times plotted in n t^2-x^2.

Alternatively, one can apply different NMO corrections to a CMP gather using a range of constant velocity values and choose the velocity that best flattens each event.

Another velocity analysis technique is based on computing the velocity spectrum, which displays a measure of signal coherency as a function of velocity and two-way zero-offset time. Signal coherency is computed in small time gates that follow a hyperbola trajectory. RMS velocities are interpreted from velocity spectra by choosing the velocity function that produces the highest coherency at times of significant event amplitudes. In particular, this technique is based on the cross correlation of the traces in a CMP gather.

For an estimation of the velocity of the top layer, one can often utilise the direct or diffracted waves. The travel time curve of the direct waves in a CMP gather is a straight line. The stacked section is utilised to calculate the velocity from diffracted waves. Namely, the hyperbolic travel time equation is linear in the n t^2-x^2 plane. The slope of this line is equal to

$$slope = \frac{4}{V_{RMS}^2}.$$

5 Migration

Migration is a method which moves events to their correct time or spatial locations and collapses energy from diffractions back to their scattering points. Until three decades ago, migration was the final processing step for seismics (Gray et al. 2001). Post-stack migration is always simpler and is usually enough for straightforward geology settings. Today, migrated seismic data almost always provide input for several steps, such as detailed analyses of attributes and signal processing methods like deconvolution. Pre-stack migration has gained popularity due to the increased computational power. The similarity between GPR and seismic reflection methods led to the application of processing techniques on the GPR data used in seismic data processing.

Still, the non-stationarity of the GPR traces makes the application of such methods insufficient. An improved Stolt migration algorithm for GPR data for relatively homogeneous velocity media was presented in Zhou et al. (2011).

A split-step migration technique was implemented in Sena et al. (2006) in order to account for dispersion effects, while the authors in Di and Wang (2004) implemented a finite element migration method which incorporates attenuation. Still, the problem of the rapidly varying velocity causes high degrees of distortion on the outcome. The insufficient application of migration on GPR data is mainly due to the inability to estimate an accurate velocity model. This is often the problem for seismics (Gray et al. 2001).

For the successful application of migration on the GPR data, multi-fold data must be acquired in order to build a high resolution velocity model (Pipan et al. 1999, 2001, 2003; Bradford 2006, 2008; Brown et al. 2009; Booth et al. 2008). Multi-fold GPR data acquisition requires multi-channel GPR devices. Processing multi-channel GPR data involves corrections and velocity analysis prior stacking.

The complex EM response of the subsurface is a wide research area and the authors of this chapter can only direct the reader to the first problem referred above which also connects with resolution of the data (Economou and Vafidis 2010; Irving and Knight 2003; Van der Baan 2012).

Migration can be considered as a smoothing technique which alters the amplitude spectrum by lowering the dominant frequency. Thus, the migrated section is characterised by lower resolution (Bancroft et al. 2012). So, consideration should be taken in order to apply deconvolution methods to the migrated sections (Vafidis et al. 2012).

6 Noise Suppression

By discarding background noise, the visual quality of the reflections of interest is enhanced (Cassidy 2009). Noise affects both deconvolution and migration techniques. There are many types of noise that can be observed in GPR data. These types can be divided into two major categories: coherent and incoherent noise. Incoherent noise consists of noise that has no discernible pattern from trace to trace, while coherent noise does display regularity from trace to trace. Coherent noise is mainly attributed to multiples, direct air and ground waves, and side reflections. As long as coherent noise can be separated in the frequency domain, single channel techniques such as band-pass filtering can be utilised. Predictive deconvolution can also be used, by which multiple patterns are identified and removed from the data. Still, the most effective techniques for suppressing coherent noise have turned out to be multichannel in nature. Multichannel filtering includes the weighted trace mix (background noise removal), f-k filtering, Karhunen-Loeve (KL) transform, the tau-p transform, the generalised Radon transform, and the radial trace transform (Nuzzo and Quarta 2004; Jeng et al. 2011; Kim et al. 2007; Carpentier et al. 2010).

A limited number of filtering techniques have been proposed to remove incoherent noise (Economou and Vafidis 2010; Baili et al. 2009). The simplest method of suppressing random noise is stacking. Simple stacking does not always result in a satisfactory improvement of the signal due to amplitude fluctuations between the stacked signals, mostly caused by DC quantisation errors and imperfect alignment related to synchronisation errors when sampling (Lahouar 2003). The authors in Economou and Vafidis (2010) utilised user-defined time-varying band-pass filters. The seismic industry offers a variety of algorithms for the suppression of random noise but many of these methods (referred in seismic data) are related to common source or stacked data (e.g. F-x deconvolution). Random noise suppression can be realised by:

1. Stacking, using local correlation, can be automated and become part of the GPR devices (Liu et al. 2009).
2. Empirical mode decomposition (EMD) was introduced by NASA researchers in order to reduce random noise in non-stationary signals (Huang et al. 1998). Compared with wavelet or Fourier transform methods, this method has no pre-defined basis, while it can capture non-stationary and nonlinear variations in the signal, acting like a time varying band pass filter (Flandrin et al. 2005).
3. Smoothing of the spectrum denoises the signal (Berkhout 1977; Taner et al. 2003). Parameters, such as degree of smoothness and time window of the relevant application, can be predefined.

7 Attributes

Attribute analysis and classification are useful for the interpretation of GPR data. Attribute analysis enhances signal characteristics otherwise not perceptible and contributes to interpretation. There are five main categories of attributes:

1. Instantaneous attributes provide useful local information and are calculated via Hilbert transform. The basic attributes are instantaneous envelope and phase. From these, other attributes can be calculated including instantaneous frequency and instantaneous gain.
2. Wavelet attributes are computed at the peak of the envelope (Zhao et al. 2012).
3. Texture attributes describe the data samples, through gray-level co-occurrence matrices (GLCM). These matrices are calculated by rescaling the GPR records to a limited number of gray levels. Then, the co-occurrence or the joint probability of the gray values is calculated within a spatiotemporal sliding window. Various attributes can be extracted from the GLCM including energy, entropy, homogeneity and contrast (Gao 2008; McClymont et al. 2008).
4. Geometrical statistical attributes are calculated directly from the data, within a spatiotemporal sliding window. Local range, variance and standard deviation are the most common.
5. Coherency-semblance attributes provide spatiotemporal relations of the dataset (Lizzi et al. 2009).

Apart from interpreting specific attribute sections, several attributes can be combined for classification. The classification reduces the amount of multidimensional data by categorising or grouping similar data. One of the motivations for using classification algorithms is to aid the construction of categories by minimising human intervention. The authors in Spanoudakis (2007) used attributes and self organising maps to interpret GPR data. The authors in Economou et al. (2007) utilised the same tools to map a three arched roman cistern and the ducts supplying water to the cistern, (Pipan et al. 2001) which focused on archaeological investigations, while (McClymont et al. 2008) utilised geometrical attributes for the visualisation of active faults.

Other classification techniques are based on the exploitation of learning-by-example (LBE) strategies for the detection and classification of buried objects by using support vector machines (Lizzi et al. 2009) or support vector regression (LeBastard et al. 2013). Great efforts have also been devoted towards the development of inversion techniques, able to profitably combine global and local search approaches with multi-focusing strategies, such as the iterative multi scaling approach (Salucci et al. 2013) or imaging approaches based on Bayesian compressive sensing (Oliveri et al. 2011; Poli et al. 2013) and interval analysis (Manica et al. 2013). Finally, specific processing has recently been developed, including the Jonscher parameterisation of the medium (Ihamouten et al. 2011) and the so-called transparent 3D half bird's-eye view of the GPR data volume or its sub-volumes (Kadioglu 2008, 2010, 2013; Kadioglu and Daniels 2008).

8 Applications

In this section, three interesting applications of GPR data processing are reviewed.

The first application (the Omalos survey) highlights the usefulness of classification techniques to the interpretation of 3D GPR data.

The second application (the Sandbox experiment) illustrates a controlled experiment conducted in a sandbox at the University of Louis Pasteur in Strasbourg, France. This is a noteworthy example of time varying deconvolution and migration before visualization and interpretation. As a controlled experiment, it demonstrates what the GPR user should expect from the GPR data before and after data processing. Finally, the third application depicts the importance of GPR data frequency domain analysis. In particular, a controlled experiment was carried out in the laboratory of civil engineering at the University of Rome "Roma TRE", Italy. The GPR data processing in the frequency domain has evidenced that a shift of the frequency GPR spectrum towards lower frequencies is expected, as the clay content increases in the soil under investigation.

9 Omalos Survey

A GPR survey was conducted at Omalos Highlands, Western Crete, Greece for doline and karst system delineation. The mean height of the area is approximately 1500 m and consists of a typical karst terrain with subsidence phenomena.

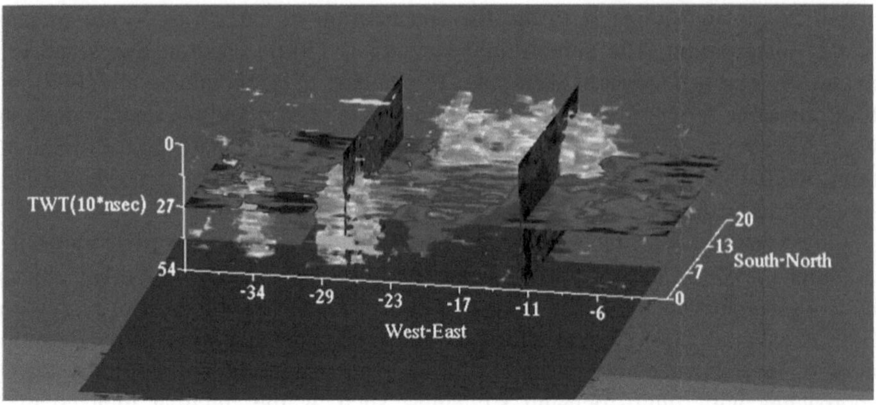

Fig. 2 The classified geovolume (run 1–20 classes). The transparent mode emphasises the doline system (20th class in *red color*) together with a slice at 270 ns and sections at 11 and 26 m (Vafidis et al. 2004)

Dolomitic limestone is covered by alluvium, whose maximum thickness is 100 m. A 110 MHz common-offset 3D GPR survey was performed on a rectangular grid of 41 approximately north–south oriented lines with line spacing of 1 m. Each GPR line contained 81 traces, equally spaced at 0.25 m. Antennas were oriented parallel to each other and perpendicular to the in-line direction. The separation between antennas was 1 m.

Processing of the 3D GPR dataset initially included dewow and high-cut filtering (Spanoudakis 2007). The direct air and ground waves were effectively removed by subtracting the average of the traces across a horizontally moving window. Finally, the data were smoothed using a 3D moving average window.

The next processing step included attribute analysis. A set of four physical (instantaneous envelope, dB instantaneous envelope, instantaneous phase and instantaneous frequency) and two geometrical (semblance of pre-processed data and semblance of instantaneous phase) attributes was calculated. The final processing step prior to classification was the extraction of 30 horizontal slices from each attribute volume within a window of 18 ns (equal to one pulse width). The classification was performed using the following steps (see Fig. 2):

- K-means cluster analysis to identify 50 cluster centres at each slice. The result was 1.500 cluster centres within the whole volume.
- SOM classification with these samples into 20 groups.

10 Sandbox Experiment

A controlled experiment was conducted in a sandbox at the University of Louis Pasteur in Strasbourg, France, using a Pulse-Ekko1000 GPR system with 1200 MHz shielded antennas (Loeffler and Bano 2004; Bano et al. 2009). The water level was

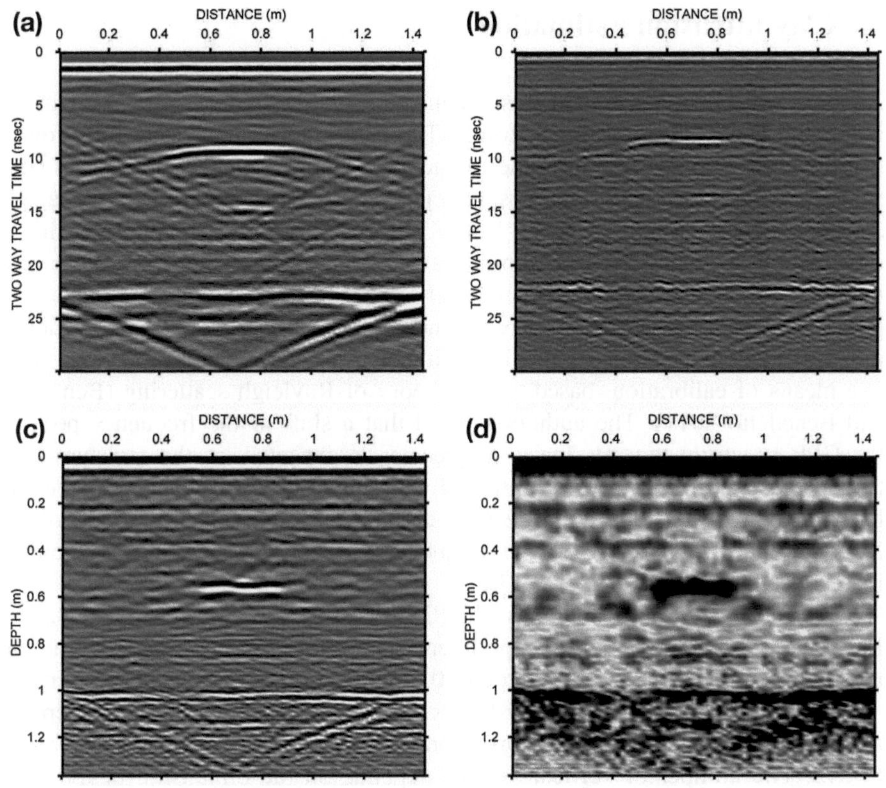

Fig. 3 A GPR section from the sandbox experiment (Loeffler and Bano 2004). **a** A GPR section above a metal pipe buried in the sand, where water surface is at 72 cm depth, **b** after time varying deconvolution and kurtosis corrections, **c** is (**b**) after time migration and time-to-depth conversion and **d** is the instantaneous amplitude of (**c**)

72 cm below surface. The high amplitude reflection in the centre of the section (recording time around 10 ns) is caused by a steel pipe (see Fig. 3). Dewow and inverse amplitude decay gain were applied to the raw data (Fig. 3a). Time varying deconvolution and maximum kurtosis phase corrections successfully increased temporal resolution for reflected waves traveling in both the unsaturated and the saturated sand (Fig. 3b).

Time migration was followed by time to depth conversion, using velocities of 0.116 m/ns for the first 72 cm (where the water table is) and 0.075 m/ns for the rest as proposed by Loeffler and Bano (2004); (Fig. 3c). The instantaneous amplitude gave a very close to reality depth section of the sandbox (Fig. 3d). Note the different nature of the recordings below the water table (>0.72 m) and up to the base of the sandbox (~1 m).

11 Clay Intrusion Estimation

Another controlled experiment was conducted in the laboratory of civil engineering at the University of Rome "Roma TRE", Italy, to evaluate clay intrusion in soil, by GPR data processing in the frequency domain. This experiment was considered necessary as one of the most prevalent causes of road pavement damage is clay intrusion in structural layers. Hence, the development of methods for inferring reliable estimates of clay content is a crucial research field involving economic, social and strategic aspects in road safety for a great number of public and private agencies. In particular, a recent new approach was proposed to evaluate moisture content in a porous medium without using core sampling as a destructive means of calibration, based on the theory of Rayleigh scattering (Benedetto and Benedetto 2011). The authors showed that a shift of the frequency peak of the GPR spectrum towards lower frequencies is expected, as the moisture content increases in the soil. The controlled experiment at the Roma Tre University has highlighted the same trend also for clay intrusion in soil, i.e. a shift of the frequency spectrum towards lower frequencies is expected, as the clay content increases (Benedetto and Tosti 2013).

Ground-coupled radar antennas (the RIS/MF system produced by IDS S.p.A., Italy) were used for this GPR experiment. The GPR operates using two antennas with central frequencies of around 0.6 and 1.6 GHz. The received signal is sampled in the time domain at 7.8125×10^{-2} ns. The vertical radar sweep has a duration of 39.936 ns which corresponds to a number of sampled data equal to 512. The number of vertical sweeps depends on the dimension of the GPR scanning on the horizontal axis. After collecting the GPR data, we have performed the frequency-based analysis in post processing and only the signal at the monostatic channel at 0.6 GHz has been post-processed. The GPR has been calibrated in standard conditions extracting the spectra of reflected signals over a metal plate (total reflection) and over a sample of known dry material previously classified and without any clay. Moreover, the end-reflection of the antenna (or system reverberation) was measured from a free-space experiment and removed prior to processing. The sample is located in a box 0.40 m $\times$ 0.40 m and 0.13 m high, so that the tested sample of soil is 0.33 m $\times$ 0.33 m and 0.10 m thick. The box is impervious and electrically isolated. The laboratory configuration remains absolutely fixed during all the tests. In particular, the so-called Misto della Magliana soil (i.e. an alluvial soil with low ultrafine soil fractions) has been used for the tests, increasing the clay content in sequential steps. Figure 4 shows the GPR spectra of the tested soil in the presence of clay intrusion from 7 % (Fig. 4a), 14 % (Fig. 4b) and 22 % (Fig. 4c), and finally with a percentage of clay equal to 30 % (Fig. 4d).

These results experimentally show that a shift of the frequency spectrum towards lower frequencies is observed, as the clay content increases in the soil under investigation. This frequency shift is not only due to scattering but also to

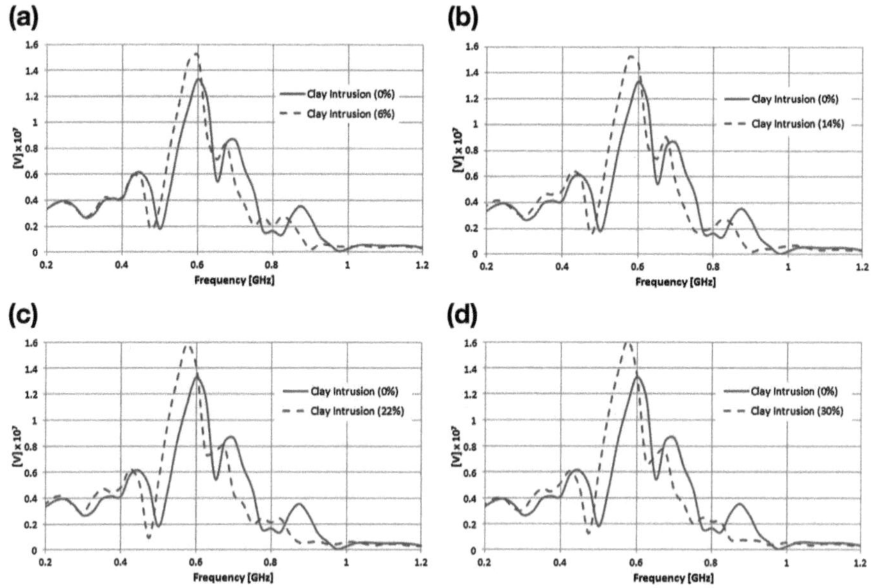

Fig. 4 GPR spectra of the tested soil in the presence of clay intrusion for: **a** 7 % of clay; **b** 14 % of clay; **c** 22 % of clay; **d** 30 % of clay intrusion (Benedetto and Tosti 2013)

attenuation which increases in the presence of clay. This makes higher frequency harmonics attenuate faster. Note that the spectrum corresponds to the trace and not to a part of it. Thus, this method could be used for the lateral (e.g. not in-depth) study of clay content, representing a qualitative estimation of the amount of clay in the soil.

12 Conclusions

This chapter has presented a review of the GPR data processing techniques required for geotechnical applications. We have focused mainly on the enhancement of the GPR data by signal processing methods such as denoising, deconvolution, migration and attribute analysis. Finally, some applications of GPR in the civil engineering field have been introduced and discussed.

As a final conclusion, it is the opinion of the authors that, although this research area has already been well studied by many researchers, more research is needed and, indeed, expected from automatic GPR data analysis.

Acknowledgments The authors acknowledge the COST Action TU1208 "Civil Engineering Applications of Ground Penetrating Radar", in support of this chapter.

References

Alani, A.M., Aboutalebi, M., Kilic, G.: Applications of ground penetrating radar (GPR) in bridge deck monitoring and assessment. J. Appl. Geophys. **97**, 45–54 (2013). doi:10.1016/j.jappgeo.2013.04.009

Annan, P.: GPR methods for hydrogeological studies. In: Rubin, Y., Hubbard, S.S. (eds.) Hydrogeophysics, vol. 50, pp. 185–213. Springer, Berlin (2005)

Baili, J., Lahouar, S., Hergli, M., Al-Qadi, I., Besbes, K.: GPR signal denoising by discrete wavelet transform. NDT E Int. **42**, 696–703 (2009)

Bancroft, J., Guirigay, T., Isaac, H.:Enhancing a seismic image after migration using deconvolution, SEG Technical Program Expanded Abstracts, pp. 1–5 (2012)

Bano, M.: Modelling of GPR waves for lossy media obeying a complex power law of frequency for dielectric permittivity. Geophys. Prospect. **52**, 11–26 (2004)

Bano, M., Loeffler, O., Girard, J.F.: Ground penetrating radar imaging and time-domain modelling of the infiltration of diesel fuel in a sandbox experiment. C.R. Geosci. **341**, 846–858 (2009)

Battista, B., Addison, A., Knapp, C.: Empirical mode decomposition operator for dewowing GPR data. J. Environ. Eng. Geophys. **14**(4), 163–169 (2009)

Belina, F.A., Dafflon, B., Tronicke, J., Holliger, K.: Enhancing the vertical resolution of surface georadar data. J. Appl. Geophys. **68**(1), 26–35 (2008)

Benedetto, A., Benedetto, F.: Remote sensing of soil moisture content by GPR signal processing in the frequency domain. IEEE Sens. J. **11**(10), 2432–2441 (2011)

Benedetto, F., Tosti, F.: GPR spectral analysis for clay content evaluation by the frequency shift method. J. Appl. Geophys. **97**, 87–96 (2013)

Benedetto, A., Benedetto, F., De Blasiis, M.R., Giunta, G.: Reliability of signal processing technique for pavement damages detection and classification using ground penetrating radar. IEEE Sens. J. **5**(3), 471–480 (2005)

Berkhout, A.J.: Least-squares inverse filtering and wavelet deconvolution. Geophysics **42**, 1369–1383 (1977)

Booth, A., Linford, N., Clark, R., Murray, T.: Three-dimensional, multi-offset ground-penetrating radar imaging of archaeological targets. Arch. Prosp. **15**, 93–112 (2008)

Bradford, J.H.: Applying reflection tomography in the post-migration domain to multi-fold ground-penetrating radar data. Geophysics **71**, K1–K8 (2006)

Bradford, J.H.: Measuring lateral and vertical heterogeneity in vadose zone water content using multi-fold GPR with reflection tomography. Vadose Zone J. **7**, 184–193 (2008)

Bradford, J., Wu, Y.: Instantaneous spectral analysis: time-frequency mapping via wavelet matching with application to contaminated-site characterization by 3D GPR. Lead. Edge **26**, 1018–1023 (2007)

Brooke, A., Maillol, J.: Multi-offset ground penetrating radar data for improved imaging in areas of lateral complexity—application at a native American site. J. Appl. Geophys. **62**, 167–277 (2007)

Brown, J., Nichols, J., Steinbronn, L., Bradford, J.: Improved GPR interpretation through resolution of lateral velocity heterogeneity: example from an archaeological site investigation. J. Appl. Geophys. **68**, 3–8 (2009)

Carpentier, S.F.A., Horstmeyer, H., Green, A.G., Doetsch, J., Coscia, I.: Semiautomated suppression of above-surface diffractions in GPR data. Geophysics **75**, J43–J50 (2010)

Cassidy, N.J.: Ground penetrating radar data processing, modelling and analysis. In: Jol, H.M. (ed.) Ground Penetrating Radar: Theory and Applications, pp. 141–176. Elsevier, Sydney (2009). ISBN 978-0-444-53348-7

Chahine, K., Baltazarta, V., Wang, Y., Déroberta, X.: Blind deconvolution via sparsity maximization applied to GPR data. Eur. J. Environ. Civ. Eng. **15**(4), 575–586 (2011)

Di, Q.Y., Wang, M.Y.: Migration of ground-penetrating radar data method with a finite element and dispersion. Geophysics **69**(2), 472–477 (2004)

Diamanti, N., Redman, D., Field observations and numerical models of GPR response from vertical pavement cracks. J. Appl. Geophys. (2012, in press)

Economou, N., Vafidis, A., Spanoudakis, N.S., Hamdan, H.A., Niniou-Kindeli, V.: Application of classification methods on geophysical data from the archaeological site of Aptera, Chania, Greece. In: Near Surface Geophysics. EAGE, Istanbul (2007)
Economou, N.: Development of GPR data processing techniques using S-transform. Ph.D dissertation at the School of Mineral Resources Engineering, Environmental Geotechnology Post-Graduate Program (2010)
Economou, N., Vafidis, A.: Spectral balancing GPR data using time variant band-width in t-f domain. Geophysics **75**(3), J19–J27 (2010)
Economou, N., Vafidis, A.: Deterministic deconvolution for GPR data in t-f domain. Near Surf. Geophys. **9**(5), 427–433 (2011)
Economou, N., Vafidis, A.: GPR data time varying deconvolution by kurtosis maximization. J. Appl. Geophys. **81**, 117–121 (2012)
Economou, N., Vafidis, A., Hamdan, H., Kritikakis, G., Andronikidis, N., Dimitriadis, K.: Time varying deconvolution of GPR data in civil engineering. Nondestr. Test. Eval. **27**(3), 285–292 (2012)
Flandrin, P., Rilling, G., Goncalves, P.: Empirical mode decomposition as a filter bank. IEEE Sign. Proc. Lett. **11**, 112–114 (2005)
Gao, D.: Application of seismic texture model regression to seismic facies characterization and interpretation. Lead. Edge **27**, 394–397 (2008)
Gray, S., Etgen, J., Dellinger, J., Whitmore, D.: Seismic migration problems and solutions. Geophysics **66**(5), 1622–1640 (2001)
Huang, N.E., Shen, Z., Long, S.R., Wu, M.L., Shih, H.H., Zheng, Q., Yen, N.C., Tung, C.C., Liu, H.H.: The empirical mode decomposition and Hilbert spectrum for nonlinear and nonstationary time series analysis. Roy. Soc. Lond. Ser. A Math. Phys. Eng. Sci. **454**, 903–995 (1998)
Ihamouten, A., Chahine, K., Baltazart, V., Villain, G., Dérobert, X.: On the variants of Jonscher's model for the electromagnetic characterization of concrete. IEEE Trans. Instr. Meas. **60**(11), 3658–3668 (2011)
Irving, J.D., Knight, R.J.: Removal of wavelet dispersion from ground-penetrating radar data. Geophysics **68**, 960–970 (2003)
Jeng, Y., Lin, C.H., Li, Y.W., Chen, C.S., Yu, H.M.: Application of sub-image multiresolution analysis of ground penetrating radar data in a study of shallow structure. J. Appl. Geophys. **73**, 251–260 (2011)
Jol, H.M.: Ground Penetrating Radar: Theory and Applications. Elsevier Science, Sydney (2009). ISBN 978-0-444-53348-7
Kadioglu, S.: Photographing layer thicknesses and discontinuities in a marble quarry with 3D GPR visualization. J. Appl. Geophys. **64**, 109–114 (2008)
Kadioglu, S.: Definition of buried archaeological remains with a new 3D visualization technique of ground penetrating radar data set in temple Augustus in Ankara-Turkey. Near Surf. Geophys. Spec. Issue GPR Archaeol. **8**(5), 397–406 (2010)
Kadioglu, S.: Transparent 2d/3d half bird's-eye view of ground penetrating radar data set in archaeology and cultural heritage, chapter 5. In: Kharfi, F. (ed.) Imaging and Radioanalytical Techniques in Interdisciplinary Research-Fundamentals and Cutting Edge Applications. InTech, Croatia (2013)
Kadioglu, S., Daniels, J.J.: 3D visualization of integrated GPR data and EM-61 data to determine buried objects and their characteristics. J. Geophys. Eng. **5**, 448–456 (2008)
Kim, J.H., Cho, S.J., Yi, M.J.: Removal of ringing noise in GPR data by signal processing. J. Geosci. **11**, 75–81 (2007)
Lahouar, S.: Development of data analysis algorithms for interpretation of ground penetrating radar data. Ph.D. Dissertation, Virginia Polytechnic Institute and State University, Blacksburg, VA (2003)
Le Bastard, C., Baltazart, V., Wang, Y., Saillard, J.: Thin pavement thickness estimation using GPR with high and super resolution methods. IEEE Trans. Geosci. Rem. Sens. **45**(8), 2511–2519 (2007)

LeBastard, C., Wang, Y., Baltazart, V., Dérobert, X.: Time delay and permittivity estimation by ground penetrating radar with support vector regression. IEEE Geos. Rem. Sens. Lett. **11**(4), 349–353 (2013)

Levy, S., Oldenburg, D.W.: The deconvolution of phase shifted wavelets. Geophysics **47**, 1285–1294 (1987)

Li, L., Tan, A.E.-C., Jhamb, K., Rambabu, K.: Buried object characterization using ultra-wide-band ground penetrating radar. IEEE Trans. Microw. Theory Tech. **60**(8), 2654–2664 (2012)

Liu, G., Fomel, S., Jin, L., Chen, X.: Seismic data stacking using local correlation. Geophysics **74**, V43–V48 (2009)

Lizzi, L., Viani, F., Rocca, P., Oliveri, G., Benedetti, M., Massa, A.: Three-dimensional real-time localization of subsurface objects—from theory to experimental validation. IEEE Int. Geoscience and Remote Sensing Symp., vol. 2, pp. II-121–II-124 (2009)

Loeffler, O., Bano, M.: Ground penetrating radar measurements in a controlled Vadose Zone: influence of the water content. Vadose Zone J. **3**, 1082–1092 (2004)

Longbottom, J., Walden, A.T., White, R.E.: Principles and application of maximum kurtosis phase estimation. Geophys. Prospect. **36**, 115–138 (1998)

Manica, L., Rocca, P., Salucci, M., Carlin, M., Massa, A.: Scattering data inversion through interval analysis under Rytov approximation. 7th European Conf. on Antennas Propag., Gothenburg, Sweden, 2013

Massa, A., Boni, A., Donelli, M.: A classification approach based on SVM for electromagnetic subsurface sensing. IEEE Trans. Geosci. Remote Sens. **43**(9), 2084–2093 (2005)

McClymont, A.F., Green, A.G., Streich, R., Horstmeyer, H., Tronicke, J., Nobes, D.C., Pettinga, J., Campbell, J., Langridge, R.: Visualization of active faults using geometric attributes of 3D GPR data: an example from the Alpine Fault Zone, New Zealand. Geophysics **73**, 11–23 (2008)

Neto, P., Medeiros, W.: A practical approach to correct attenuation effects in GPR data. J. Appl. Geophys. **59**, 140–151 (2006)

Nuzzo, L., Quarta, T.: Improvement in GPR coherent noise attenuation using t-p and wavelet transforms. Geophysics **69**(3), 789–802 (2004)

Oliveri, G., Rocca, P., Massa, A.: A Bayesian compressive sampling based inversion for imaging sparse scatterers. IEEE Trans. Geosci. Remote Sens. **49**(10), 3993–4006 (2011)

Pipan, M., Baradello, L., Forte, E., Prizzon, A., Finetti, I.: 2-D and 3-D processing and interpretation of multi-fold ground penetrating radar data: a case history from an archaeological site. J. Appl. Geophys. **41**, 271–292 (1999)

Pipan, M., Baradello, L., Forte, E., Finetti, I.: Ground penetrating radar study of iron age tombs in southeastern Kazakhstan. Arch. Prosp. **8**, 141–155 (2001)

Pipan, M., Forte, E., Dal Moro, G., Sugan, M., Finetti, I.: Multifold ground-penetrating radar and resistivity to study the stratigraphy of shallow unconsolidated sediments. Lead. Edge **22**, 876–880 (2003)

Poli, L., Oliveri, G., Rocca, P., Massa, A.: Bayesian compressive sensing approaches for the reconstruction of two-dimensional sparse scatterers under TE illumination. IEEE Trans. Geosci. Remote Sens. **51**(5), 2920–2936 (2013)

Saarenketo, T., Scullion, T.: Road evaluation with ground penetrating radar. J. Appl. Geophys. **43**, 119–138 (2000)

Salucci, M., Sartori, D., Anselmi, N., Randazzo, A., Oliveri, G., Massa, A.: Imaging buried objects within the second-order born approximation through a multiresolution regularized inexact-newton method, Electromagnetic Theory EMTS), Proceedings of 2013 URSI International Symposium on, 116–118 (2013).

Schimmel, M., Gallart, J.: Frequency-dependent phase coherence for noise suppression in seismic array data. J. Geophys. Res. **112**, B04303 (2007)

Schmelzbach, C., Scherbaum, F., Tronicke, J., Dietrich, P.: Bayesian frequency-domain blind deconvolution of ground-penetrating radar data. J. Appl. Geophys. **75**(4), 615–630 (2011)

Sena, A., Stoffa, P., Sen, M.: Split-step Fourier migration of GPR data in lossy media. Geophysics **71**, K77–K91 (2006)
Shao, W., Bouzerdoum, A., Lam Phung, S., Lijun, S., Indraratna, B., Rujikiatkamjorn, C.: Automatic classification of ground penetrating radar signals for railway ballast assessment. IEEE Trans. Geosci. Remote Sens. **49**(10), 3961–3972 (2011)
Spanoudakis, N.,Vafidis, A.: GPR data interpretation using self organizing maps. In: Near Surface Geophysics. EAGE, Istanbul (2007)
Stockwell, R.G., Mansinha, L., Lowe, R.P.: Localization of the complex spectrum: the S-transform. IEEE Trans. Sign. Proc. **44**, 998–1001 (1996)
Taner, T., Luo, Y., Kelamis, P., Kellogg, S., Craigie, E.:Frequency domain smoothing for enhanced seismic resolution, SEG Expanded Abstracts (2003)
Todoeschuck, J.P., LaFleche, P.T., Jensen, O.G., Judge, A.S., Pilon, J.A.: Deconvolution of ground probing radar data. In: Pilon, J. (ed.) Ground Penetrating Radar, Geological Survey of Canada, pp. 227– 230 (1992)
Turner, G.: Subsurface radar propagation deconvolution. Geophysics **59**, 215–223 (1994)
Vafidis, A., Manoutsoglou, M., Hamdan, H., Andronikidis, N., Koukadaki, M., Kritikakis, G., Oikonomou, N., Spanoudakis, N.: Geophysical survey at the Omalos plateau, Chania, Crete. In Proc. of the 10th Int. Congress Bulletin of the Geological Society of Greece, vol. XXXVI, pp. 1204–1213. Thessaloniki, 2004
Vafidis, A., Andronikidis, N., Economou, N., Panagopoulos, G., Zelilidis, A., Manoutsoglou, E.: Reprocessing and interpretation of seismic reflection data at Messara Basin, Crete, Greece. J. Balkan Geophys. Soc. **15**(2), 31–40 (2012)
Van der Baan, M.: Time-varying wavelet estimation and deconvolution by kurtosis maximization. Geophysics **73**, V11–V18 (2008)
Van der Baan, M.: Bandwidth enhancement: inverse Q filtering or time-varying Wiener deconvolution? Geophysics **77**(4), V133–V142 (2012)
White, R.E.: Maximum kurtosis phase correction. Geophys. Int. J. **95**, 371–389 (1988)
Xia, J., Franseen, E.K., Miller, R.D., Weis, T.V.: Application of deterministic deconvolution of ground-penetrating radar data in a study of carbonate strata. J. Appl. Geophys. **56**, 213–229 (2004)
Zhao, W., Tian, G., Wang, B., Shi, Z., Lin, J.: Application of 3D GPR attribute technology in archaeological investigations. Appl. Geophys. **9**(3), 261–269 (2012)
Zheng, Y.F., Li, Z., Zhou, L.C., Lv, D., Ying, M., Men, Y.: Increase the accuracy of GPR in tunnel detection. Adv. Mater. Res. **183–185**, 1529–1533 (2011)
Zhou, H., Wan, X., Duan, R., Li, W.: Improved stolt migration algorithm for GPR imaging using segmentation velocity model. J. Comp. Inf. Syst. **7**(16), 5829–5836 (2011)

Part IV
Different Applications of GPR and Other NDT Technologies in CE

terrestrial radar, which later developed together with the airborne radar, becoming an effective military defensive system against enemy planes during World War II. Later, in the 1940s, an airplane incident ("LostSquadron", 1942), caused by not detecting an ice layer, aroused the interest of scientists for the application of the radar as a subsoil prospective method. After that, the method fell into disuse, and its popularity increased considerably since the 1970s. There were therefore carried out the first studies in Lunar, geological and glaciological investigations (Campbell and Orange 1974; Annan and Davis 1976) and then, the technique was introduced into the civil engineering field (1970–1980s). After that, the applications of radar have been broadened in a number of fields. A comprehensive history of development of GPR has been compiled by Daniels et al. (1988).

Since its origins, the GPR has been commonly used as a prospecting method for security applications. Humanitarian security includes mining and UXOs (unexploded ordnance), recognition of underground defensive spaces, human remains detection, as well as detecting and locating trapped people in disaster areas.

In the last decades, many countries have been affected by wars that have left the territory plagued by Explosive Remnants of War (EWR), like mines, cluster bombs and unexploded objects: Afghanistan, Angola, Burundi, Cambodia, Democratic Republic of Congo, Iran, Iraq, the former Yugoslavia, Pakistan, and Rwanda are only a few examples. Clearing terrains from ERWs, is a difficult and dangerous task, that is made harder by variability of environmental variables (e.g., earth composition and humidity, climate, and vegetation) and by modifications happening over time (displacement and covering of objects). Cost of clearance (typically 300–1000 € per item removed) is a limiting factor that prevents substantial progress (200,000 mines removed each year, compared to some 100 million on the ground). Such cost, as well as slowness and danger of clearing operations, is largely determined by the difficulty of locating buried ERW (Balsi et al. 2009).

Landmines are a humanitarian challenge because they indiscriminately kill and maim people because they are weapons that cannot distinguish between a soldier and a civilian, and they remain active for decades. As a result, most of the victims of mines are innocent men, women and children. Landmines are used in so many conflicts, in such large numbers and so indiscriminately that during war they are a cause of displacement, and after hostilities they endanger the lives of returnees and humanitarian aid workers, delay return and impede reintegration and reconstruction. Landmines are also being used in current conflicts to block humanitarian access. In this frame, the GPR technique supposes a promising method to assist in landmine clearance, as many authors propose (Gader et al. 2001; Daniels 2006).

It was in the mid-1970s that GPR appeared with archaeological purposes but the first contributions related to 3D GPR imaging were in the 1990s (van Deen and de Feijter 1992; Grasmueck and Green 1996). In fact, the 3D GPR surveying continuous to flourish during the last decades, and the technique has demonstrated its potential to produce more realistic images of the underground space, which allows not only for the location, but also for the 3D reconstruction of the buried targets (Leckebusch 2003; Lin et al. 2011; Novo et al. 2012a).

There are still a large number of remaining defensive underground constructions, mainly built during the World War I, World War II, and the Cold War, that could be reconstructed using 3D GPR methodologies. The most of them were initially constructed as defensive military installations, but then they served also to protect civilians from falling bombs and other attacks, as well as from natural catastrophes such in the case of tornadoes. Some examples of defensive spaces are: trench bunkers, blockhouses, artillery installations, underground tunnel-linked (as the case of the Maginot Line in France), and air-raid shelters, or bomb shelters, as protection from aerial bombardments (as in the case of the Spanish Civil War). In UK, for example, the subways were extensively used as bomb shelters during the World War II.

Other applications of GPR for rescue, surveillance and security operations are focused on searching and rescuing human remains and trapped people in disaster areas. During the last decades, government and law enforcement agencies have promote the use of GPR in forensic investigations. Forensic anthropology includes the study of war crime and homicide victims, as well as mass disasters. Since the 1980s, GPR has been successfully applied to assist in forensic investigations, and interesting information can be provided about the dimensions and limits of graves, as well as to detect buried bodies or bone remains (Freeland et al. 2003; Schultz and Martin 2011). These results allow saving time-consuming and efforts since smaller areas are highlighted for further testing with traditional searching methods such as excavations. In addition, the technique is non-intrusive, preserving the scene and avoiding the destruction of forensic evidences that allows also reconstructing the events.

Disasters like earthquakes, floods, avalanches, explosions or similar occurrences result in structural collapses and then, probable trapped people under building debris, mudflows or avalanche snow. If people are buried alive, time for rescue is critical, and the GPR technology has shown its effectiveness to locate victims within minutes. The location of live victims hidden by walls or rubble was possible with GPR by detecting motion or even breathing and heartbeat when person is unconscious (Sachs et al. 2008).

In the next Sections, some of the published works in the frame of the mentioned security applications and field operations have been compiled to show the potential of the GPR. The purpose is to demonstrate how the method significantly benefits the procedures for inspection and also, successfully solved some of the limitations of traditional methods. On the other hand, the most important limitations of the technique are also presented in this Chapter.

2 Landmine and UXO Detection

Mine detection is an ongoing concern that lasts in time because of the danger they represent for civil population. This is a real problem that involves millions of human beings all around the world. Due to the long lifetime of these artefacts, the victims of landmines often have no relation with the original motivation for which they were settled. Nowadays, there still remain millions of anti-personnel mines in different countries around the world (Table 1). These mines kill or wound someone every 20 min, which supposes 70 people per day and more than 20,000 people per year (Kowalenko 2004). Currently, in the arsenals of Governments and under control of armed groups, there are hundreds of million mines in use (Banks 1997). The cost of producing one single anti-personnel mine is not elevated and it can be rapidly placed in large quantities by unqualified personnel. On the contrary, demining must be carried out by highly qualified personnel, and results considerably more expensive. United Nations (UN) statistics indicate the decease of almost two deminers every 1000 removed mines (Joynt 1998). All these negative statistics could be decreased in favour of human life by using an effective method to detect and remove landmines and unexploded ordnance (UXO), which still remain in minefields and former battle camps over the world.

As different authors recommend (Daniels 2006; Tesfamariam and Mali 2012), the ideal landmine detection equipment has to be designed to work in different environmental conditions such as, in a wide temperature range −20 to 60 °C, from arid deserts to overgrown jungle, from rocks to swampy areas, from plain roads to sloppy hill sides, rain, dust, humidity; at different soil types such as sand, clay soils, magnetic soils, loom soils, saline soils etc. All the above mentioned environmental conditions should be considered in the design of a landmine detector (Bruschini et al. 1998). According to that, many methods and combinations thereof have been tried for the detection of mines. These include: Acoustic Sensors, Infrared Imaging Systems, Nuclear Magnetic Resonance (NMR), Nuclear Quadrupole Resonance (NQR), X-Ray Backscatter, Microwave Techniques and Biological Detector (dogs, pigs and rats). Despite of disposing of all that amount of technology, the most usual techniques applied for mine detection are manual, concretely prodding. It consists of using a rigid stick (prodder) as unique supporting tool to scan the soil at a shallow angle of typically 30°. Though effective, this technique is slow and hugely dangerous for deminers. Besides prodding, another close-in hand-held technology currently used for landmine and unexploded ordnance detection is the metal detector (MD). The basic MD used for mine detection, based on the phenomenon of

Table 1 Estimated world distribution of landmines and unexploded ordnance (UXO)

Country	Landmines	UXO
Afghanistan	4 million	Large
Bosnia and Herzegovina	1 million	Large
Cambodia	300,000–1 million	2.5 million
Croatia	1–1.2 million	0
Ecuador	50,000–60,000	Small
Egypt	5–7.5 million	15–15.5 million
Ethiopia	1.5–2 million	Large
Vietnam	3.5 million	Large
Zimbabwe	2.5 million	Unknown

Source United States commitment to humanitarian demining

Electro Magnetic Interference (EMI), measures the disturbance of an emitted electromagnetic field caused by the presence of metallic objects in the soil. The major problem is that many modern mines are made of plastic and contain little or have no metal parts, except for the small striker pin. In addition, there is another handicap concerning this detection technique. The usual presence of metal waste all over the minefields may lead to an increase of the False Alarm Rate (FAR), therefore, supposes an evident decrease of efficiency of the method.

The GPR method, which has proven its suitability for subsoil prospection, is a well-recognized technique by the scientific community for mine and UXOs detection (Gader et al. 2001; Sato et al. 2005; Daniels 2006; González-Huici 2012). In this sense, the GPR postulates as an ideal solution for mine clearance and unexploded ordnance removal. The technique suitability is based on its rapidity and safety as non-destructive technique (Maierhofer 2003) compared to other more invasive procedures, such as excavations or traditional tastings methods, which may be obviously dangerous. Moreover, the GPR technique has the ability to detect either, metallic or non-metallic landmines, as far as there is a difference in electromagnetic contrast between the ground and the target. It must be also taken into account that the performance of the signal processors depend on the target type and size, the soil environment, the moisture and the roughness of the ground surface. Furthermore, the detection performance of a signal processor depend on the type of casing of the target (plastic, wooden, metallic, ceramic) and on the amount of metal that the target. There is no signal processing technique with better performance regarding all these named conditions and for all kinds of targets than GPR technology (Daniels 2006).

Thus, the GPR System meets the criteria for the ideal humanitarian demining system (Capineri et al. 2008), namely:

1. Low cost construction.
2. Easy transportation.
3. Possibility of operation and interpretation by non-experts.
4. Rapid deployment in emergencies.
5. High probability of detection for landmine.
6. Low false alarm rate.
7. Identification of landmine type.
8. No degradation in performance in the rugged conditions of actual minefields.

An example of application of GPR in mine detection is presented in Fig. 1. An experimental minefield scenario was simulated within the dependences of the Spanish Naval Academy (Marín, Spain). The aim of the work was to calibrate the GPR signal (central frequencies ranging from 2.3 GHz to 500 MHz) to better detect landmines and UXOs, as well as to characterize the patterns of reflections produced. As described in Fig. 2, different types of landmines (anti-personnel and anti-tank mines, as well as mortar and hand grenades) were assumed in testing. The whole scenario was designed with the targets disposed at several depths and dispositions to analyse their influence in detection. Figure 1 illustrates the partial results obtained with a central frequency 2.3 GHz, in which all the targets were successfully detected.

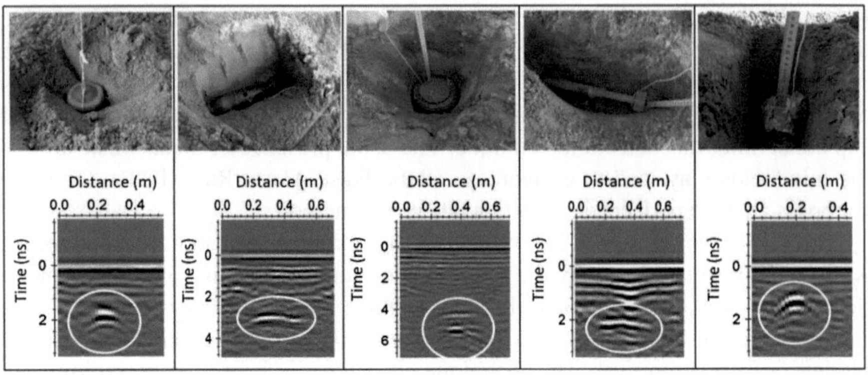

Fig. 1 GPR data acquired with a 2.3 GHz central frequency in an experimental minefield scenario, showing the pattern of reflection produced by each landmine or grenade (Data provided by the Applied Geotechnologies research group of the University of Vigo)

Object	Designation	Image	Dimensions	Casing	Depth
Landmine	AP-SB33		t = 3.23 cm, Ø = 8.46 cm	Plastic	10 cm
Mortar Grenade	GM-ECIA		L = 38 cm, Ø = 7.86 cm	Metal	20 cm
Anti-Tank Mine	AT-SB81	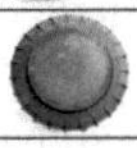	t = 10 cm, Ø = 22 cm	Plastic	25 cm
Mortar Grenade	INSTALAZA II M-63		L = 34.5 cm, Ø = 3.56 cm	Metal	15 cm
Hand Grenade	ALHAMBRA-EJ		L = 7.54 cm, Ø = 6.37 cm	Plastic	10 cm

Fig. 2 Description of the landmines and grenades in Fig. 1: designations, dimensions, casing materials and depths of burial

Although promising, this technology has limitations. In particular, the resolution needed to detect small objects involves GHz frequencies, which decreases soil penetration and increases image clutter. A major difficulty for operation of GPR systems is the presence of clutter within or on the surface of the material, or in the side and back lobes of the antenna and sources of surface clutter (Daniels 2006; Tesfamariam and Mali 2012). This has been modelled by Firoozabadi et al. (2007). Clutter is defined as sources of unwanted reflections that occur within the effective bandwidth and search window of the radar and are present as spatially coherent reflectors. Its presence can completely obscure the buried landmine and a proper understanding of its source and impact on the radar is essential.

The GPR sensors should not provide indications on the following sources of clutter (Tesfamariam and Mali 2012):

- Small metal fragments.
- Shrapnel.
- Spent bullet and cartridge cases.

- Ground Topographical variations.
- Puddles of water.
- Tufts of grass.
- Rocks and stones.
- Animal borrows.
- Cracks and fissures in ground.

The advanced available signal processing techniques may be used to reduce the clutter components, in order to reduce the false alarm rates and increase the probability of detection. In this sense, Abujarad (2007) exposed different signal processing techniques to reduce clutter in GPR, namely: Singular Value Decomposition (SVD), Factor Analysis (FA), Principal Component Analysis (PCA), FastICA, Infomax, SOBI, JADE, Wavelet Transform and Wavelet Packets Combined with Higher-Order-Satatistics. The comparison of these techniques was based on SNR and EPC curve. By using SNR as criterion to compare above techniques, it could be stated that the JADE based technique appears to have the best overall performance by a significant margin. Many authors are researching on the ways to improve the detection capability of GPR and improve the efficiency of landmine clearance process, Zoubir et al. (2002) used many techniques, giving emphasis to Kalman filtering for clutter reduction, Barakat et al. (2000) and Sun and Li (2003) have used time-frequency analysis techniques for plastic landmine detection, Carevic (2000) used wavelet packet decomposition techniques, Carevic (1999) applied Kalman filtering approach, Van der Merwe and Gupta (2000) used novel signal processing techniques for clutter reduction in GPR measurements. Sezgin (2011) used fuzzy based techniques for background calculation and target energy moment calculation to suppress the clutter. The above mentioned techniques have shown that the performance of GPR is improved after the clutter components are suppressed. However, some of the techniques such as Kalman filtering approach are computationally expensive.

The influence of the homogeneity of the soil on GPR for landmine detection is another usual handicap that must be taken into account (Igel and Preetz 2009). In this frame, Takahashi et al. (2012) explain how landmine detection by GPR becomes challenging when soil is inhomogeneous. Soil inhomogeneity causes unwanted reflections (clutter) which disturb reflections from landmines and this scattering is governed by Mie scattering. Numerical simulation is currently being applied in some works, to facilitate the complex understanding of the electromagnetic waves propagation phenomena through the media. Thus, some authors (Balsi et al. 2009; Deiana and Anitori 2010; Nabelek and Ho 2013) use the Finite-Difference Time-Domain (FDTD) method to assist in the characterization of the reflections patterns produced in the real data.

As the interpretation of the 2D GPR images (radargrams) is not trivial, the use of 3D processing and visualization techniques in landmine detection has been widely increased (Ligthart et al. 2004). Using 3D visualization produces more realistic images of the underground space, improving location and allowing for the 3D reconstruction of the buried targets (Dyana et al. 2012). Therefore, 3D visualization provides better interpretation of the results (Eide and Hjelmstad 2004; Capineri et al. 2008).

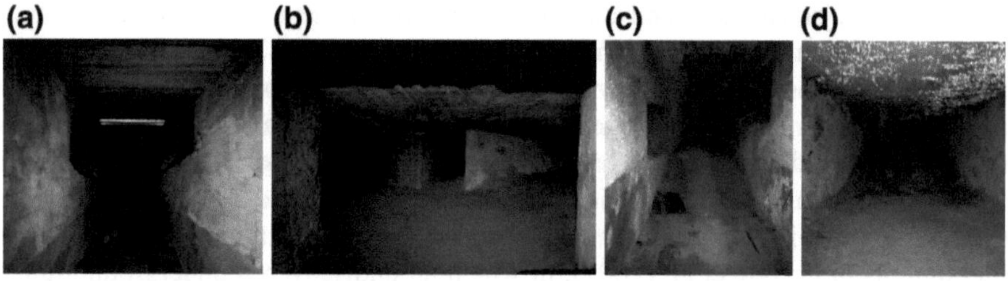

Fig. 4 Detailed images from the interior of the tunnel: **a** main corridor and lateral branches, **b** final trifurcation of galleries at the end of the main corridor, **c** ramp of the principal corridor, and **d** evidences of condensation inside one of the lateral branches

branches and the final trifurcation at the end of the main corridor (Fig. 4a, b). Although the tunnel is symmetric with respect to its central corridor (Fig. 4a), the lateral branches at one side of the central axis were not easily distinguished from the radar data, and only the one corresponding to the final trifurcation was partially detected. The most probable cause for this lack of information is due to an improper isolation of the tunnel lining, and subsequent attenuation of the radar-wave signal because of the presence of moisture as deduced from a visual inspection in the interior of the structure, where gravitational exudations were observed in a lateral branch (Fig. 4e).

The extraction of a 3D volume, which represents the entire tunnelling space, was possible by using the iso-surface rendering technique (Fig. 3b). It reproduced the hidden underground, showing an intuitive overview picture with an estimation of the dimensions of the target. This volumetric reconstruction visually confirms the existence of the two levels in the principal corridor of the tunnel (Fig. 4c). Moreover, it is clearly appreciated the gaps of data and consequent loss of information at the location of the existent ramp that separates these two levels (from 8 to approximately 13 m in the x-line of the iso-surface render).

Otherwise, precise positioning of the GPR antenna during data acquisition is crucial. The combination of GPR with Global Positioning Systems (GPS) provides accurate trace positioning in a global reference coordinate system. Grasmueck and Viggiano (2007) have integrated a novel Rotary Laser Positioning System (RLPS) technology with GPR into a highly efficient and simple to use 3D imaging system. A simpler but ingenious solution has been proposed by Lualdi and Zanzi (2003) who have invented the PSG (Pad System for Geo-radar). RTK-GPS was also implemented by other authors for positioning (Leckebusch 2005; Trinks et al. 2008).

The extra time needed for collecting 3D data makes it very costly for large areas. More recently, the use of new antenna arrays are able to cover areas of 2000 m^2 in only one hour preserving 3D dense acquisition (Trinks et al. 2008). Innovative GPR systems are based on massive antenna arrays which provide full-resolution imaging while speeding-up data acquisition (Linford et al. 2010; Trinks et al. 2010; Novo et al. 2012b). Current GPR equipment used for 3D data acquisition are Terravision (GSSI), RIS MF Hi-Mod (IDS), Stream X (IDS), MIRA array (Malå Geoscience), NogginPlus (Sensors and Software), Geoscope (3d-Radar), GroundVue (Utsi electronics), and Zond-12e (Radar Systems Inc.).

4 Human Remains and Life Detection

The main role of forensic anthropology traditionally has been to study and identify human remains, in single cases and in the study of war crime victims or mass disasters. Invariably, the search is expensive and time-consuming because of the variety of locations that need to be investigated. In recent decades, there has been an increase in the use of geophysics in forensic investigations for the narrowing down of sites to be excavated (France et al. 1992). As a result, investigators can then focus their efforts on searching and excavating smaller areas instead of digging up the entire scene. The advantages of using geophysical methods are the preservation of the crime scene, minimized destruction of forensic evidence, and a consequent possibility of reconstructing events at the scene (Bruschini et al. 1998; Schultz 2007; Pringle et al. 2008; Schultz 2008).

The GPR is a well-accepted technique for the detection of clandestine graves since it was first successfully utilized in 1986 by Vaughn (Schultz 2003). Despite the fact that GPR can help as part of a forensic investigation sequence (Ruffel et al. 2009), it has not been commonly standardized by national law enforcements because specialize training is necessary for acquiring, processing and interpreting the data. Although GPR can provide noteworthy information for forensics, the effectiveness of the techniques is highly dependent on environmental background, and there were some difficulties in terms of collection, analysis and interpretation of the GPR data in this context. In complex heterogeneous environments covered by trees, bushes, stumps and stones, data acquisition was a difficult task. The interpretation of GPR data can also be difficult when soils contain high concentrations of rocks, gravel and roots that result in false anomalies or clutter (Congram 2008; Novo et al. 2011). Other authors have noted that field soil properties greatly influence the appropriate transmission of the radar wave. The presence of conductive media, such as clay-rich subsoil, can attenuate the electromagnetic signal, and detecting underground features becomes difficult because the signal is not able to penetrate (Mellet 1992; Freeland et al. 2003; Schultz et al. 2006). The use of GPR in forensic applications is a quite recently implemented practice, as it has only been used since the 1990s (Daniels 2004), but in recent years, the amount of published works with successful results in this context has increased (Ruffel and McKinley 2005; Pringle et al. 2012a).

Pringle et al. (2012b) have evaluated the potential of the method for establishing relevant forensic search methodologies with the aim of identifying the optimal configuration. The central frequencies of 225, 450 and 900 MHz were evaluated in this work, and the GPR data showed 450 MHz frequency antenna to be optimal for burial target detection. Other authors, as Freeland et al. (2003), analysed the capabilities of the 400 and 900 MHz to assist in criminal investigations. They concluded that 400 MHz was appropriate to produce detailed data, discerning the grave walls and folded tarp covering the lower body. Schultz and Martin (2011) present the comparison of data produced by 500 and 250 MHz antenna for surveying a controlled grave scenario containing a pig carcass. The 500 MHz data provided more information within the grave, while the 250 MHz data displayed excellent imagery. In both cases, the detection of the grave was possible by a disruption of the soil horizon, which guarantees the location even when no response is obtained from the body. This work also demonstrated how the survey orientation may also affect detection. The pig carcass, as well as the floor of the grave, was detected when surveying in perpendicular direction.

The state of decomposition that includes complete skeletonisation, and subsequent compactness of the backfill, is another crucial factor for GPR analysis in forensic applications (Congram 2008). In Schultz et al. (2006), a 500 MHz antenna was used to monitor pig burials for durations of both 12 and 21 months in order to correlate the decomposition state of the cadaver with the data produced. Additionally, two different natural soil conditions were considered: burials in sand and burials in which cadavers were in contact with the upper surface of a clay horizon. As result, cadavers in sand were easily detected for the duration of the study (21 months), even when completely skeletonized. On the other hand, the cadavers in clay became increasingly difficult for identification over the first year of burial, even when they still retained extensive soft tissue structures.

The use of three-dimensional (3D) processing and visualization produces more realistic images of the underground space, which gives an absolutely accurate and intuitive display of the underground reality (Pringle et al. 2008). Nevertheless, 3D strategies for GPR are rarely applied in real forensic cases because they require special training for both data acquisition and 3D data processing. In addition, rough surfaces and time are often strong limitations in homicide investigations, and therefore the most of the surveys are still carried out in only two dimensions. Furthermore, these investigations are commonly involved in complex environments covered by trees, stumps and stones, which difficult data acquisition and produce a variety of non-target anomalies. Pringle et al. (2012c) conducted a GPR survey to systematically assess the changing geophysical response of simulated clandestine graves during the first 3 years after burial. The graves were excavated ranging at <1 m depth and two end-member scenarios were considered for this study: namely a naked and wrapped burial. Data was acquired at 3-monthly intervals after burial, and the central frequencies of 110, 225, 450 and 900 MHz were investigated to be used in forensic search cases effectively. 3D data acquisition and processing were also considered in this work. The results shown that GPR is not optimal to detect targets if there is an advanced state of decomposition, over 18 months. Although 3D data were generated to locate more subtle features that may be missed from the 2D data, a variety of non-target anomalies were also present in time-slices, making more difficult to locate the forensic targets.

In Novo et al. (2011), a forensic case study was carried out in a mountainous environment. Main objective was to locate a clandestine grave which is around

10–20 years old and contains human remains of one individual and a metallic tool, probably a pick. The use of GPR for detecting the skeletal remains of an individual buried in a clandestine grave many years ago presents a considerable challenge due to the small size of the remains in relation to the scale of the survey area, the low geophysical contrast of the body and the strong dependence on the soil effects on skeletonisation. The efforts of this study focused on detecting the ferrous part of the pick. Data were collected using 3D ultra-dense methodologies with the 250 MHz antenna to have a record of the reflection of the pick. Unfortunately, the presence of many metal rubbish items buried in the first 20–30 cm led to useless data due to many false-positive indicators and metal detectors became unreliable when metal objects were located deeper. Moreover, the 3D imaging results became unreliable in terms of distinguishing similar anomalies with the real target (Fig. 5).

More recently, there has also been research to numerically model the expected GPR responses from buried human remains. Hammon et al. (2000) employed Finite-difference time-domain (FDTD) simulations to model the GPR response for various body cross-sections with different depths of burial, soil types, soil moisture contents, survey frequencies and antenna separations. Accurate information was provided on the number, dimensions, locations and orientations of body elements. In addition, results revealed the attenuation in the surrounding soil and within the body (biological tissues) as the main limitation on image quality.

Solla et al. (2012) present the use of FDTD modelling and GPR signal characterization in forensics. Experimental scenes that mimic the most frequent real forensic cases were built by considering several buried objects: bone remains, clothes, active and inactive mobile phones, drug caches, guns and bullet shells, etc. Additionally, the geometric characterization of the scenes was made using photogrammetric methods. The 3D models of the experimental grids were provided,

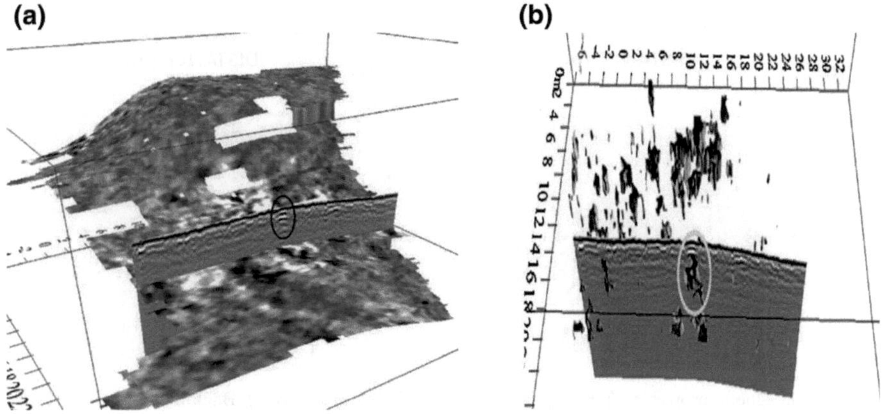

Fig. 5 **a** Topo corrected slice at 1 m depth and 250 MHz radargram, highlighting the presence of a metallic target that could correspond to the pick. *Blank areas* represent no data due to the presences of obstacles: trees, bushes and tree stumps (*small squares*). **b** Iso-volume (95 %) associated to the simulated grave containing the pick does not differ much from other anomalies present in the cube (Data available from Novo et al. 2011)

and the 2D sections obtained from these models were used as inputs in creating the synthetic models. Both synthetic and field data were compared to assist in the interpretation. As example, Fig. 6 illustrates the real scene (a) and the 2D ortho-image generated (b), as well as both synthetic and field data produced (c and d, respectively) for the case of the scene containing bone remains at 0.5 m depth. Although the field data showed more complex reflections produced by the heterogeneous backfill (Fig. 6d), the interpreted reflection patter produced by the bone remains is very similar to that modelled in the synthetic radargram (Fig. 6c).

A 3D FDTD modelling approach over a clandestine human burial target was performed by Millington et al. (2011). Based on the Born approximation, an innovative frequency-domain, micro-wave tomographic inversion technique was used that is based on the solution to the inverse-scattering problem of electromagnetic waves in a 2D plane. The results provided useful information on the true form, depth, location and spatial interrelationships of the buried features and, therefore, improved interpretations can be obtained in a 3D context.

The GPR method has been also developed for life detection as supporting tool in rescue missions. Wars and natural catastrophes result in structural collapses, and people could be trapped under building debris. In such circumstances, time is critical since an early detection can potentially reduce the mortality rate, so the

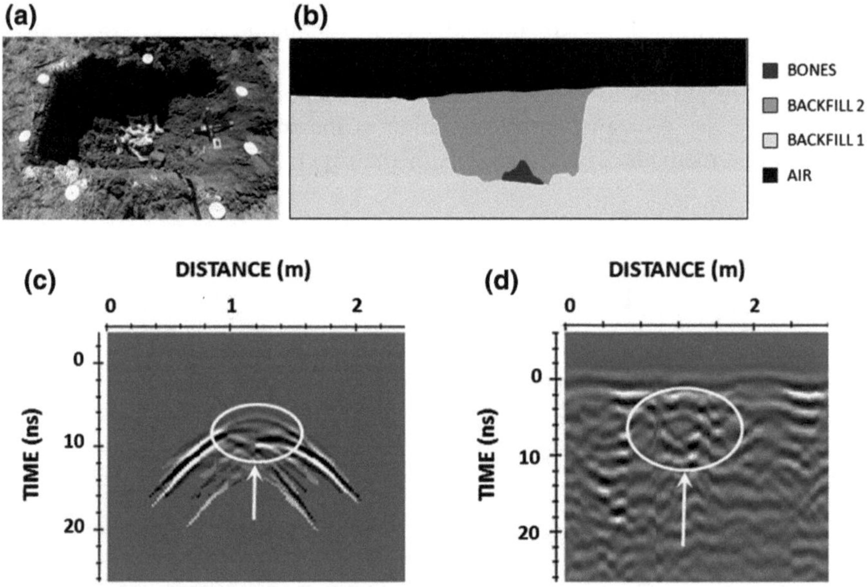

[Electromagnetic properties (permittivity and conductivity, respectively): Backfill 1: 0.033 S/m, 9.0; Backfill 2: 0.0033 S/m, 9.0; Bones: 0.13 S/m, 12.0]

Fig. 6 Experimental scene: bone remains at 0.5 m depth. **a** Real scene, **b** synthetic model built from the longitudinal section provided by photogrammetric methods, **c** synthetic data generated from FDTD modelling in GprMax software, and **d** 500 MHz field data collected (Data available from Solla et al. 2012) development of life detection systems is desirable. The effectiveness of the GPR for life location was demonstrated by detecting motion or vital signs like breathing and heartbeat. A review on GPR for the detection of buried or trapped victims in disaster scenarios, such as avalanches or earthquakes is presented in Crocco and Ferrara (2014).

Different works in the use of GPR for the detection of humans being trapped under avalanches are included in the specialised literature, as Jaedicke (2003) and Instantes et al. (2004). Searching for victims under avalanches is commonly carried out using the standard UHF GPR systems. Buried persons can be identified as a perturbation of the backscattered signal, but the method is not able to distinguish is the person is alive or not. As supported by Modroon and Olhoeft (2004), the electromagnetic waves have excellent propagation in snow and the detection of buried human bodies is possible due to high electrical contrast. The relative dielectric permittivity of snow varies from 1.2 to 12.0 depending on several factors, including the amount of moisture content, density, grain size and shape, temperature, and frequency. The average human body permittivity of 50 was determined using a weighted average of the five main components of the human body: skin, bone, blood, organs, and fat. The permittivity of the range of other materials that might be present in avalanche debris (rocks, trees, boots, etc.) ranges from 2 to 12. Results also showed proper imaging and identification of freezing bodies. The detection of vital signals and, subsequently, the location of human alive bodies was possible by using continuous-wave (CW) microwave transceiver systems (Pieraccini et al. 2008). In this work, operating with 2.4 GHz, both breathing and heartbeat frequencies have been detected through a 1.8 m-thick snow barrier.

Regarding the detection of victims trapped in debris, there are different techniques that allow searching survivors under rubble by measuring vital signals such as motion or even breathing and heartbeat. They are recognized by inducting a low-frequency perturbation of the backscattered radar signals. CW systems have been first used for detection of even a non-moving person, in which the vital signals cause an amplitude modulation in the phase of the radar signal (Doppler spectrum). It is possible using temporal and spectral characteristics of the detected signals (Lukin and Konovalov 2004). Nevertheless, the most of the published studies have considered the ultra-wide band (UWB) radar to be capable of detecting breathing and even heartbeat signals (Sachs et al. 2008; Zaikov et al. 2008; Wu et al. 2012). The use of UWB radar technology can guarantee higher time stability with respect to pulsed radar.

Scattering, or clutter, is the main inconvenience when surveying in complex environments, and the GPR signal might be enormously affected by inhomogeneous building debris mainly composed by rubble and metal fragments. To overcome this limitation, the specialised literature has published different algorithms that are used to remove background noise, or clutter, and to improve the recognition of vital signals. In Grazzini et al. (2010), an enhanced UWB, Continuous Wave Stepped Frequency (CW-SF), GPR radar has been tested for buried victims' detection. The CW-SF technique uses a signal generation process that enhance the overall time stability of the system, improving the capability of this technique to detect low-frequency movements. A real test site is presented, in which a volunteer was hidden inside a concrete pipe, with diameter about one meter, covered by a debris layer about one meter composed by chunks of walls, columns, rocks and bricks. The breathing frequency from the buried volunteer was identified in the B-scan after background removal, B-scan after horizontal pass-band filtering and range-doppler data processing. Another study (Cist 2009) has demonstrated the potential of the method to detect breathing frequencies under concrete slabs in a debris pile. Local cell phones were also considered to create background radio noise and reduce system sensitivity. Horizontal Fourier transform was used to look for these signs of life. A series of test with concrete slabs below 25 cm thicknesses, and from 1.5 to 4 m away, shown a detection success rate of about 90 %. The detection limit was found to be at 4 m through 1 m of rebar reinforced concrete. Similar results were found by other authors (Li et al. 2012a) when using a practical adaptive line enhancer (ALE) algorithm controlled by the least mean square (LMS) algorithm. The experimental results obtained in this work demonstrated that ALE method can monitor, with more than 90 % accuracy, the respiratory signal through a brick wall and with a distance of 5 m between the radar and the target. Successful results were also obtained using the discrete wavelet transform (DWT), which is a wavelet-based strong clutter removal technique for UWB life detection radar (Li et al. 2012b). This transform allows for the analysis of non-stationary signals like GPR signals and to remove strong clutters reflected by ruins.

There are different systems specifically developed to detect the movement of victims buried in soil, rock, rubble or snow. The Rescue Radar GPR sensor (Sensors and Software Inc. 2014) was designed for rapid deployment by search and rescue teams. This system is sensitive to small movements of arms, legs or even breathings. Exploration depth can vary from a meter or two in high-loss environments like rock, building materials and other debris to more than 10 m in highly transparent materials like snow. Other existing system is the LifeLocator® (Geophysical Survey Systems Inc. 2014), which uses the principals of ultra-wide band (UWB) GPR to detect movement in up to 10 m of debris and breathing in up to 8 m.

The most important in rescue missions is to detect buried victims being found alive for which time is crucial. In relation to this, another issue in this framework is the adoption of faster techniques to be used during emergencies. The use of array systems can reduce time in rescue because of the massive acquisition data they provide. An array antenna composed by five 1.2 GHz antennas was developed by Akiyama et al. (2007), which incorporate UWB pulse radar to detect moving and/or vital signals. Advances in the array system are shown in Takeuchi et al. (2008).

On the other hand, to reach inaccessible or dangerous areas, rescue radar systems were also designed and tested to operate mounted on-board of unmanned aerial vehicles (UAVs). Donelli (2011) presents a light-weight microwave system to locate victims trapped under rubble. The system, based on continuous wave X-band radar, enables to detect breathing and heartbeat signals. To remove noise and clutter, the independent component analysis (ICA) algorithm is used. A real test was performed, and the system detected and identified the life signals with a reasonable degree of accuracy. Studies were also carried out in such framework to develop a fast and reliable method to locate avalanche victims. In Heilig et al. (2008), a 400 MHz GPR system was mounted on an aerial railway system, approximately 6 m high, to simulate a helicopter flight. Water bags were used to simulate the victim, and the experiments were performed in both dry and moist snow. Although in the first case the target was easily detected until a maximum transversal distance of a victim at 1.5 m from the direction of the flight however, the detection was not possible when surveying in moist snow. For this approach, a real-time detection is presented in Fruehauf et al. (2009), showing an algorithm that automatically detects hyperbolas in radargrams. The localization algorithm yields a hit rate of 100 % without false detections.

5 Conclusion

The GPR method has shown its potential as a suitable tool for landmine and unexploded ordnance detection. Although promising, this technology has limitations, as the resolution needed to detect small objects that involves GHz frequencies, which decreases soil penetration and increases image clutter. Thinking about the future, the main challenge in landmine and UXO detection for hand-held radar, is the further reduction in the rate of false alarm. The current generation of dual sensor detectors reduces the ratio but is not still enough. There may be developed robust techniques that reduce the ratio down so that the efficiency improvement in humanitarian operations will be greater. Moreover, vehicle-based radar has to achieve orders of magnitude performance improvement to enable route clearing military operation to proceed at speed. Humanitarian clearance may tolerate speed reduction but still requires high detection rates. This applies to both stand off and close in GPR systems. Besides, airborne radar supposes an enormous technical challenge. However, a new generation of unmanned airborne vehicles may provide suitable platforms for the close in GPR systems if ground skimming can be achieved. This would allow reconnaissance vehicles to run ahead of convoys and would reduce the need to mine protect vehicles. The most successful developments in the field of landmine detection have been where GPR is used in conjunction with other sensors, primarily the metal detector (MD). The market for humanitarian mine detectors is fragmented and very cost sensitive and will be difficult to develop, but it is to be hoped that parallel developments and procurements for military applications will result in products being used to meet the humanitarian need.

On the other hand, there has been an increase in the use of non-destructive testing technologies for prospecting investigations, reflecting security and societal concerns for the conservation and maintenance of underground spaces. Among these techniques, the GPR method has been quite commonly applied for high-resolution imaging in many archaeological and underground applications. GPR is an effective tool to locate shallow man-made constructions, or cavities, and their interpretations in terms of structural shapes, dimensions, estimation of building materials and fillings, as well as the analysis of their state of conservation. However, the understanding of GPR results among non-geophysicists has been a long term challenge because the interpretation of radargram images is not straightforward. In this frame, the use of 3D imaging techniques and processing software has produced more realistic images of the underground, which permits not only finding but also obtaining 3D reconstructions of buried objects. On the other side, the extra time needed for collecting 3D data makes it very costly for large areas. Otherwise, precise positioning of the GPR antenna during data acquisition presents as crucial in this field of work. The combination of GPR with Global Positioning Systems (GPS) provides accurate trace positioning in a global reference coordinate system.

Otherwise, there are several advantages of using geophysical methods for human remains detection applied in forensic anthropology, namely: the preservation of the crime scene, minimized destruction of forensic evidence, and a consequent possibility of reconstructing events at the scene. Despite the fact that GPR can help as part of a forensic investigation, it has not been commonly standardized by national law enforcements because specialize training is necessary for acquiring, processing and interpreting the data. Although GPR can provide noteworthy information for forensics, the effectiveness of the techniques is highly dependent on environmental background, and there may appear some difficulties in terms of collection, analysis and interpretation of the GPR data. In complex heterogeneous environments covered by trees, bushes, stumps and stones, data acquisition becomes a difficult task. The interpretation of GPR data can also be difficult when soils contain high concentrations of rocks, gravel and roots that result in false anomalies or clutter. The use of three-dimensional (3D) processing and visualization produces more realistic images of the underground space, which gives an accurate and intuitive display of the underground reality. Nevertheless, 3D strategies for GPR are rarely applied in real forensic cases because they require special training for both data acquisition and 3D data processing. In addition, rough surfaces and time are often strong limitations in homicide investigations, and therefore the most of the surveys are still carried out in only two dimensions. Furthermore, these investigations are commonly involved in complex environments covered by trees, stumps and stones, which difficult, as said before, data acquisition and produce a variety of non-target anomalies.

The GPR method has been also developed for life detection as supporting tool in rescue missions. War and natural catastrophes result in structural collapses, and people may be trapped under building debris. In such circumstances, time is critical since an early detection can potentially reduce the mortality rate, so the development of life detection systems is desirable. The effectiveness of the GPR for life location has been demonstrated by detecting motion or vital signs like breathing and heartbeat. Otherwise, scattering or clutter, is the main inconvenience when surveying in complex environments, and the GPR signal might be enormously affected by inhomogeneous building debris mainly composed by rubble and metal fragments. The most important issue in rescue missions is to detect buried victims that still are alive and for whom time is crucial. Accordingly, the use of faster techniques during such kind of emergencies results in a primary need. In this sense, the use of array systems can reduce the rescue time because of the massive acquisition data they provide. On the other hand, to reach inaccessible or dangerous areas, rescue radar systems are also designed and tested to operate mounted onboard of unmanned aerial vehicles (UAVs).

Finally, we can conclude that GPR has demonstrated its potential as a promising technique for humanitarian assistance and security applications. The main problem related to this technology is that the returns from shallow buried objects can be obscured by ground returns and presence of clutter. Therefore, further investigation is required focused on advanced signal processing techniques to suppress the clutter components in order to decrease the false alarm ratio.

Acknowledgments The authors acknowledge the COST Action TU1208 "Civil Engineering Applications of Ground Penetrating Radar", supporting this work.

References

Abujarad, F., Omar, A.S.: Factor and principle component analysis for automatic landmine detection based on ground penetrating radar. In: Proceedings of German Microwave Conference (GeMIC'06), Karlsruhe, Mar 2006

Abujarad, F: Ground penetrating radar signal processing for landmine detection. Dissertation, Otto von Guericke University Magdeburg (2007)

Akiyama, I., Ohya, A., Aoki, Y., Matsuno, F.: Search for survivors buried in rubble by rescue radar with array antennas—extraction of respiratory fluctuation. In: Proceedings of the 2007 IEEE International Workshop on Safety, Security and Rescue Robotics, Rome, 27–29 Sept 2007

Annan, P., Davis, J.: Impulse radar sounding in permafrost. Radio Sci. **11**, 383–394 (1976)

Balsi, M., Esposito, S., Frezza, F., Nocito, P., Pajewski, L., Porrini, L., Schettini, G., Twizere, C.: FDTD simulation of GPR measurements in a laboratory sandbox for landmine detection. In: Proceedings of the 5th International Workshop on Advanced Ground Penetrating Radar (IWAGPR'09), Granada, 27–29 May 2009

Banks, E.: Brassey's Essential Guide to Anti-Personnel Landmines; Recognising and Disarming. Brassey's, London (1997)

Barkat, B., Zoubir, A.M., Brown, C.L.: Application of time-frequency techniques for the detection of anti-personnel landmines. In: Proceedings of the 10th IEEE Workshop on Statistical Signal and Array, Pocono Manor, 14–16 Aug 2000

Bruschini, C., Gros, B., Guerne, F., Pièce, P.Y., Carmona, O.: Ground penetrating radar and imaging radar detector for antipersonnel mine detection. J. Appl. Geophys. **40**, 59–71 (1998)

Campbell, K., Orange, A.: A continuous profile of sea ice and freshwater ice thickness by impulse radar. Polar Record **17**(106), 31–41 (1974)

Capineri, L., Ivashov, S., Bechtel, T., Zhuravlev, A., Falorni, P., Windsor, C., Borgioli, G., Vasiliev, I., Sheyko, A.: Comparison of GPR sensor types for landmine detection and classification. In: Proceedings of the 12th International Conference on Ground Penetrating Radar, University of Birmingham, Birmingham, 16–19 June 2008

Carevic, D.: A Kalman Filter-Based Approach to Target Detection and Target-Background Separation in Ground Penetrating Radar Data. DSTO Electronics and Surveillance Research Laboratory, Salisbury (1999)

Carevic, D.: Clutter reduction and detection of mine-like objects in ground penetrating radar data using wavelets. Subsurf. Sens. Technol. Appl. **1**(1), 101–118 (2000)

Chamberlain, A.T.: Cave detection in limestone using ground penetrating radar. J. Archaeol. Sci. **27**, 957–964 (2000)

Chan, Y.K., Koo, V.C.: An introduction to synthetic aperture radar (SAR). Prog. Electromagnet. Res. B **2**, 27–60 (2008)

Cist, D.B.: Non-destructive evaluation after destruction: using ground penetrating radar for search and rescue. In: Non-destructive Testing in Civil Engineering (NDTCE'09), Nantes, 30 June–3 July 2009

Congram, D.R.: A clandestine burial in Costa Rica: prospection and excavation. J. Forensic Sci. **53**(4), 793–796 (2008)

Crocco, L., Ferrara, V.: A review of ground penetrating radar technology for the detection of buried or trapped victims. In: Proceedings of the 2014 International Conference on Collaboration Technologies and Systems (CTS 2014), Minnesota, 19–23 May 2014

Curlander, J.C., McDounough, R.N.: Synthetic Aperture Radar, Systems and Signal Processing. Wiley, New York (1991)

Daniels, D.J.: Locating caves, tunnels and mines. Geophy. Lead. Edge Explor. **7**, 32–52 (1988)

Daniels, D.J.: Ground Penetrating Radar. The Institution of Electrical Engineering, London (2004)

Daniels, D.J.: A review of GPR for landmine detection. Sens. Imaging Int. J. **7**(3), 90–123 (2006)

Daniels, D.J., Curtis, P., Amin, R., Hunt, N.: MINEHOUND™ production development. In: Proceedings of Detection and Remediation Technologies for Mines and Minelike Targets X, vol. 5794, pp. 488–494 (2005)

Daniels, D.J., Gunton, D.J., Scott, H.F.: Introduction to subsurface radar. Radar Sig. Proc. IEE Proc. F **135**(4), 278–320 (1988)

Deiana, D., Anitori, L.: Detection and classification of landmines using AR modeling of GPR data. In: Proceedings of the 13th International Conference on Ground Penetrating Radar, Lecce, 21–25 June 2010

Doheny, R., Burke, S., Cresci, R., Ngan, P., Walls, R.: Handheld standoff mine detection system (HSTAMIDS) field evaluation in Thailand. In: Proceedings of Detection and Remediation Technologies for Mines and Minelike Targets X, vol. 5794, pp. 889–900 (2005)

Donelli, M.: A rescue radar system for the detection of victims trapped under rubble based on the independent component analysis algorithm. Prog. Electromagnet. Res. M **19**, 173–181 (2011)

Dyana, A., Rao, C.H., Kuloor, R.: 3D segmentation of ground penetrating radar data for landmine detection. In: Proceedings of the 14th International Conference on Ground Penetrating Radar, Shanghai, 4–8 June 2012

Eide, E., Hjelmstad, J.: UXO and landmine detection using 3-dimensional ground penetrating radar system in a network centric environment. In: Proceedings of ISTMP 2004 (2004)

El-Qady, G., Hafer, M., Abdalla, M.A., Ushijima, K.: Imaging subsurface cavities using geoelectric tomography and ground-penetrating radar. J. Cave Karst Stud. **67**(3), 174–181 (2005)

Feng, X., Fujiwara, J., Zhou, Z., Kobayashi, T., Sato, M.: Imaging algorithm of a hand-held GPR MD sensor (ALIS). In: Proceedings of Detection and Remediation Technologies for Mines and Minelike Targets X, SPIE, vol. 5794, pp. 1192–1199 (2005)

Firoozabadi, R., Miller, E.L., Rappaport, C.M., Morgenthaler, A.W.: Sub-surface sensing of buried objects under a randomly rough surface using scattered electromagnetic field data. IEEE Trans. Geosci. Remote Sens. **45**(1), 93–104 (2007)

France, D.L., Griffin, T.J., Swanburg, J.G., Lindemann, J.W., Davenport, G.C., Trammell, V., et al.: A multidisciplinary approach to the detection of clandestine graves. J. Forensic Sci. **37**(6), 1445–1458 (1992)

Freeland, R.S., Miller, M.L., Yoder, R.E., Koppenjan, S.K.: Forensic application of FMCW and pulse radar. J. Environ. Eng. Geophys. **8**(2), 97–103 (2003)

Frigui, H., Ho, K.C., Gader, P.: Real-time landmine detection with ground-penetrating radar using discriminative and adaptive hidden Markov models. EURASIP J. Adv. Sig. Process. **2005**, 419248 (2005)

Fruehauf, F., Heilig, A., Schneebeli, M., Fellin, W., Scherzer, O.: Experiments and algorithms to detect snow avalanche victims using airborne ground-penetrating radar. IEEE Trans. Geosci. Remote Sens. **47**(7), 2240–2251 (2009)

Gader, P.D., Mystkowski, M., Zhao, Y.: Landmine detection with ground penetrating radar using hidden Markov models. IEEE Trans. Geosci. Remote Sens. **36**(6), 1231–1244 (2001)

Gao, P., Collins, L.: A two-dimensional generalized likelihood ratio test for land mine and small unexploded ordnance detection. Sig. Process. **80**, 1669–1686 (2000)

Garreau, P., Cottard, G., Berthaud, P., Beaumont, E., Bolomey, J-Ch: Potentials of microwave tomographic imaging for on line detection of land mines. In: Proceedings of the EUREL International Conference on the Detection of Abandoned Land Mines, Edinburgh, 7–9 Oct 1996

Geophysical Survey Systems Inc.: LifeLocator®. http://www.gssilifelocator.com/. Accessed 24 May 2014

González-Huici, M.A.: Strategy for landmine detection and recognition using simulated GPR responses. In: Proceedings of the 14th International Conference on Ground Penetrating Radar, Tongji University, Shanghai, 4–8 June 2012

Grasmueck, M., Green, A.G.: 3-D georadar mapping: looking into the subsurface. Environ. Eng. Geosci. **2**(2), 195–200 (1996)

Grasmueck, M., Viggiano, D.A.: Integration of ground-penetrating radar and laser positioning sensors for real-time 3-D data fusion. IEEE Trans. Geosci. Remote Sens. **45**(1), 130–137 (2007)

Grazzini, G., Pieraccini, M., Parrini, F., Spinetti, A., Macaluso, G., Dei, D., Atzeni, C.: An ultra-wide band high-dynamic range GPR for detecting buried people after collapse of buildings. In: Proceedings of the 13th International Conference on Ground Penetrating Radar, Lecce, 21–25 June 2010

Hammon, W.S., McMechan, G.A., Zeng, X.: Forensic GPR: finite-difference simulations of responses from buried human remains. J. Appl. Geophys. **45**(3), 171–186 (2000)

Heilig, A., Schneebeli, M., Fellin, W.: Feasibility study of a system for airborne detection of avalanche victims with ground penetrating radar and a possible automatic location algorithm. Cold Reg. Sci. Technol. **51**(2/3), 178–190 (2008)

Igel, J., Preetz, H.: Small-scale variability of electromagnetic soil properties and their influence on landmine detection: How to measure, how to analyse, and how to interpret? In: Proceedings of SPIE Detection and Sensing of Mines, Explosive Objects, and Obscured Targets XIV, vol. 7303, p. 730312 (2009)

Instanes, A., Lønne, I., Sandaker, K.: Location of avalanche victims with ground penetrating radar. Cold Reg. Sci. Technol. **38**, 55–61 (2004)

Jaedicke, C.: Snow mass quantification and avalanches victim search by ground penetrating radar. Surv. Geophy. **24**(5/6), 431–445 (2003)

Jin, T., Zhou, Z.: Ultrawideband synthetic aperture radar lanmine detection. IEEE Trans. Geosci. Remote Sens. **45**(11), 3561–3573 (2007)

Jin, T., Zhou, Z.: Ultrawideband synthetic aperture radar unexploded ordnance detection. IEEE Trans. Aerosp. Electron. Syst. **46**(3), 1201–1213 (2010)

Joynt, V.P.: Mobile metal detection: a field perspective. In: Proceedings of the 2nd International Conference on the Detection of Abandoned Land Mines, Edinburgh, 12–14 Oct 1998

Kowalenko, K.: Saving lives, one land mine at a time. IEEE Inst. **28**, 10–11 (2004)

Lazăr, C., Ene, D., Parnic, V., Popovici, D.N., Florea, M.: Ground penetrating radar prospections in Romania. Măriuţa-la movilă necropolis, a case study. Mediterr. Archaeol. Archaeometry 11(2):79–89 (2011)

Leckebusch, J.: Ground-penetrating radar: a modern three-dimensional prospection method. Archaeol. Prospection **10**, 213–240 (2003)

Leckebusch, J.: Precision real-time positioning for fast geophysical prospection. Archaeol. Prospection **12**(3), 199–202 (2005)

Li, W., Jing, X., Li, Z., Wang, J.: A new algorithm for through wall human respiration monitoring using GPR. In: Proceedings of the 14th International Conference on Ground Penetrating Radar, Tongji University, Shanghai, 4–8 June 2012 (2012a)

Li, Z., Jing, X., Li, W., Wang, J.: A wavelet-based strong clutter removal technique for UWB life detection. In: Proceedings of the 14th International Conference on Ground Penetrating Radar, Tongji University, Shanghai, 4–8 June 2012 (2012b)
Ligthart, E.E., Yarovoy, A.G., Roth, F., Ligthart, L.P.: Landmine detection in high resolution 3-D GPR images. In: Proceedings of MIKON 2004
Lin, A., Novo, A., Har-Noy, S., Ricklin, N., Stamatiou, K.: Combining GeoEye-1 satellite remote sensing, UAV aerial imaging, and geophysical surveys in anomaly detection applied to archaeology. IEEE J. Sel. Top. Appl. Earth Observations Remote Sens. **4**, 870–876 (2011)
Linford, N., Linford, P., Martin, L., Payne, A.: Stepped frequency ground-penetrating radar survey with a multi-element array antenna: results from field application on archaeological sites. Archaeol. Prospection **17**, 187–198 (2010)
Lorenzo, H., Hernández, M.C., Cuéllar, V.: Selected radar images of man-made underground galleries. Archaeol. Prospection **9**(1), 1–7 (2002)
Lualdi, M., Zanzi, L.: 3D GPR investigations on building elements using the PSG. In: Proceedings of Symposium on the Application of Geophysics to Engineering and Environmental Problems (SAGEEP), San Antonio, Texas, 6–10 Apr 2003
Lukin, K., Konovalov, V.: Through wall detection and recognition of human beings using noise radar sensor. In: Proceedings of the RTO SET Symposium on Target Identification and Recognition Using RF Systems, Oslo, 11–13 Oct 2004
Maierhofer, C.: Nondestructive evaluation of concrete infrastructure with ground penetrating radar. J. Mater. Civ. Eng. **15**(3), 287–297 (2003)
Martinaud, M., Frappa, M., Chapoulie, R.: GPR signal for the understanding of the shape and filling of manmade underground masonry. In: Proceedings of the 10th International Conference on Ground Penetrating Radar, Delft University of Technology, Delft, 21–24 June 2004
Mellet, J.S.: Location of human remains with ground penetrating radar. In: Hanninen, P., Autio, S. (eds.) Proceedings of the 4th International Conference on Ground Penetrating Radar, Finland, 2012 (1992)
Millington, T.M., Cassidy, N.J., Nuzzo, L., Crocco, L., Soldovieri, F., Pringle, J.K.: Interpreting complex, three-dimensional, near-surface GPR surveys: an integrated modelling and inversion approach. Near Surf. Geophy. **9**, 297–304 (2011)
Modroo, J.J., Olhoeft, G.R.: Avalanche rescue using ground penetrating radar. In: Proceedings of the 10th International Conference on Ground Penetrating Radar, Delft University of Technology, Delft, 21–24 June 2004
Nabelek, D.P., Ho, K.C.: Detection of shallow buried objects using an autoregressive model on the ground penetrating radar signal. In: Proceedings of SPIE 8709, Detection and Sensing of Mines, Explosive Objects, and Obscured Targets XVIII, 87091I, 7 June 2013
Novo, A., Solla, M., Montero-Fenollós, J.L., Lorenzo, H.: Searching for the remains of an Early Bronze Age city at Tell Qubr Abual-'Atiq (Syria) through archaeological investigations and GPR imaging. J. Cult. Heritage (2013). http://dx.doi.org/10.1016/j.culher.2013.10.006
Novo, A., Lorenzo, H., Rial, F.I., Solla, M.: 3D GPR in forensics: finding a clandestine grave in a mountainous environment. Forensic Sci. Int. **204**, 134–138 (2011)
Novo, A., Lorenzo, H., Rial, F.I., Solla, M.: From pseudo-3D to full-resolution GPR imaging of a complex roman site. Near Surf. Geophy. **10**, 11–15 (2012a)
Novo, A., Dabas, M., Morelli, G.: The STREAM X multi-channel GPR system: first test at Vieil-Evreux (France) and comparison with other geophysical data. Archaeol. Prospection **19**(3), 179–189 (2012b)
Pérez-Gracia, V., Canas, J.A., Pujades, L.G., Clapes, J., Caselles, O., Garcia, F., Osorio, R.: GPR survey to confirm the location of ancient structures under the Valencian Cathedral (Spain). J. Appl. Geophys. **43**(2–4), 167–174 (2000)
Pieraccini, M., Luzi, G., Dei, D.: Detection of breathing and heartbeat through snow using a microwave transceiver. IEEE Geosci. Remote Sens. Lett. **5**(1), 57–59 (2008)

Pringle, J.K., Doyle, P., Babits, L.E.: Multidisciplinary investigations at stalag luft III allied prisoner-of-war camp: the site of the 1944 great escape, Zagan, Western Poland. Geoarchaeology Int. J. **22**(7), 729–746 (2007)

Pringle, J.K., Jervis, J., Cassella, J.P.N., Cassidy, J.: Time-lapse geophysical investigations over a simulated urban clandestine grave. J. Forensic Sci. **53**(6), 1405–1416 (2008)

Pringle, J.K., Ruffell, A., Jervis, J.R., Donnelly, L., McKinley, J., Hansen, J., Morgan, R., et al.: The use of geoscience methods for terrestrial forensic searches. Earth Sci. Rev. **114**(1–4), 108–123 (2012a)

Pringle, J.K., Holland, C., Szkornik, K., Harrison, M.: Establishing forensic search methodologies and geophysical surveying for the detection of clandestine graves in coastal beach environments. Forensic Sci. Int. **219**, e29–e36 (2012b)

Pringle, J.K., Jervis, J.R., Hansen, J.D., Glenda, M.J., Cassidy, N.J., Cassella, J.P.: Geophysical monitoring of simulated clandestine graves using electrical and ground-penetrating radar methods: 0–3 years after burial. J. Forensic Sci. **57**(6), 1467–1486 (2012c)

Ruffell, A., McKinley, J.: Forensic geoscience: applications of geology, geomorphology and geophysics to criminal investigations. Earth Sci. Rev. **69**, 235–247 (2005)

Ruffell, A., Donnelly, C., Carver, N., Murphy, E., Murray, E., McCambridge, J.: Suspect burial excavation procedure: a cautionary tale. Forensic Sci. Int. **183**, 11–16 (2009)

Sachs, J., Aftanas, M., Crabbe, S., Drutarovsky, M., Klukas, R., Kocur, D. et al.: Detection and tracking of moving or trapped people hidden by obstacles using ultra-wideband pseudo-noise radar. In: Proceedings of the European Radar Conference, Amsterdam, 30–31 Oct 2008

Sato, M., Fujiwara, J., Feng, Z., Zhou, Z., Kobayashi, T.: Development of a hand-held GPR MD sensor system (ALIS). In: Proceedings of SPIE, Detection and Remediation Technologies for Mines and Minelike Targets X vol. 5794, pp. 1192–1199, 10 June 2005

Sato, M., Kobayashi, T., Takahashi, K., Fujiwara, J., Feng, X.: Vehicle-mounted SAR-GPR and its evaluation. In: Proceedings of SPIE 6217, Detection and Remediation Technologies for Mines and Minelike Targets XI, 62172H, 18 May 2006

Schultz, J.J.: Detecting buried remains in Florida using ground-penetrating radar. Dissertation, University of Florida (2003)

Schultz, J.J.: Using ground-penetrating radar to locate clandestine graves of homicide victims. Homicide Stud. **11**(1), 15–29 (2007)

Schultz, J.J.: Sequential monitoring of burial containing small pig cadavers using ground-penetrating radar. J. Forensic Sci. **53**(2), 279–287 (2008)

Schultz, J.J., Martin, M.M.: Controlled GPR grave research: comparison of reflection profiles between 500 and 250 MHz antennae. Forensic Sci. Int. **209**(1–3), 64–69 (2011)

Schultz, J.J., Collins, M.E., Falsetti, A.B.: Sequential monitoring of burials containing large pig cadavers using ground-penetrating radar. J. Forensic Sci. **51**(3), 607–616 (2006)

Sensors and Software Inc.: Rescue radar. https://sensoft.ca/Products/Rescue-Radar/Overview.aspx. Accessed 24 May 2014

Sezgin, M.: Simultaneous buried object detection and imaging technique utilizing fuzzy weighted background calculation and target energy moments on ground penetrating radar data. EURASIP J. Adv. Sig. Process. **2011**, 55 (2011). doi:10.1186/1687-6180-2011-55

Solla, M., Lorenzo, H., Novo, A., Rial, F.I.: Ground-Penetrating Radar Assessment of the Medieval Arch Bridge of San Anton, Galicia, Spain. Archaeol. Prospection **17**(4), 223–232 (2010)

Solla, M., Riveiro, B., Álvarez, M.X., Arias, P.: Experimental forensic scenes for the characterization of ground-penetrating radar wave response. Forensic Sci. Int. **220**, 50–58 (2012)

Solla, M., Núñez-Nieto, X., Novo, A., Lorenzo, H.: Uso del Georradar en Aplicaciones Militares: Caso Particular de Detección de Túneles Subterráneos. Revista General de Marina, Ministerio de Defensa. Editorial MIC, León (2014)

Sun, Y., Li, J.: Time-frequency analysis for plastic landmine detection via forward-looking ground penetrating radar. IEE Proc. Radar Sonar Navig. **150**(4), 253–261 (2003)

Takahashi, K., Preetz, H., Igel, J.: The influence of soil properties on landmine detection, 10–12. SPIE Newsroom. doi: 10.1117/2.1201206.004265. (2012)

Takeuchi, T., Uematsu, Y., Saito, H., Aoki, Y.: Measurement of survivor location for rescue radar system by using two dimensional array antenna. In: Proceedings of the 2008 IEEE International Workshop on Safety, Security and Rescue Robotics, Sendai, 21–24 Oct 2008

Testafariam, G.T., Mali, D.: GPR technologies for landmine detection. Int. J. Comput. Sci. Commun. Technol. **5**(1), 768–774 (2012)

Trinks, I., Nissen, J., Johansson, B., Emilsson, J., Gustafsson, C., Friborg, J., Gustafsson, J.: Pilot study of the new multichannel GPR system MIRA for large scale, high-resolution archaeological prospection at the site of the Viking town Birka in Sweden. ISAP News **16**, 4–7 (2008)

Trinks, I., Johansson, B., Gustafsson, J., Emilsson, J., Friborg, J., Gustafsson, C., Nissen, J., Hinterleitner, A.: Efficient, large-scale archaeological prospection using a true three-dimensional ground-penetrating radar array system. Archaeol. Prospection **17**, 175–186 (2010)

Valová, P., Glisníková, V.: Using geophysical survey to investigate underground passages in and around Matky Boží chapel. Interdisciplinaria Archaeologica Nat. Sci. Archaeol.II **2**, 175–179 (2011)

Van Deen, J.K., de Feijter, J.W.: Three-dimensional ground probing radar. Geol. Surv. Finland Spec. Pap. **16**, 35–40 (1992)

Van der Merwe, A., Gupta, I.: A novel signal processing technique for clutter reduction in GPR measurements of small, shallow land mines. IEEE Trans. Geosci. Remote Sens. **38**, 2627–2637 (2000)

Vickers, R.S.: Design and applications of airborne radars in the VHF/UHF Band. In: IEEE AES Magazine, 26–29 June 2002

Wang, J., Li, Y., Zhou, Z. et al.: Image formation techniques for vehicle-mounted forward-looking ground penetrating SAR. In: Proceedings of the International Conference on Information and Automation, Changsha, 20–23 June 2008

Wu, S., Tan, K., Xu, Y., Chen, J., Meng, S., Fang, G.: A simple strategy for moving target imaging via an experimental UWB through-wall radar. In: Proceedings of the 14th International Conference on Ground Penetrating Radar, Tongji University, Shanghai, 4–8 June 2012

Zaikov, E., Sachs, J., Aftanas, M., Rovnakova, J.: Detection of trapped people by UWB radar. In: Proceedings of the German Microwave Conference (GeMiC'08), Hamburg, 10–12 March 2008

Zoubir, A.M., Chant, I.J., Brown, C.L., Barkat, B., Abeynayake, C.: Signal processing techniques for landmine detection using impulse ground penetrating radar. IEEE Sens. J. **2**, 41–51 (2002)

Applications of GPR in Association with Other Non-destructive Testing Methods in Surveying of Transport Infrastructures

Mercedes Solla, Henrique Lorenzo, Joaquin Martínez-Sánchez and Vega Pérez-Gracia

Abstract Preservation and maintenance of transport infrastructure is a global concern that affects social and economic development in all countries. During the last decades, there has been a continuous increase in the use of non-destructive testing (NDT) applied to many aspects related to civil engineering field. Ground Penetrating Radar (GPR) has become an established method of inspection. This paper presents a compilation of works in the frame of the applications of GPR and other NDT methods in the evaluation of transport infrastructures. Published works in roads and pavements, concrete and masonry structures, and tunnel testing are mentioned. It has been demonstrated that such methods have significantly benefited the procedures for inspection and also, successfully solved some of the limitations of traditional methods.

1 Introduction

The deterioration and distress mechanisms that are active under the surface cannot be assessed by traditional visual and optical inspection. Alternative methods are therefore required for inspection. GPR has been established as one of the most

M. Solla (✉)
Defense Center University (University of Vigo), Spanish Naval Academy, Pontevedra, Spain
e-mail: merchisolla@uvigo.es

H. Lorenzo · J. Martínez-Sánchez
Department of Natural Resources and Environment Engineering,
University of Vigo, Vigo, Spain
e-mail: hlorenzo@uvigo.es

J. Martínez-Sánchez
e-mail: joaquin.martinez@uvigo.es

V. Pérez-Gracia
Department of Strengthen of Materials and Structures, EUETIB,
Technical University of Catalonia, Catalonia, Spain
e-mail: vega.perez@upc.edu

A. Benedetto and L. Pajewski (eds.), *Civil Engineering Applications of Ground Penetrating Radar*, Springer Transactions in Civil and Environmental Engineering,
DOI 10.1007/978-3-319-04813-0_13

recommended NDT methods for routine sub-surface inspections. The use of GPR in civil engineering applications began to appear in the mid-1970s and the 1980s. Some of its main applications include services such as pavements, utilities and voids detection, as well as different structures associated with the transport infrastructure, such as bridge decks, retaining walls, masonry structures and tunnel inspections.

In the next Section, "Applications of GPR", some of the published works has been compiled to show the potential of the method. A brief review regarding the use of other NDT methods was also included.

2 Applications of GPR

This section is divided into four sub-sections, concerning different types of transport infrastructure and construction materials: roads and pavements (Sect. 2.1), concrete structures (Sect. 2.2), masonry structures (Sect. 2.3), and tunnels (Sect. 2.4).

2.1 Roads and Pavements

The pavement life-cycle is not only affected by the number of heavy loads but also layer thickness is a vital factor defining the quality of pavements. Deficiencies in thickness reduce their lives, and periodical rehabilitation is therefore necessary in any country's road management program to maintain roads in optimal conditions of use or monitor quality control. Road inspections imply the evaluation of different parameters such as roughness of the pavement, skid resistance, and presence and condition of cracks, voids and delamination. Also, the study of sub-layers under the pavement is an important task to assure the maintenance of the infrastructure. The loss of the local structural support of the pavement is one of the main causes of fissures and cracks. These damages produce discontinuities in the supporting loads structure. Therefore, the base of the pavement cannot absorb the loads in a homogeneous manner. Two are the main causes of this problem. One of them is the existence of changes in the water content caused by a non-appropriate drainage system. The second one is the presence of variations in the porosity and the changes in the cohesion between particles.

GPR technology is rapid, cost effective, and allows field surveys to be conducted without disturbing the pavement structure and the normal traffic flow (Lorenzo et al. 2011; Plati and Loizos 2012). GPR has been successfully used to find voids and cracks under pavement (Stryk 2008; Diamanti and Redman 2012; Krysiński and Sudyka 2013) as well as to monitor quality control on new asphalt overlays or to evaluate base course quality (Saarenketo and Scullion 2000). But above all, measuring pavement layer thickness is one of the most known applications of GPR (Gordon et al. 1998; Fauchard et al. 2003; Loizos and Plati 2007a; Vafidis et al. 2011; Grégoire and Van Geem 2013). Detection of changes in water content in sub-layers and sub-asphalt soil is other important application. Grote et al. (2005) applied GPR in an experimental infiltration test in pavements, using this methodology to monitor the soil water content during the experiment. Pedret et al. (2012) determine seasonal changes in water content in pavements, as a consequence of variations in radar images. Chen et al. (2006) evaluated also the moisture en pavements with GPR to identify the most possible causes of pavement failure in the main lanes of a highway. Radar data was used combined with falling weight deflectometer in order to determine areas with low density, voids and damp, as well as layers thicknesses.

There are complementary NDT technologies to the measuring of different road characteristics. Some examples are mobile LiDAR for geometric measurements and laser profilers for the evaluation of the pavement surface roughness (International Roughness Index–IRI) (Suksawat 2011). In Puente et al. (2013), a novel method consisting of mobile LiDAR technology is presented to evaluate layer thicknesses and volumes for newly constructed pavements. The method was favourably validated by GPR. Figure 1 presents both GPR and LiDAR data acquired through a road section (A and B, respectively) in order to validate the laser scanning approach for obtaining thickness measurements with appropriate accuracy. The GPR data was collected with a 2.3 GHz antenna and the radargram produced allowed for the identification of reflections at the air/ground, intermediate/base and base/soil-cement interfaces. To compare both data (C in Fig. 1) and to calibrate the LiDAR, the

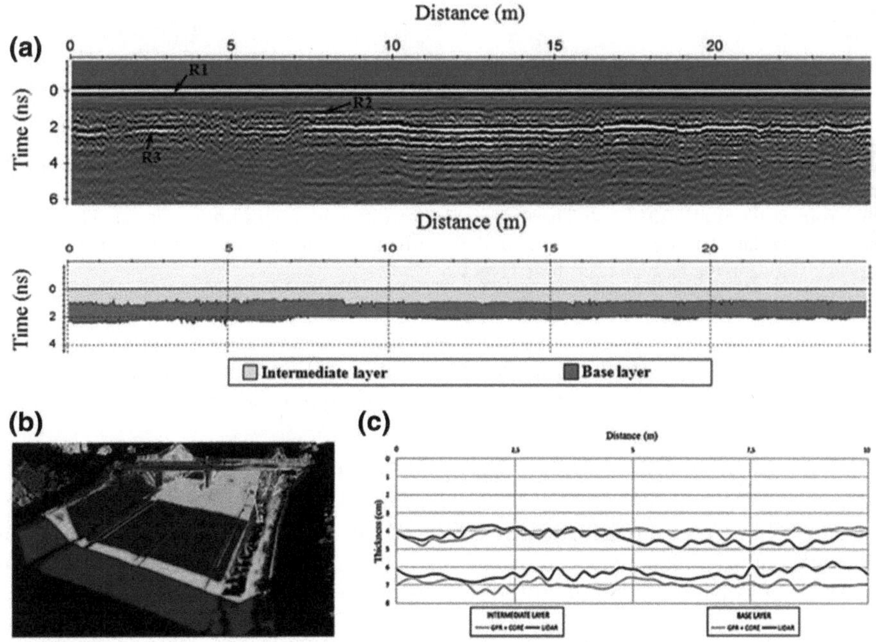

Fig. 1 The use of both GPR and LiDAR approaches for thicknesses measurements of pavement: **a** 2.3 GHz data collected showing the reflections at different interfaces: air/ground (R1), intermediate/base (R2) and base/soil-cement (R3), **b** general overview of the pavement layer point clouds from laser scanning approach after the registration, and **c** thickness measurement by both GPR and laser scanning for the intermediate and base layers [Data available from Puente et al. (2013)]

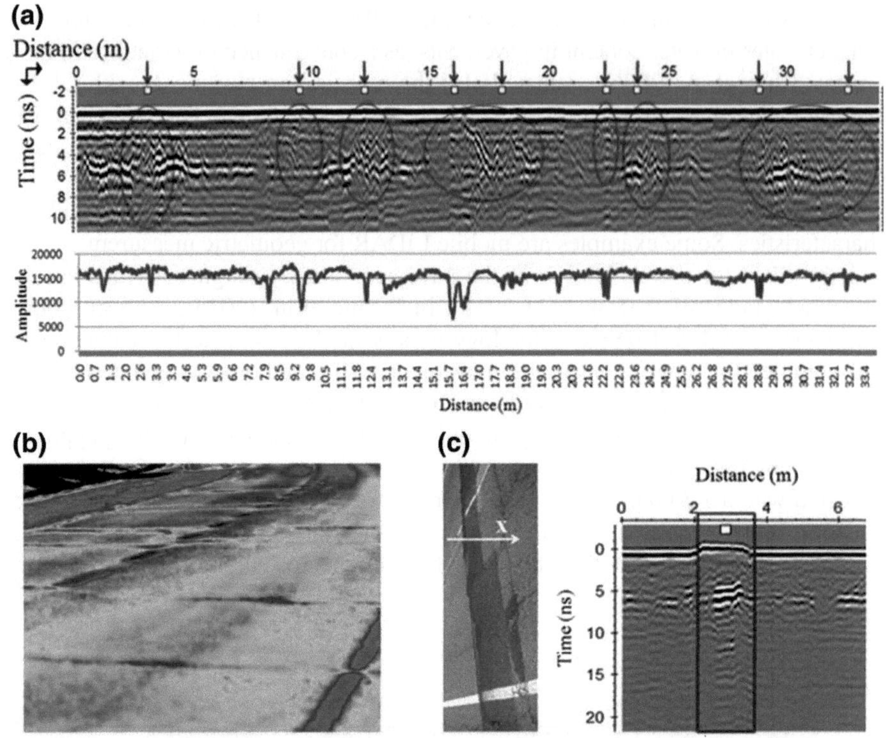

Fig. 2 The combination of GPR and thermography to analyze cracking in pavement: **a** 1 GHz data acquired showing the amplitude profile where the peaks in value indicate the presence of cracks in the surface road (possible pavement failures in depth are also illustrated) [Data available from Solla et al. (2014)], **b** Thermographies of different cracks present in an actual road, and **c** 1 GHz data acquired through a natural bump in a road surface, showing an anomaly at the interface between the base and sub-base layers

absolute errors were calculated as the difference between the LiDAR and the GPR data (ground truth results), and the average percentage error was 9 % at maximum, being slightly smaller for the intermediate layer (7 %).

Dumoulin et al. (2010) apply infrared thermography for the detection of pavement cracks or moisture content. Solla et al. (2014) present the combined application of GPR and infrared thermography to the detection and characterization of cracks in pavement. GPR demonstrated its capabilities to detect cracking in road surface thanks to the amplitude variations appearing where cracks are visible in surface (Fig. 2a). Nevertheless, it was not possible to establish a relationship between the measured amplitude and the depth of crack. Regarding infrared thermography, this technique has been proved as an adequate method for crack detection (Fig. 2b), and different tendencies or equations were established to estimate the depth of crack based on the pixels in image (thermograph) and the temperature difference between the surface and the bottom of the crack. On the other hand, the GPR provided the advantage for the detection of the origins of cracking in depth (reflections in red circles in Fig. 2a), as well as for the qualitative evaluation of the flaw: layers separation, levelling faults, etc. Additionally, Fig. 2c illustrates the GPR profile recorded, in the longitudinal direction, over a section of the road presenting a natural bump to analyze the most probable cause of failure. Observing the radargram, an anomaly was identified in the continuous reflection (at 5 ns in depth) produced by the interface between the base and the sub-base layers. It seems to indicate the existence of plastic deformation on the grading and the numbness of its materials, most probably because of water absorption. It is most likely that the soil-cement layers composing the sub-base were not treated with pre-cracking techniques during the construction of the road.

The structural evaluation (bearing capacity and layer stability) is quantitatively evaluated with the Falling Weight Deflectometer (FWD) that measures the deformations of the pavement in response to heavy loads (Grégoire and Van Geem 2013). Sybilski et al. (2012) present a case study of premature cracking where combination of GPR with FWD methods and laboratory measurements allowed the identification of severe water content as the reason for cracking. Domitrović and Rukavina (2013) deal with the comparison of pavement layer moduli calculated from FWD deflection data using layer thickness obtained by GPR and coring. Furthermore, this paper concluded that there is a tendency in reinforcement projects to apply recycling methods and use recycled materials so, knowing the continuous thickness of asphalt layers by GPR is essential in order to determine the optimum thickness available for milling and thus achieve optimization of recycling process.

Pedret and Pérez-Gracia (2011) evaluate correlations between measurements with a rolling wheel deflectometer and GPR wave velocity, obtaining layers thicknesses from coring. Figure 3 shows data from a combined method using GPR and rolling wheel deflectometer.

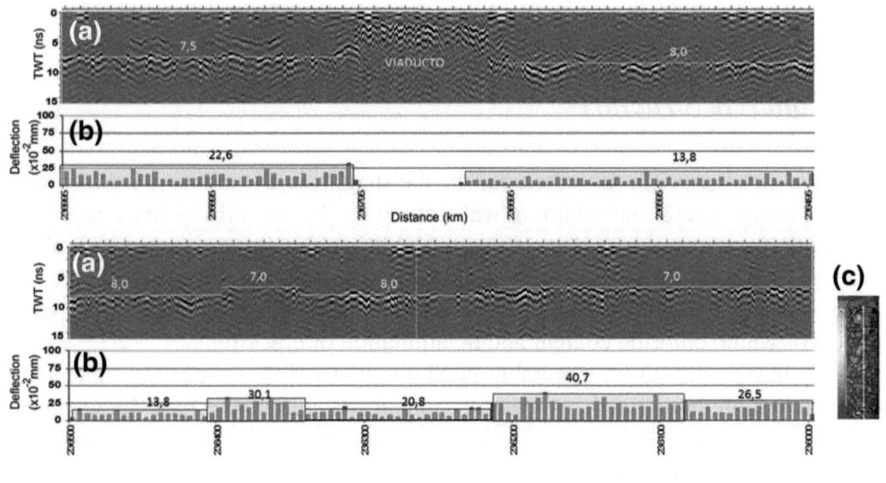

Fig. 3 GPR data (**a**) compared to results of a rolling wheel deflectometer (**b**). Thickness from GPR TWT conversion into depth could be compared with coring (**c**) (Data available from Pedret and Pérez-Gracia 2011)

Currently, there are different automatic inspection systems developed for road surveying and quality control that can integrate many sensors, such as global position systems (GPS/GLONASS) aided with inertial measurement units (IMU) for navigation and positioning, light detection and ranging (LiDAR) for geometric data, RGB cameras or video for imaging, infrared thermography for cracking, optical profilers for the measurement of pavement surface roughness and GPR for subsurface evaluation and pavement thicknesses. Examples of recent automatic systems that combines some of the related sensors are VISUALISE (Papi et al. 2012), road-scanner SITECO (SITECO 2014), Road-scanner Road Doctor (Road Doctor 2014), and SITEGI (Martínez-Sánchez et al. 2013). The objective is to provide users with effective tools for quickly diagnosis of infrastructure at highway speed and to reduce costs in maintenance (Barber et al. 2008; Jaakkola et al. 2008).

In addition to road/highway pavements, this sub-section includes airport runways and railroad ballast. Delamination, cracks and voids are the most common diseases in airport runway, which are particularly more severe due to the higher traffic loads that pavements are supporting. GPR is commonly combined with the Heavy Weigh Deflectometer (HWD) for airfield pavements inspection (Szynkiewicz and Grabowski 2004).

GPR also provides noteworthy information of the ballast quality and the track bed condition (Loizos and Plati 2007b), in addition to the geotechnical properties or subgrade and subsoil materials (Hugenschmidt 2000; Santos-Assunçao et al. 2013). Three-dimensional images are usual tools to determine discontinuities in the layers that could be possible fragile areas of the infrastructure (Fig. 4). Sandoval et al. (2011) combine different geophysical techniques (3D GPR, ERT, and microgravity) to analyze the stability of railways, which allowed the determination of ballast thickness, collapse, deformation of the layering of the railway, and other anomalies such as voids (Fig. 5).

2.2 Concrete Structures

Concrete structures are also included in the diagnosis of transport infrastructure such as bridge decks and retaining walls along roads and railway lines, as well as over/underpasses to ensure the passage of wildlife animals, person and agricultural machinery. The diagnosis of concrete includes: estimation of thicknesses, location of reinforcing bars and metallic ducts, estimation of bar size, location of voids, effects of water, chloride content and delamination or cracking.

GPR has proven to be a suitable NDT method for the inspection of concrete structures as demonstrated by several publications within the past decades (Utsi and Birtwisle 2012). Bala et al. (2011) applied pseudo 3D surveys to determine the location of rebars in reinforced concrete pavements. Detecting water content is an important phase for the diagnosis of concrete. Klysz et al. (2007) observed a linear

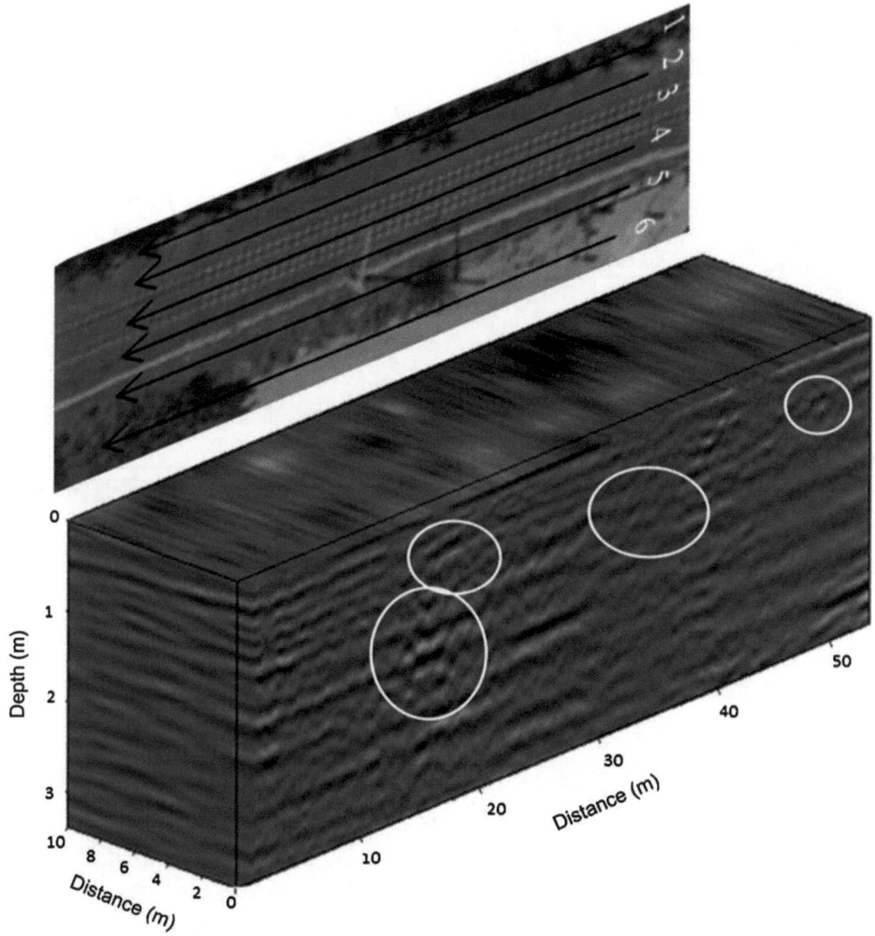

Fig. 4 3D GPR data in a railway ballast study. Different bed layers are visible, and possible voids and anomalous zones are detected (Data available from Santos-Assunçao et al. 2013)

relation between the velocity of propagation of the direct wave and moisture. In addition, GPR can be successfully used for the location of salt ingress due to the influence of chloride content on the permittivity of the concrete (Sbartaï et al. 2006). What is more, Lai et al. (2010) show its capabilities to analyze the corrosion of reinforcing induced by chloride content.

As example, Fig. 6 presents a radargram showing the corrosion of the reinforcing bars produced by high chloride content (zone into the red square). The expected reflections from the bars are in the form of consecutive hyperbolas, and as seen in Fig. 6, the radar wave signal is partially lost in this portion of the concrete specimen affected by corrosion.

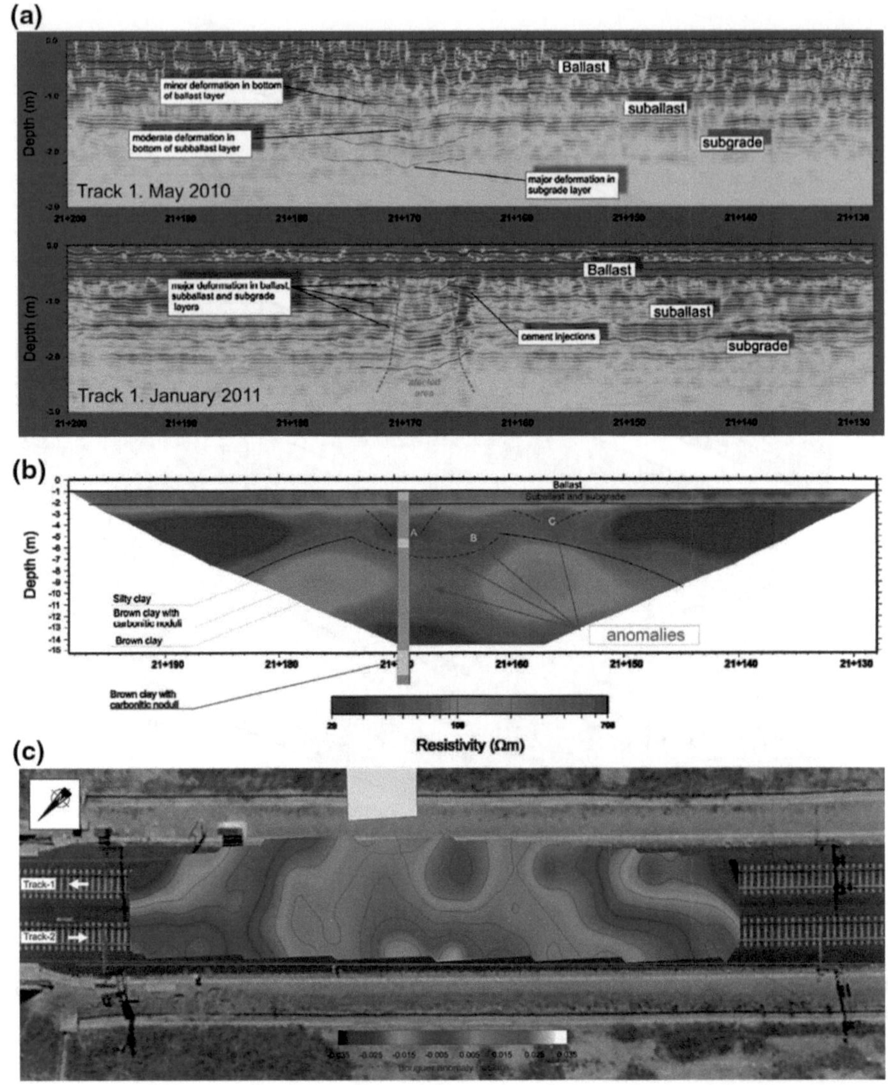

Fig. 5 The combination of geophysical techniques to evaluate railway platforms. **a** Radargrams from a 3D multifrequency survey, showing a small deformation of the ballast layer (*up*) and a collapsed area (*down*). **b** ERT model obtained, presenting different anomalies detected. **c** Bouguer anomaly map derived from the microgravity survey (Data provided by the company Geofísica Consultores, S.L.)

In other published works, the method enables inspectors to obtain information about reinforcing bars, defining depth of rebar, and location of tendon-ducts on bridge decks (Hugenschmidt et al. 2010). The reflection coefficient of a thin layer into concrete was evaluated in Van der Wielen et al. (2012) in order to estimate the detection limit of the GPR antenna. The reflection coefficient revealed to be linearly dependent on the thickness-on-wavelength ratio for thicknesses less than λ/11.

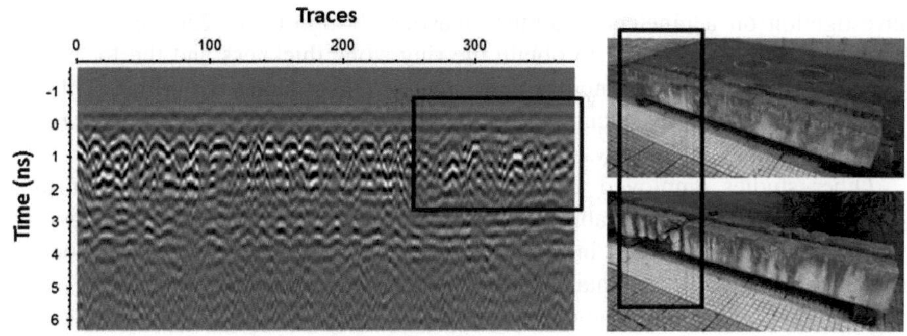

Fig. 6 The influence of chloride content in rebar detection: radargram collected with a 2.3 MHz antenna showing the corrosion of the reinforcing by the presence of chloride content (zone in *red square*) in the concrete specimen. The GPR profile was conducted from right to left (Data provided by the Applied Geotechnologies research group of the University of Vigo)

Other NDT measurements (such as electrical, sonic, seismic and infrared thermography) combined with GPR have earned the interest of researchers on the assessment of water content and alkalini-aggregate reaction on concrete, as well as concrete quality and delamination (Breysse et al. 2008; Sbartaï et al. 2012). Villain et al. (2012) present a combination of GPR, capacitive and impact-echo measurements to analyze porosity and water and chloride contents. Other possibility which is currently promoted is combination of GPR with laser scanning that provides exact information about the surface of roads and structures, and their suburb in combination to GPR data of the sub-surface (Solla et al. 2013a).

2.3 Masonry Structures

The most of the masonry bridges are the oldest structures still in use in the transport infrastructure. These constructions are subjected to special tension conditions because of the increase in traffic loads and ageing, which produce material degradation and structural damage. Some of the most typical damages in masonry bridges are: moisture, differential settlements, and thermal deformation with subsequent bulging of spandrels, damages in wing walls, cracking, as well as arch mechanism failure and loss of ashlar.

During the last few decades, GPR has been demonstrated its capabilities for bridge inspection. Solla et al. (2012a) successfully applied GPR for assessing masonry bridges, which determined the effectiveness of the method in obtaining relevant structural information concerning the presence of cavities and faults, and reinforcement elements, in addition to ring stone thickness and foundation conditions. Pérez-Gracia (2001) applies GPR for assessing historical bridges to reach their foundations. In addition, the results obtained with 100 and 500 MHz antennas provided information about the water table and sub-bottom geological structure near riverbed level. In another study, Fernandes (2006) performs a GPR investigation on a nineteenth century masonry bridge using 250 and 500 MHz antennas. The purpose was to obtain the ring stone thickness and the location of drainage channels and drainpipes under the pavement along the bridge. This aim was achieved with good accuracy and the presence of significant moisture content within the whole structure was also determined.

Other studies employed GPR together with other NDT methods for more exhaustive evaluation of stability. Orbán and Gutermann (2009) have shown the joint effectiveness of GPR, infrared thermography, and sonic methods for obtaining unknown geometric data and finding hidden characteristics such as voids, moisture content, and inhomogeneous filling. Many structural problems can be best analyzed by a particular NDT method owing to the physical properties of their construction materials. Colla et al. (1997) employs radar, sonic and conductivity methods for assessing an historical stone masonry bridge. These tests have demonstrated the ability to achieve inhomogeneity identification and layering within the bridge together with moisture-drainage problems. Furthermore, Flint et al. (1999) combine seismic, radar and electrical resistivity tomography methods to evaluate changes in the internal condition of a masonry viaduct. The response of these techniques has shown to be useful in monitoring the presence of voids into the structure as well as changes in internal moisture content. Other authors determine vibration modes combined with GPR information to model the dynamic behaviour of the structure (Pérez-Gracia et al. 2011). Radar data provides valuable information to define most accurate models.

In addition to geophysical inspection, other NDT optical methods were also for appropriate combination. The metric information obtained by photogrammetry or laser scanning allows for the characterization of the stonework and filling, which provides better understanding of the GPR propagation phenomena (Arias et al. 2007). This combined approach can be also used to define a hypothesis for structural analysis, which describes structural behaviour of the structure or structural stability of arches (Solla et al. 2012b). The complex external geometry of the structure is elucidated using the photogrammetry or laser scanning data, and the internal composition is defined by the GPR data (Lubowiecka et al. 2011).

Masonry structures are built using heterogeneous filling that often complicates the interpretation and analysis of field GPR data. FDTD modelling of the GPR signal is therefore typically used as additional interpretational tool. Some authors (Diamanti and Giannopoulos 2011) have employed novel FDTD numerical modelling sub-gridding scheme to simulate GPR responses from delamination or ring separation in brick masonry arch bridges, in which different aspects were considered such as, the effect of varying the thickness of faults, their location, and the effect of water ingress in hairline delamination on GPR signals. More sophisticated and realistic modelling can be obtained when a combination of different NDT, such as infrared thermography and GPR, is considered to create models (Solla et al. 2013b).

Figure 7 shows the methodology approach assumed to determine the presence of water content in masonry by field GPR data and FDTD modelling based on the interpretation provided by thermography methods. GPR data was collected

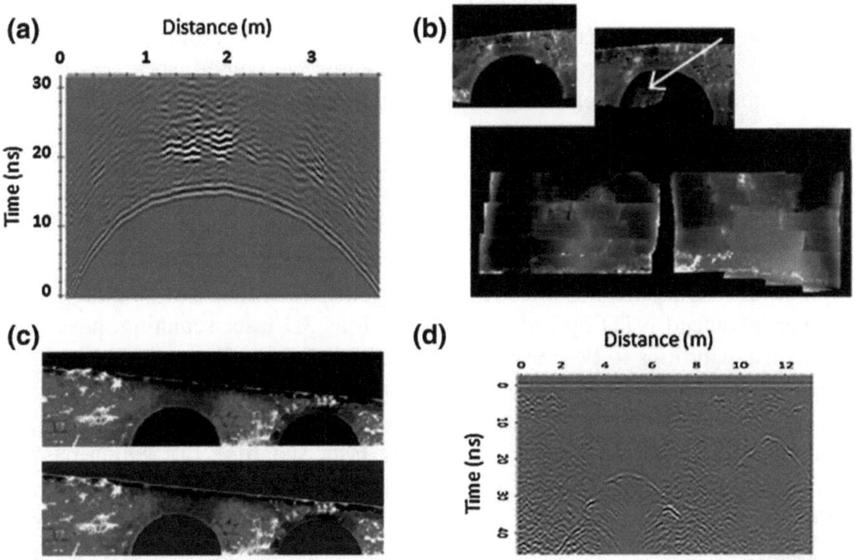

Fig. 7 The analysis of water content in masonry by GPR, thermography and more realistic FDTD modelling based on orthothermograms: **a** 1 GHz GPR data collected over the internal surface of the arch vault showing the zone affected by high water content, **b** 3D model of the arch vault textured with thermograms and orthothermograms created, **c** Orthothermogram (*up*) and synthetic model built for FDTD modelling (*down*), and **d** synthetic data obtained from the model simulated (Data provided by the Applied Geotechnologies research group of the University of Vigo)

with the 1 GHz antenna over the internal surface of the vault to avoid the complex pattern of reflections that occurs at the fill/ring stones interface because of the irregular shape of the ring stones (A). Additionally, infrared thermography was applied to map moisture areas in masonry and to assist in the interpretation of the GPR data. First, a 3D model of the structure was created by photogrammetry and, then, this model was textured with the thermograms acquired, resulting in a thermographic 3D model (B). Furthermore, FDTD modelling was used to improve the interpretation of the field GPR data. More realistic synthetic models were built from the orthothermograms generated (C), and the synthetic radargram obtained (D) allowed for more thoroughly understand radar-wave propagation phenomena. The combination of all these techniques provided a more accurate and complete vision of the real state of the structure.

2.4 Tunnels

GPR has proved a valuable method for tunnel quality management and detection. There are successful works in detecting thicknesses of both lining and backfill grouting layers behind the concrete lining (Liu et al. 2012), as well as to identify defects in lining such as voids and cracks (Karlovšek et al. 2012). Additional information can be obtained to define rebar geometry, including corrosion detection (Xiang et al. 2013). Variations in water content, and other aspects, as the presence of reinforcement elements embedded in lining (Parkinson and Ékes 2008), can be also obtained by GPR.

Integration of different geophysical methods was also considered for evaluating tunnel stability. There is published works which integrate GPR with seismic method in order to provide additional mechanical characteristics of the damage (Abraham and Dérobert 2003) in addition to elastics characteristics of discontinuities in order to examine possible areas of instability (Cardarelli et al. 2003).

Other advanced NTD optical methods, such as 3D laser scanning, have shown also their capabilities to be integrated with GPR (Xie and Zeng 2012). Such technology allows for the detection of defects on tunnel surface such as water exudation, which could be directly related with inner faults detected by GPR.

Figure 8 presents the results obtained by laser scanning (a), which has demonstrated its capabilities for the testing and documentation of the surface situation of the tunnel, such as faults in surface and moist zones. GPR results are also

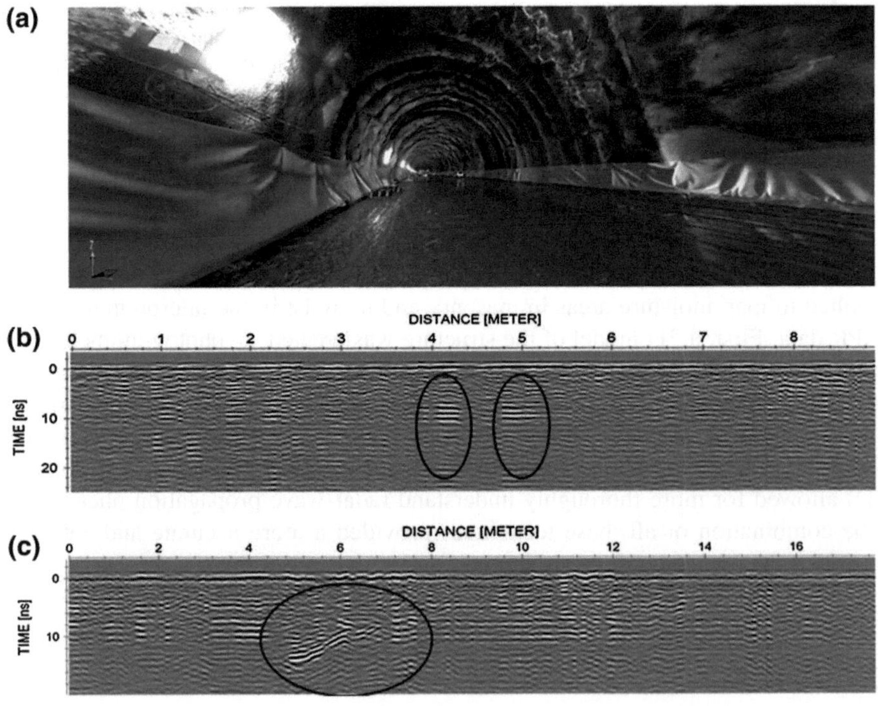

Fig. 8 Application of GPR and laser scanning to evaluate tunnels: **a** 3D model of the tunnel provided by laser scanning, **b** 1 GHz data showing the reflections produced by the metal trusses used for reinforcement, and **c** 1 GHz data presenting a lining cavity (Data provided by the Applied Geotechnologies research group of the University of Vigo)

illustrated showing the presence of internal metal elements, or trusses, used for structural reinforcement (b), and the existence of a fissure in lining at the concrete lining and grouting layer interface (c).

3 Conclusion

Although GPR have demonstrated its capabilities for roads and transport infrastructure inspection, alone or in combination with other NDT, further research is required in developing more cost-effective acquisition methodologies and processing.

The presence of defects in pavements, as well as concrete and masonry structures is difficult to detect using standard procedures. Combined approaches by considering different NDT methods are needed because there are many different aspects influencing the detection of such defects. For example, infrared thermography is useful in combination with GPR to map shallower moisture content, in addition to the use of LiDAR or even multispectral cameras that delimit moist areas from the intensity attribute.

Further development is also demanded concerning to affordable systems for road/highways inspection, as well as to use in airport runways and railways diagnosis. The tendency is to integrate complementary sensors such as LiDAR, RGB cameras, profilometer or video camera, thermography, and GPR, all mounted in a moving vehicle and spatially related to a common trajectory defined by global position systems (GPS/GLONASS) aided with inertial measurement unit (IMU) Examples of recent automatic systems that combine some of related sensors are Road Doctor (Roadscanner), SITEGI (Uvigo), and SITECO.

The large amount of data collected when surveying relevant infrastructure makes difficult to process all the data and manage in a reasonable amount of time. More powerful and versatile processing tools are therefore demanded to optimize time and resources invested. Advances are also required in developing new software in order to visualize and integrate additional data acquired from different sensors in moving vehicle systems for road inspection.

Acknowledgments The authors acknowledge the COST Action TU1208 "Civil Engineering Applications of Ground Penetrating Radar", supporting this work.

References

Abraham, O., Dérobert, X.: Non-destructive testing of fired tunnel walls: the Mont-Blanc Tunnel case study. NDT&E Int. **36**, 411–418 (2003)

Arias, P., Armesto, J., Di-Capua, D., González-Drigo, R., Lorenzo, H., Pérez-Gracia, V.: Digital photogrammetry, GPR and computational analysis of structural damages in a medieval bridge. Eng. Fail. Anal. **14**, 1444–1457 (2007)

Bala, D.C., Garg, R.D., Jain, S.S.: Rebar detection using GPR: an emerging non-destructive QC approach. Int. J. Eng. Res. Appl. (IJERA) **1**(4), 2111–2117 (2011)

Barber, D., Mills, J., Smith-Voysey, S.: Geometric validation of a ground-based mobile laser scanning system. ISPRS J. Photogram. Remote Sens. **63**(1), 128–141 (2008)
Breysse, D., Klysz, G., Déboret, X., Sirieix, C., Lataste, J.F.: How to combine several non-destructive techniques for a better assessment of concrete structures. Cem. Concr. Res. **38**, 783–793 (2008)
Cardarelli, E., Marrone, C., Orlando, L.: Evaluation of tunnel stability using integrated geophysical methods. J. Appl. Geophys. **52**, 93–102 (2003)
Chen, D.H., Chen, T.-T., Scullion, T., Bilyeu, J.: Integration of field and laboratory testing to determine the causes of a premature pavement failure. Can. J. Civ. Eng. **33**, 1345–1358 (2006)
Colla, C., Das, P.C., McCann, D., Forde, M.: Sonic, electromagnetic and impulse radar investigation of stone masonry bridges. NDT&E Int. **30**(4), 249–254 (1997)
Diamanti, N., Giannopoulos, A.: Employing ADI-FDTD subgrids for GPR numerical modeling and their application to study ring separation in brick masonry arch bridges. Near Surf. Geophys. **9**, 245–256 (2011)
Diamanti, N., Redman, D.: Field observations and numerical models of GPR response from vertical pavement cracks. J. Appl. Geophys. **81**, 106–116 (2012)
Domitrović, J., Rukavina, T.: Application of GPR and FWD in assessing pavement bearing capacity. In: Proceedings of International Scientific Conference on Road Research and Administration, Bucharest, 4–5 July 2013
Dumoulin, J., Ibos, L., Ibarra-Castanedo, C., Mazioud, A., Marchetti, M., Maldague, X., Bendada, A.: Active infrared thermography applied to defect detection and characterization on asphalt pavement samples: comparison between experiments and numerical simulations. J. Mod. Opt. **57**(8), 1759–1769 (2010)
Fauchard, C., Dérobert, X., Cariou, J., Côte, Ph: GPR performances for thickness calibration on road test sites. NDT&E Int. **36**, 67–75 (2003)
Fernandes, F.: Evaluation of two novel NDT techniques: microdrilling of clay bricks and ground penetrating radar in masonry. Ph.D. thesis, Universidade do Minho
Flint, R.C., Jackson, P.D., McCann, D.M.: Geophysical imaging inside masonry structures. NDT&E Int. **32**, 469–479 (1999)
Gordon, M.O., Broughton, K., Hardy, M.S.A.: The assessment of the value of GPR imaging of flexible pavements. NDT&E Int. **31**(6), 429–438 (1998)
Grégoire, C., Van Geem, C.: Use of radar in road investigation—BRRC experience. In: Proceedings of 7th International Workshop on Advanced GPR, 6 pp, Nantes, France, 2–5 July 2013
Grote, K., Hubbardb, S., Harvey, J., Rubin, Y.: Evaluation of infiltration in layered pavements using surface GPR reflection techniques. J. Appl. Geophys. **57**, 129–153 (2005)
Hugenschmidt, J.: Railway track inspection using GPR. J. Appl. Geophys. **43**, 147–155 (2000)
Hugenschmidt, J., Kalogeropoulos, A., Soldovieri, F., Prisco, G.: Processing strategies for high-resolution GPR concrete inspections. NDT&E Int. **43**, 334–342 (2010)
Jaakkola, A., Hyyppä, J., Hyyppä, H., Kukko, A.: Retrieval algorithms for road surface modelling using laser—based mobile mapping. Sensors **8**(9), 5238–5249 (2008)
Karlovšek, J., Scheuermann, A., Willimas, D.J.: Investigation of voids and cavities in Bored Tunnels using GPR. In: Proceedings of 14th International Conference on Ground Penetrating Radar, pp. 496–501, Shanghai, China, 4–8 June 2012
Klysz, G., Balayssac, J.P.: Determination of volumetric water content of concrete using ground-penetrating radar. Cem. Concr. Res. **37**, 1164–1171 (2007)
Krysiński, L., Sudyka, J.: GPR abilities in investigation of the pavement transversal cracks. J. Appl. Geophys. **97**, 27–36 (2013)
Lai, W.L., Kind, T., Wiggenhauser, H.: Detection of accelerated reinforcement corrosion in concrete by ground penetrating radar. In: Proceedings of 13th International Conference on GPR, 5 pp, Lecce, Italy, 21–25 June 2010
Liu, H., Xie, X., Sato, M.: Accurate thickness estimation of a backfill grouting layer behind shield tunnel lining by CMP measurement using GPR. In: Proceedings of 14th International Conference on GPR, pp. 137–142, Shanghai, China, 4–8 June 2012

Loizos, A., Plati, C.: Accuracy of pavement thicknesses estimation using different ground penetrating radar analysis approaches. NDT&E Int. **40**, 147–157 (2007)

Loizos, A., Plati, C.: Ground penetrating radar: a smart sensor for the evaluation of the railway trackbed. In: Proceedings of Conference on Instrumentation and Measurement Technology Conference, 6 pp, Warsaw, Poland, 1–3 May 2007b

Lorenzo, H., Rial, F.I., Pereira, M., Solla, M.: A full non-metallic trailer for GPR road surveys. J. Appl. Geophys. **75**, 490–497 (2011)

Lubowiecka, I., Arias, P., Riveiro, B., Solla, M.: A multidisciplinary approach to assess historic bridges using photogrammetry, ground penetrating radar and finite elements analysis. Comput. Struct. **89**, 1615–1627 (2011)

Martínez-Sánchez, J., Nogueira, M., González-Jorge, H., Solla, M., Arias, P.: SITEGI Project: applying geotechnologies to road inspection. Sensor integration and software processing. ISPRS Annals of the Photogrammetry, Remote Sensing and Spatial Information Sciences, vol. II-5/W2, pp. 181–186. ISPRS Workshop Laser Scanning 2013, Antalya, Turkey, 11–13 Nov 2013

Orbán, Z., Gutermann, M.: Assessment of masonry arch railway bridges using nondestructive in-situ testing methods. Eng. Struct. **31**(10), 2287–2298 (2009)

Papí, J., Esteban, M.: Artificial vision real results. Thinking Highways, pp. 60–62 (2012)

Parkinson, G., Ékes, C.: Ground penetrating radar evaluation of concrete tunnel linings. In: Proceedings of 12th International Conference on GPR, 11 pp, Birmingham, UK, 16–19 June 2008

Pedret, J., Pérez-Gracia, V.: Study of ground penetrating radar sensitivity to asphalt mixtures void content in road pavements. Revista Ingeniería de Obras Civiles-RIOC, vol. 1, pp. 5–18 (2011)

Pedret, J., Pérez-Gracia, V., Valdés, G.A.: Fundamentos y ejemplos de aplicación del ensayo no destructivo de reflexión electromagnética mediante georradar en pavimentos asfálticos. In: resúmenes del congreso 10° PROVIAL, Innovación Tecnológica para la Gestión Vial 2012, 8–10 October, Santiago de Chile, Chile (2012)

Pérez-Gracia, V.: Radar de subsuelo. Evaluación para aplicaciones en arqueología y en patrimonio histórico-artístico. PhD thesis. Universidad Politécnica de Cataluña (2001)

Pérez-Gracia, V., Di Capua, D., Caselles, O., Rial, F., Lorenzo, H., González-Drigo, R., Armesto, J.: Characterization of a Romanesque Bridge in Galicia (Spain). Int. J. Arch. Herit. **5**(3), 251–263 (2011)

Plati, C., Loizos A.: Using ground-penetrating radar for assessing the structural needs of asphalt pavements. Non Destruct. Test Eval. **27**(3), 273–284 (2012)

Puente, I., Solla, M., González-Jorge, H., Arias P.: Validation of mobile LiDAR surveying for measuring pavement layer thicknesses and volumes. NDT&E Int. **60**, 70–76 (2013)

Road Doctor—Roadscanner: Available from http://www.roadscanners.com/index.php/consulting/laser-scanning-services Accessed 14 Jan 2014

Saarenketo, T., Scullion, T.: Road Evaluation with ground penetrating radar. J. Appl. Geophys. **43**, 119–138 (2000)

Sandoval, S., Mínguez, R., Nestares, E., Carbó, A.: Multidisciplinary study of a ballast collapse in a high-speed railway track in Spain. In: Proc. 6th Int. Conf. on applied geophysics for environmental and territorial system engineering, Sardinia, Italy, April 28-30 (2011)

Santos-Assunçao, S., Pedret Rodés, J., Pérez-Gracia, V.: Ground Penetrating Radar Railways Inspection. In: Proc. 75th EAGE Conference & Exhibition incorporating SPE EUROPEC 2013, (paper Tu P09 02), London, UK, June 10–13 (2013)

Sbartaï, Z.M., Laurens, S., Balayssac, J.P., Arliguie G., Ballivy G.: Ability of the direct wave of radar ground-coupled antenna for NDT of concrete structures. NDT&E Int. **39**, 400–407 (2006)

Sbartaï, Z.M., Breysse, D., Larget, M., Balayssac, J.P.: Combining NDT techniques for improved evaluation of concrete properties. Cem. Concr. Compo. **34**, 725–733 (2012)

Siteco—Road Scanner Mapping Mobile System: Available from http://www.sitecoinf.it/siteco_eng/eng/roadscanner.html Accessed 14 Jan 2014

Solla, M., Lorenzo, H., Rial, F.I., Novo, A.: Ground-penetrating radar for the structural evaluation of masonry bridges: Results and interpretational tools. Constr. Build. Mat. **29**, 458–465 (2012a)

Solla, M., Caamaño, J.C., Riveiro, B., Arias, P.: A novel methodology for the structural assessment of stone arches based on geometric data by the integration of photogrammetry and ground-penetration radar. Eng. Struct. **35**, 296–306 (2012b)

Solla, M., González-Jorge, H., Varela, M., Lorenzo H.: Ground-penetrating radar for inspection of in-road structures and data interpretation by numerical modeling. J. Constr. Eng. Manag. **139**(6), 749–753 (2013a)

Solla, M., Lagüela, S., Riveiro, B., Lorenzo, H.: Non-destructive testing for the analysis of moisture in the masonry arch bridge of Lubians (Spain). Struct. Control Health Monit. **20**, 1366–1376 (2013b)

Solla, M., Lagüela, S., González-Jorge, H., Arias, P.: Approach to identify cracking in asphalt pavement using GPR and infrared thermographic methods: Preliminary findings. NDT&E Int. **62**, 55–65 (2014)

Stryk, J.: Road diagnostics—ground penetrating radar possibilities. Art. n°5, Intersections Test Methods **5**(1), 48–57 (2008)

Suksawat, B.: Development of a multifunction international roughness index and profile measuring device. In: Proc. Int. Conf. Control Autom Systems, pp. 795–799 (2011)

Sybilski, D., Bańkowski, W., Sudyka, J., Krysiński, L.: Reasons of premature cracking pavement deterioration—a case study. In: Proc. 7th RILEM Int. Conf. on Cracking in Pavements, RILEM vol. 4, pp. 1029–1038 (2012)

Szynkiewicz, A., Grabowski, P.: GPR Monitoring of pavements on airfield. In: Proc. 10th Int. Conf. on Ground Penetrating Radar, Delft, The Netherlands, June 21–24, pp. 803–806 (2004)

Utsi, V., Birtwisle, A.: Evaluation of bridge decks using ground probing radar (GPR). In: Mechtcherine, Schenck (eds.) Concrete Solutions, Grantham, Taylor & Francis group, London (2012)

Vafidis, A., Economou, N., Dimitriadis, K.: Time varying zero-phase filtering of GPR data for imaging pavement layers. In: Proc. 73rd European Association of Geoscientists and Engineers Conference and Exhibition, Vienna, Austria, May 23–26, vol. 1, pp. 2794–2798 (2011)

Van der Wielen, A., Courard, L., Nguyen, F.: Static detection of thin layers into concrete with ground penetrating radar. Restor. Build. Monuments **18**(3/4), 247–254 (2012)

Van Geen, C., Grégoire, C.: Rehabilitation of roads containing of cobblestone pavements covered with a bituminous layer. In: Proc., 9th Int. Conf. on the Bearing Capacity of Roads, Railways and Airfields (BCRRA), Trondheim, 25–27 June, 10 pp (2013)

Villain, G., Sbartaï, Z.M., Dérobert, X., Garnier, V., Balayssac, J.P.: Durability diagnosis of a concrete structure in a tidal zone by combining NDT methods: Laboratory tests and case study. Constr. Build. Mat. **37**, 893–903 (2012)

Xiang, L., Zhou, H., Shu, Z., Tan, S., Liang, G., Zhu, J.: GPR evaluation of the Damaoshan highway tunnel: A case study. NDT&E Int. **59**, 68–76 (2013)

Xie, X., Zeng, C.: Non-destructive evaluation of shield tunnel condition using GPR and 3D laser scanning. In: Proc., 14th Int. Conf. on GPR, Shanghai, China, June 4–8, pp. 479–484 (2012)

Advanced Electric and Electromagnetic Methods for the Characterization of Soil

Marc Van Meirvenne

1 Introduction

Soil is the topmost layer of the Earth's crust and operates as the interface between the atmosphere, hydrosphere and geosphere and on the continents, soil is the major host for the biosphere. Consequently, soil fulfils several functions and delivers essential services to our environment. These were identified by Blum (2005) as:

- Biomass production, ensuring the delivery of food, fodder, renewable energy and raw materials.
- Protection against environmental threats, including solid, liquid or gaseous, inorganic or organic depositions affecting global cycles such as the carbon cycle.
- Gene reservoir. Soils are a primary source of biodiversity, supporting biotechnological and bioengineering processes.
- Physical basis of human activities, i.e. the support for buildings and all types of infrastructures (urban soils, sport fields, brownfields…).
- Source of raw materials such as peat, clay, sand, gravel, water etc. Often these form the basis for industrial and socio-economic development.
- Safeguarding the geogenic and cultural heritage. Soils, as a part of present or past landscapes conceal and protect paleo-geological features and archaeological objects. They are of high value for the understanding of man's history and that of the earth.

The detailed spatial characterization of soil properties is essential for the management of soil to provide all these functions.

Soil is considered to be the result of the interactions between several soil forming factors. Jenny (1941) identified these as: climate, parent material (rock type), topography, organisms and time. These interactions result in soil being an intimate

M. Van Meirvenne (✉)
Research Unit Soil Spatial Inventory Techniques, Department Soil Management,
Faculty of Bioscience Engineering, Ghent University, Coupure 653, 9000 Ghent, Belgium
e-mail: marc.vanmeirvenne@ugent.be

A. Benedetto and L. Pajewski (eds.), *Civil Engineering Applications of Ground Penetrating Radar*, Springer Transactions in Civil and Environmental Engineering,
DOI 10.1007/978-3-319-04813-0_14

mix of many types of material: a variety of minerals (with quartz and clay mineral being the most dominant), organic matter (being the heterogeneous decay-result of fresh biomass), water, ions and air. A "healthy" soil has ideally a pore volume of around 50 % which is ideally partly filled with water (say 25 %) and partly filled with air (say 25 %). The composition of both soil-water and soil-air is different from the above ground equivalents: in soil water contains more ions due to the interactions with the charged soil minerals and organic matter, and the air is enriched by CO_2 and contains less O_2 due to the respiration activities of roots and other living organisms. In general, fluxes of matter and energy are very slow in soil compared to aquatic or aerial environments. As a consequence, soil can be very heterogeneous in composition and very strong gradients in differences can be maintained over long periods of time. As many soil forming factors tend to increase in variability by increasing spatial dimensions, the variability of soil tends to increase as well. However, it is a frequent observation that the order of soil variability encountered at a regional scale may be already present within individual fields (e.g. Van Meirvenne 2003). Moreover, besides a horizontal variation, soil genesis may create a vertical differentiation known as soil horizons. The identification of the sequence and properties of soil horizons forms the basis of soil classification or taxonomy.

To investigate and characterise the complex composition and highly variable nature of soil, soil scientists had no alternative to digging or augering to obtain soil samples. These were identified in the field or taken to the laboratory for analysis. Since this procedure is time and labour intensive, soil sampling is kept to a strict minimum. To obtain soil spatial continuous maps, soil surveyors often rely on information provided by more easily observable land properties, such as topography and vegetation. Yet soil mapping is largely a predictive exercise. If quantitative predictions are aimed, statistical procedures such as geostatistical (kriging) algorithms are required.

To overcome the limitations of punctual sampling and analysis, combined with spatial prediction, soil scientists welcomed the introduction of proximal soil sensors. These allow the rapid and non-invasive sampling of thousands of locations with a field of a few ha, contrasting strongly with the density of classical sampling schemes.

2 Proximal Soil Sensing

Proximal soil sensing (PSS) was defined as the use of field-based sensors to obtain information from the soil when the sensor is in direct contact with or close to (within 2 m) the soil (Viscarra Rossel et al. 2010). In recognition of its importance, the International Union of Soil Sciences recently established a Working Group on Proximal Soil Sensing (www.proximalsoilsensing.org), which aims to provide a framework for greater interaction between soil scientists and engineers for the development of proximal soil sensing technologies.

Viscarra Rossel et al. (2011) provide an overview of the wide variety of available technologies for PSS. Soil sensors can be invasive (i.e. they require soil-to-sensor contact, either at the soil surface or they need to be inserted into the soil) and non-invasive sensors. They can also be active (i.e. they have their own source of energy) or non-active, mobile or stationary and allow direct or indirect inference of the targeted soil properties.

It is beyond the scope of this chapter to provide an overview of all available proximal soil sensor. Here we will focus on sensor relying on electric and electromagnetic principles.

3 Electric and Electromagnetic Soil Properties

The key electric and electromagnetic properties of geomaterials, such as soil, are:

- electrical conductivity σ (or EC) being a measure of the availability and mobility of electrical charges within a medium; its reciprocal is the electrical resistivity ρ.
- magnetic susceptibility χ (or MS) being the ability of a medium to respond to a magnetic field.
- dielectric permittivity ε being the ability of a medium to allow passage of electromagnetic energy.

3.1 Electrical Conductivity

If an electrical potential difference ΔV (SI unit: voltage, V) exists between the ends of a wire then a current with strength I (Ampère, A) is induced according to *Ohm's law*:

$$R = \frac{\Delta V}{I} \tag{1}$$

with R (Ohm, Ω) being the electrical resistance of the wire. R depends on the cross-sectional area A of the wire and its length g, so it depends both on the type of material and how much of it is present. To characterize the material itself, the electrical resistivity ρ (Ω m) is used:

$$\rho = R\frac{A}{g} \tag{2}$$

R is measured, but it has to be converted into ρ to characterize the material. The reciprocal of ρ is the electrical conductivity σ (Siemens per m or S m^{-1}).

Ohm's law can be generalized to any material, so both ρ and σ can be used to characterize different types of material. Some typical values of σ (mS m^{-1}) are: quartz: 10^{-8}–10^{-11}; sandy soil (moist): 5–25; clayey soil (moist): 40–100; sea water: 5000; iron: 10,000.

Since most particles of the sand and silt fractions are quartz, this material can be considered as electrical neutral and are inert to the passage of a current. Dry clay is also electrical inert. But, since clay minerals are mostly moist and negatively charged due to isomorphic substitutions in their crystalline build-up, these minerals have a capacity to retain ions in solutions near their surface. This capacity is called the cation exchange capacity. If a soil is moist the retained ions are in equilibrium with the surrounding pore solution. Other soil colloids such as organic matter have a similar capacity. Therefore, pathways involving the liquid phase will dominate the passage of an electrical current.

If a soil contains free salts (e.g. due to a saline ground water) it will be very conductive. However, in the absence of free salts the finer pores tend to retain more ions than the often empty or only partially filled large pores. Therefore, under salt-free conditions compaction increases σ. Any porous-free material which prevents the passage of electrical currents, such as a gravel layer, a bedrock (of non-conductive minerals) or a buried wall, will strongly reduce σ. On the other hand, metal (both ferrous and non-ferrous) is an excellent conductor. Finally, temperature has a positive effect on σ because it decreases the viscosity of the soil solution and increases the ionic activity of the dissolved ions.

Understanding the mechanisms of current flow is crucial to interpret bulk measurements of electrical conductivity and resistivity. Especially changes and contrasts in their values between different locations, or over time, are of interest.

3.2 Magnetic Susceptibility

The magnetic susceptibility χ is the capacity of a material to increase the strength of an external magnetic field. When a magnetic field with strength **H** (the *magnetic field intensity*, A m^{-1}) is applied to a medium it will cause the material to become magnetized as a function of its magnetic susceptibility. The resulting *magnetic induction* **B** (Tesla, T) is given by:

$$\mathbf{B} = \mu_0(1 + \chi)\mathbf{H} \tag{3}$$

with μ_0 the *magnetic permeability of vacuum* ($=4 \times 10^{-7}$ T A^{-1} m). χ is dimensionless, but since its value is different depending on units used, often "SI" is written to indicate that the units which have been used are according to the international system of units (SI).

B can be measured passively with a magnetometer, which is a widely used type of sensor in archaeology, but since it is not based on active electromagnetic signals, this method will not be discussed here.

χ can be measured in the laboratory or in a soil profile with a handheld magnetic susceptibility sensor. Some reference values of χ of various materials are (in 10^{-3} SI): quartz: −0.01; clay minerals: 0.2; granite: 2.5; basalt: 70; hematite: 6.5; magnetite: 6000. It can be observed that mainly the ferrous minerals, and especially magnetite ($Fe^{3+}{}_2Fe^{2+}O_4$), have a strongly elevated magnetic susceptibility.

Given the key role of magnetite, its content in soil is very important because it is a naturally magnetized mineral. Moreover, magnetite is very resistant to weathering so it tends to accumulate in sediments and topsoil. Consequently, soil developed in situ on magnetite rich rocks, such as basalt, will have a much stronger anomaly than soil developed from magnetite poor rocks, such as granite or limestone. Additionally, many human soil interferences, such as digging, burying objects or building fundamentals, will cause disruptions of the distribution of the magnetic susceptibility within the soil profile and can therefore be used to trace these activities.

3.3 Dielectric Permittivity

When a direct electrical current flows through a wire it produces an associated magnetic field **B** with field lines perpendicular to the direction of the current (*Ampère's law*). When this current oscillates back and forth at a high frequency (alternating current in order of MHz) then the associated magnetic field oscillates in strength and direction producing an outgoing electromagnetic (EM) wave. The properties of such an EM wave are described by the Maxwell equations. The first of these is Ampère's law with Maxwell's correction. It states that a magnetic field is generated by the combination of an electrical field (Ampère's law) and the effects of the changes of this electric field over time. The differential form of this equation is:

$$\nabla \times \mathbf{B} = \sigma \mathbf{E} + \varepsilon \frac{\delta \mathbf{E}}{\delta t} \tag{4}$$

with ∇ the nabla-operator describing the curl of the magnetic field **B** is the magnetic field intensity (as in Eq. 4), and **E** is the electrical field intensity (V m^{-1}) and ε is the dielectric permittivity. The first term on the right hand side is the effect from a conduction current described by Ohm's laws, while the second term describes the displacement current. The frequency of the oscillating current determines the frequency of the EM field, which is for geophysical purposes in the order of kHz to GHz. When this frequency is low, i.e. in the order of kHz, then the displacement current can be neglected. In that case the electrical field just creates a magnetic *field*. However when the frequency is in the order of GHz to THz, then the field behaves as an EM *wave*.

EM waves can be imagined as a self-propagating transverse oscillating wave of electric and magnetic fields moving in phase perpendicular to each other. If either the magnetic of electrical component of the field is lost (attenuated, absorbed or conducted away) the wave will die. In air however, the wave will propagate continuoùsly until it encounters a medium that absorbs or reflects it.

The dielectric permittivity ε of a material is a measure of the capacity to transmit energy under the form of an EM wave. Put otherwise, the dielectric permittivity is a measure of the capacity of a material to slow down an EM wave according to:

$$v \approx \frac{c}{\sqrt{\varepsilon_r}} \tag{5}$$

with v the velocity of the EM wave in the material, c the speed of light in a vacuum or, by approximation, in air, which is 0.30 m ns^{-1}. ε_r is the relative dielectric permittivity defined as:

$$\varepsilon_r = \frac{\varepsilon}{\varepsilon_0} \tag{6}$$

with ε_0 being the dielectric permittivity of vacuum (=8.85 10^{-12} C^2 N^{-1} m^{-2}). Since ε_r is dimensionless, v has the same units as c. In PSS it is custom to express v in cm per ns. Some values of ε_r for some selected materials are: fresh water: 80; ice: 3–4; sand (dry-wet): 3–20; granite: 4–8; clay (dry-wet): 5–40; moist organic rich soil: 12–15. In soil, the water content is the most important factor retarding the passage of EM waves. Hence, soil textures which tend to be wet with small pores, such as clayey soils, typically have large value of ε_r.

4 Electrical Resistivity Sensors

4.1 Principle

Electrical resistivity (ER) sensors are active and invasive. Their aim is to measure the resistivity of the bulk soil and eventually use several measurement geometries to determine the resistivity of individual soil layers. These measurements can support soil or geological surveys or help to locate contrasting features, either natural or anthropogenic.

Two current electrodes are entered into the topsoil and an electrical current I is passed between them. In its simplest configuration, two other electrodes, the potential electrodes, are used to measure the potential difference ΔV created by this current. From these measurements the resistance R of the soil can be derived through Ohm's law (1). Finally, given de configuration of the setup, R is converted into resistivity ρ. Since these measurements are bulk measurements involving all layers within the volume influenced by the current, it is common to refer to the obtained property as the *apparent resistance* R_a or the *apparent electrical resistivity* ρ_a.

4.2 Instrumentation

Several configurations of electrode arrays can be used. Some are hand-held, others mobile. The hand-held systems are mainly used in detailed archaeological prospections, where the mobile variants serve more investigations at a landscape scale, often aimed at supporting precision agriculture.

The simplest configuration is the *Wenner array*, where the potential electrodes are positioned inside the current electrodes and the distance between all four electrodes is equal. When the resistance is measured it can be shown that for this configuration the resistivity is obtained from: $\rho_a = 2\pi dR_a$ with ρ_a the apparent resistivity and d the distance between two electrodes. The simplicity of the Wenner array is counterbalanced by the need to carry around a wide boom (if $d = 1$ m the boom is 3 m wide!) containing 4 electrodes which have to be pushed into the soil each time a measurement is needed.

As an alternative to the less handy Wenner configuration the *Twin Probe* configuration was developed. Here two pairs of a current and a potential electrode are used where one pair is put in a fixed remote position and the other pair is moved around. Both pairs are connected by a power cable. The distance between both pairs must be at least 30 times the distance between the paired electrodes. So if the electrodes are 0.5 m apart, the distance between both pairs must be at least 15 m. When this condition is met, we can assume that the distance between both electrode pairs is "infinite", which allows simplifying the conversion from R into ρ_a to: $\rho_a = \pi dR_a$ with d the distance between the current and potential electrodes of one pair (in m). When a is 0.5 m, the maximum depth of investigation is about 0.75 m. The major advantage of the Twin Probe configuration is the lower burden of carrying around and entering each time the electrodes. However one needs to keep an eye on the distance from the fixed pair of electrodes and be aware that a connection cable is lying around.

If one needs to investigate individually deeper layers, e.g. to identify saline groundwater, the separation distance between the electrodes needs to be increased. This is usually extended to a profile of 25 or 32 interconnected electrodes entered in the soil at regular intervals, usually 1 m for shallow investigations increasing up to 5 m for deeper studies. Using a Wenner configuration a computer addresses 4 electrodes in turn creating many combinations of equally spaced electrodes. These measurements can be visualized as a pseudosection of apparent resistivity values. Such a section can be used in inverse modeling to determine the resistivity of individual layers of the profile. Obviously for practical reasons this configuration is not very useful for spatial surveys, but it can complement such data with an improved depth resolution.

Mobile electrical resistivity measurements are usually based on the principle of a Wenner array but use rotating metal disks, sometimes even spiked, instead of pins. These disks are driven around by a terrain vehicle in a systematic way allowing producing a map of the ER of the surveyed area. So far, one company has commercialized such a system: Veris technologies (Fig. 1). The Veris 3100 system operates with 6 coulter disks which cut through the topsoil. One pair (disks 2–5) act as current electrodes, while disk pairs 1–6 and 3–4 are the potential electrodes. In this way simultaneously a topsoil (0–30 cm) and a subsoil (0–90 cm) measurement of the apparent electrical resistivity is obtained.

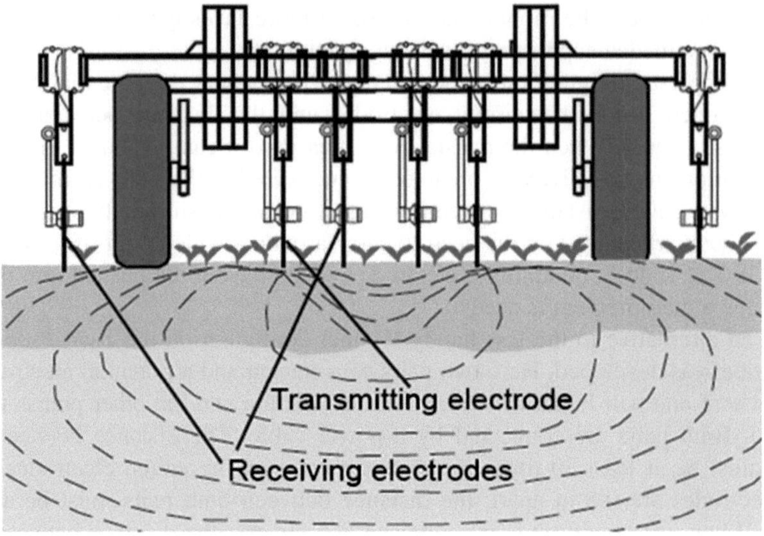

Fig. 1 The Veris 3100 soil EC mapper, an example of a mobile ER sensor (*Source* http://www.veristech.com/)

4.3 What Can Be Measured, Advantages, Limitations

A strong change in ER often indicates to presence of buried objects. Therefore these measurements are frequently used in archaeological surveys. Buried walls, stones, roads form mostly highly resistive anomalies while the infilling of ditches, pits, graves and metal objects cause low resistance measurements. Given the detailed nature of such surveys, often the manual ER-meters are used.

Vertical electrical sounding is mainly implemented in relationship to geo(hydro)logical investigations along transects. Occasionally this technique is used to supplement the spatial investigations of ρ_a.

For soil surveys, eventually supporting agricultural planning, resistivity measurements are frequently used, although for such application it is custom to express the outcome as EC. Given the large area covered during such surveys these are mostly performed with a mobile equipment.

The major *advantages* of ER measurements are:

- It is insensitive to external influences such as magnetic storms, electrical wires, high-power lines…
- It is not influenced by small metallic objects in the topsoil.
- Electrical imaging allows obtaining information about the stratigraphy and layered build-up of a soil.
- It provides stable measurements over time.
- It is a robust, simple and reliable technology.

The major *limitations* of ER measurements are:

- Bulk measurements can be dominated by a single layer or a large object which is very resistive or conductive (e.g. a perched water layer on top of a compacted horizon or a long wall).
- Soil contact can be difficult to maintain in stony terrain, in the presence of a dense crop or when yield residues are on the land, especially with the mobile systems.
- Under very dry topsoil conditions (desert, sandy soils...) the galvanic contact between the probes and the subsoil can be difficult to establish. Under such conditions ER measurements are best postponed until after rain or irrigation. Similarly, when the soil is covered with standing water, ER measurements cannot be performed.
- Not suitable to detect small metallic objects.

5 Ground Penetrating Radar

5.1 Principle

Since most chapters of this book focus on ground penetrating radar (GPR), here a brief account of the aspects relevant for soil investigations with this type of sensor will be given.

GPR is an active sensor which is mostly used with soil contact ("ground-coupled"), but it can also be used in an air-coupled configuration. The aim of GPR is to detect interfaces with contrasting dielectric properties. A short pulse of an EM-wave is sent out into the soil by a transmitting antenna. This wave will be partly reflected by the interface and partly refracted. The reflected wave is captured by a second, receiving antenna. The refracted part of the wave proceeds deeper into the soil and if it encounters another interface, this process may repeat several times until the wave has lost all its energy.

The dominant property determining the intensity (amplitude) of a reflection is the contrast in the relative dielectric permittivity of an interface. In soil, this is largely determined by variations in water content.

A key feature of GPR instruments is the frequency of the antenna. For soil investigations, the *central value* of such antennas is usually within the range of 100–500 MHz, but this value can vary widely, from 20 till 3000 MHz. The higher the frequency, the smaller its physical size, the finer its horizontal and vertical resolution and the shallower the penetration depth (Table 1).

Table 1 Frequency, penetration depth and resolution for several GPR frequencies (English Heritage 2008)

Central frequency (MHz)	Approx. maximum penetration depth in soils (m)	Wavelength λ in soil with $\varepsilon_r = 15$ (m)	Horizontal resolution at maximum depth (m)	Vertical resolution (= λ/4) (m)
1000	<1	0.08	0.2	0.02
500	2	0.16	0.4	0.04
200	3	0.39	0.8	0.10
100	5	0.77	1.4	0.19
50	7	1.55	2.4	0.39

5.2 Instrumentation

GPR instruments can be divided in time-domain and frequency-domain instruments. The first type is dominant on the commercial market. A time-domain GPR sends out a short pulse and records the reflections as a function of time. In a frequency-domain GPR, a continuous wave is transmitted at each frequency which is step-wise increased. For each frequency the characteristics of the reflected wave are compared to the transmitted signal. Frequency-domain data can be transformed into the time-domain by an inverse Fourier transformation.

In its simplest form, a time-domain GPR consist out of two antennas and a control unit, often showing in real time the reflections trace as one moves ahead (Fig. 2a). By combining more antenna-pairs the efficiency of prospection can be strongly increased. Today, full 3-D systems are available combining up to 16 GPR traces at a few cm apart. Frequency-domain GPR instruments are less common, but have some advantages. The measurement frequency can be modified more easily then with time-domain systems and the antenna-array can be lifted above the soil surface improving its manoeuvrability on uneven terrain (Fig. 2b).

Fig. 2 **a** a manually pushed time-domain GPR requiring ground-contact (GSSI, double frequency). **b** A mobile stepped frequency antenna array operated in air-coupled mode (3D-Radar)

5.3 What Can Be Measured, Advantages, Limitations

GPR instruments are excellent for positioning and delineating soil contrasts (layering) and buried objects. In hydrogeophysics, GPR is used to monitor soil water and ground water dynamics. Also non-destructive archaeology uses GPR intensively, especially when building structures need to be identified.

Some of the limitations of GPR for soil investigations are:

- Compared to other soil sensors, the optimal operational conditions for GPR are quite restrictive. In some situations the success of implementing a GPR can be limited, such as on heavy clay, saline soil, rough terrain, standing surface water, below the groundwater table...
- Post-processing requires strong insight into the physics of the method.
- Sensitive to disturbances by other EM signals such as mobile phones, radio...
- The use of GPR is prohibited near airfield and military installations.
- Due to the heavy data load, data storage and post-processing capacities can be quite system demanding, especially with the antenna array systems.

6 Electromagnetic Induction Sensors

6.1 Principle

Recall Eq. (4). When the frequency of the oscillating electrical current is in the order of kHz, the second term on the right hand side of the equation can be neglected, reducing it into: $\nabla \times \mathbf{B} = \sigma \mathbf{E}$, being Ampère's law. Under such conditions, this equation indicates that an electrical current $\mathbf{E}$ creates an associated magnetic field $\mathbf{B}$ with the electrical conductivity σ as influencing variable. In this case the associated alternating magnetic field does not behave as a propagating wave as with GPR.

The Law of Faraday, being the second Maxwell equation, states that a changing magnetic field induces an electric current in the affected material with its magnetic permeability μ as influencing property:

$$\nabla \times \mathbf{E} = -\mu \frac{\delta \mathbf{B}}{\delta t} \tag{7}$$

So, in the case of two nearby copper coils, an alternating electrical current applied to one coil creates an associated alternating magnetic field which will induce an alternating electrical current in the second coil. By measuring the second current, information on the strength of the first current is obtained without any physical contact between both coils.

In its simplest configuration, a frequency-domain electromagnetic induction (EMI) sensor consists out of a transmitter and a receiver coil positioned at a given intercoil distance and configuration (Fig. 3). An alternating current is

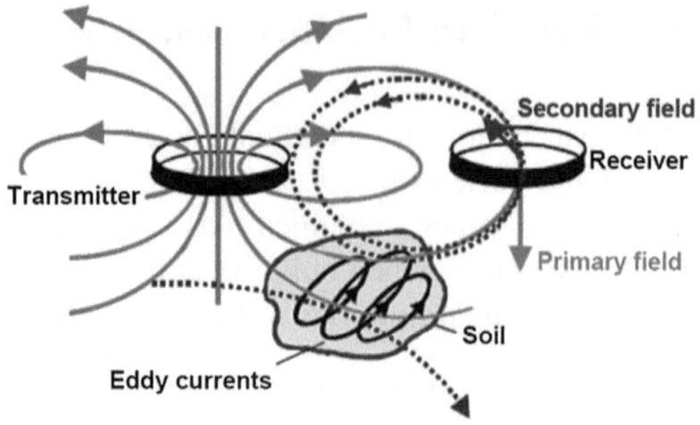

Fig. 3 Principle of electromagnetic induction with coils in a horizontal coplanar configuration. See text for explanation

passed through the transmitter coil with a fixed oscillating frequency. According to Ampère's Law, a primary magnetic field is created by the transmitter coil and this field propagates above and below the sensor. A part of the field lines of this primary field will pass through the receiver coil. According to Faraday's law, these magnetic field lines will create an electrical current through the receiver coil, which can be measured.

In the case where the sensor is near to soil, the primary magnetic field will create alternating eddy currents in the soil. The nature of these currents will depend on the electrical conductivity of this material. These currents will induce a secondary magnetic field. Some of the field lines of this secondary field will reach the receiver coil and induce an secondary electrical current, which can be measured as well.

EMI sensors are designed to separate the characteristics of the primary and secondary EM fields. The primary field generates instantaneously an alternating current in the receiver coil. On the other hand, the induced eddy currents take a finite time to generate. Consequently, there is a time gap between the primary and secondary magnetic fields at the receiver coil causing them to be out-of phase. The difference between the primary and secondary fields is expressed as the in-phase (IP) and quadrature-phase (QP) amplitudes. The IP-response is the amplitude measured when the primary field is at its maximum; the QP-response is the amplitude when the primary field is zero. Both components are divided by the maximum amplitude of the primary field so that the output is given as two ratios. For soils with a σ_a of less than about 100 mS m^{-1} (known as "*operating at low induction numbers*"), McNeill (1980) showed that the apparent electrical conductivity (EC_a) can be obtained from the QP-ratio. The same signal allows obtaining the apparent magnetic susceptibility (MS_a) of the medium from the IP-ratio. When the condition of low induction number does not hold, the two responses are no longer separable.

A recent paper by Doolittle and Brevik (2014) reviews the current state-of-the-art in respect to the application of electromagnetic induction in soil studies.

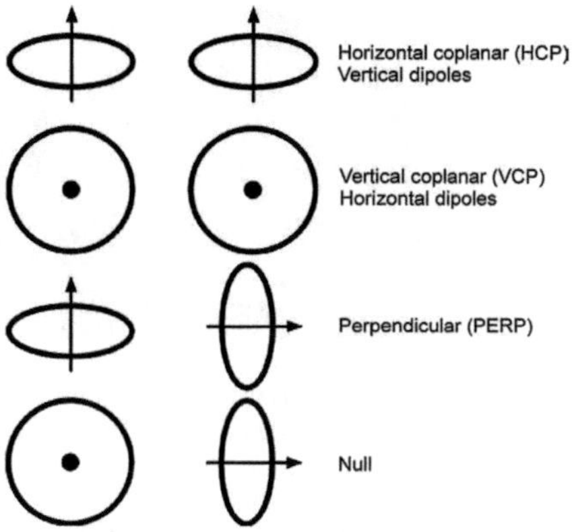

Fig. 4 Some geometries of EMI coil configurations (Simpson 2009)

6.2 Electromagnetic Induction Sensors

An electromagnetic induction sensor is an active, non-invasive sensor which uses an EM-field to characterise both and simultaneously the EC_a and the MS_a of soil material. By varying the configuration of the instrument, or by using multiple coil configurations simultaneously, different depth responses are obtained allowing converting the bulk measurements into the contribution of specific depth intervals. The geometry of the transmitter and receiver coils which can vary according to Fig. 4. The Null configuration produces only noise.

EMI sensors are available in many different configurations. The two most common EMI-systems for soil applications are the Geonics and the Dualem sensors.[1] The basic versions of both brands (the EM38 and the Dualem-1S) have an intercoil distance of 1 m. The EM38 has a single coil configuration in HCP (or VCP) orientation operating at 14.6 kHz, while the Dualem-1S has two receiver coils, one in HCP and one in PERP orientations, and operates at 9.0 kHz. More recently both companies have constructed more versatile variants of these standard models:

- Geonics EM38-MK2 has two HCP intercoil distances: 0.5 and 1 m, measuring each simultaneously EC_a and MS_a. This model is very popular for agricultural applications since the 0.5 m intercoil configuration focusses mainly on the topsoil. One of the drawbacks of these instruments is regular need for calibration. Another is the relative large noise component of the 0.5 m configuration. The depth of influence (DOI) for the two coil configurations is 0.75 and 1.5 m respectively. The EM31 has an intercoil distance of 3.66 m and operates at 9.8 kHz. In HCP configuration it has a DOI of 6 m.
- Dualem-21S has four coil configurations: both at 1 and 2 m distance a HCP geometry and at 1.1 and 2.1 m a PERP geometry, measuring each simultaneously EC_a and MS_a. A major advantage of Dualem instruments is their ability

[1] www.geonics.com and www.dualem.com.

Fig. 5 Two mobile configuration of an EMI sensors. *Left* the handheld EM38–MK2. *Right* the Dualem-21S inside a towed sled

for auto-calibration. The DOI's of these four configurations is 0.5 m (1.1 m PERP), 1 m (2.1 m PERP), 1.5 m (1 m HCP) and 3.0 m (2 m HCP). There exists also two longer versions: Dualem-421S and Dualem 642, each with six pairs of coils at 1, 2 and 4 m or 2, 4 and 6 m resp. Each distance has coils at HCP and PERP orientations. The 6 m HCP has a DOI of 9 m.

Besides these two companies, also GSSI offers an EMI sensor: the Profiler EMP-400 which has a single coil pair at 1.22 m distance and allows to modify the frequency between 1 and 15 kHz.

Since the EMI sensors have an immediate response and no physical contact is required they can easily be used in a mobile configuration, either manually or motorized (Fig. 5). In the motorized configuration several ha's can be survey per day with a measurement density of 0.2 (in-line) by 1 m (between-lines).

6.3 What Can Be Measured, Advantages, Limitations

Many soil properties are related with the EC and MS of soil. However, in practice mostly one or a few of these properties dominate the response of an EMI sensor. If salinity is present then it will be the major effect which is measured. However, in moist non-saline conditions the EC_a measurements reflect mainly the influence of the clay content (Saey et al. 2009). Changes in the moisture content generally do not affect patterns of variability, but they influence the absolute measurements (more moisture = higher EC_a). Mostly, variations in MS_a indicate human interferences (buried objects, burned or heated material) and therefore this property is very useful in archaeological studies (De Smedt et al. 2013a, b). EMI systems also allow the detection of ferrous and non-ferrous metals (Saey et al. 2011). As a consequence, EMI sensors are, together with mobile ER sensors, the most used soil sensor for general soil mapping purposes.

With multi-coil EMI systems, the simultaneous EC_a measurements, each with its own depth response curve, can be combined to isolate the contribution of a given depth interval to the overall EC_a measurement or to determine the depth of a contrasting layer (Saey 2011; Saey et al. 2012a, b). When multiple measurements are available, it is possible to reconstruct the build-up of the soil layering, similar to the profiling with electrical resistivity. Today software is available for this inverse modelling or electromagnetic tomography (e.g. http://www.emtomo.com/).

The advantages of EMI are:

- Only system producing simultaneously 2 geophysical properties: EC and MS.
- Quick to start-up, easy to operate and relatively light to transport.
- No contact is required with soil, so EMI systems can be lifted above the soil surface or even used on a floating raft under flooded conditions (e.g. Islam 2012). Hence EMI can even be used to investigate surface waters.
- Can be used under a wide range of conditions: saline–non-saline, dry–water saturated, all types of soil texture (heavy clay, pure sand, silt), peat... However, the signal-to-noise ratio is the smallest in dry and sandy (i.e. high resistive) conditions, so then the measurements are less suitable to identify subtle changes.
- Multi-receiver systems allow obtaining information about the stratigraphy and layered build-up of a soil.

Limitations of EMI as a soil sensing system are:

- Sensitive to external EM-fields caused by nearby electrical wires, fences, high-power lines...
- Measurement stability can be a problem.
- Strongly influenced by the presence of metallic objects (could be both a disadvantage or an advantage).
- Systems which are not auto-calibrating can require frequent re-calibration which makes subsequent measurements difficult to compare.
- Resolution can be somewhat coarse for very detailed investigations, due to the intrinsic nature of the sensor.

7 Conclusions

Soil is a complex and heterogeneous material fulfilling a wide range of functions and services to the environment. Its characterisation is required for efficient management and environmental health, yet this is complicated by a large variability, both horizontally and vertically, of its properties. Since physical sampling is limited to punctual locations, soil sensing is more and more used to characterise soil spatially. Among the wide variety of proximal soil sensors available, the ones operating on the basis of electrical, magnetic and electromagnetic principles are the most frequently used.

ER sensors focus on the measurement of the electrical conductivity but are limited by the requirement to insert electrodes into the soil. Mobile ER sensors use discs or spiked to reduce this limitation. ER measurements are very suitable to reconstruct the stratigraphy of soil.

GPR focusses on changes in dielectric permittivity and is very well suited to map these. Limitations to applications is soil are linked to operational difficulties under very wet and clayey conditions.

EMI is de most commonly used soil sensor because EC and MS reflect strongly the status of a number of key soil properties such as texture and the presence of anthropogenic disturbances. The recent evolution to expand these sensors to multi-receiver instruments expands further the applicability, including also the inverse modelling of the soil profile build-up.

Despite the strengths of every type of sensing system, the future will bring an increased integration of soil sensors into multi-sensor configuration allowing their fused processing. A wide range of applications will benefit from such technology.

References

Blum, W.: Functions of soil for society and the environment. Rev. Environ. Sci. Bio/Technol. **4**, 75–79 (2005)

De Smedt, P., Saey, T., Lehouck, A., Stichelbaut, B., Van De Vijver, E., Islam, M.M., Meerschman, E., Van Meirvenne, M.: Exploring the potential of multi-receiver EMI survey for archaeological prospection: a 90 ha dataset. Geoderma **199**, 30–36 (2013a)

De Smedt, P., Van Meirvenne, M., Herremans, D., De Reu, J., Saey, T., Meerschman, E., Crombé, P., De Clercq, W.: The 3-D reconstruction of medieval wetland reclamation through electromagnetic induction survey. Sci. Rep. **3**, 1517–1522 (2013b)

Doolittle, J.A., Brevik, E.C.: The use of electromagnetic induction techniques in soils studies. Geoderma **223–225**, 33–45 (2014)

English Heritage: Archaeological Field Evaluation. English Heritage, UK. Product Code 51430 (2008)

Islam, M.M.: Ability of a non-invasive sensor to improve soil management of paddy rice fields. Ph.D. thesis UGent (2012). ISBN 978-9-0598956-2-1

Jenny, H.: Factors of Soil Formation: A System of Quantitative Pedology. McGraw Hill, New York (1941)

McNeill, J.D.: Electromagnetic Terrain Conductivity Measurement at Low Induction Numbers. Technical Note TN-6. Geonics Limited, Mississauga (1980)

Saey, T.: Integrating multiple signals of an electromagnetic induction sensor to map contrasting soil layers and locate buried features. Ph.D. UGent (2011). ISBN 978-90-5989-426-6

Saey, T., Van Meirvenne, M., Vermeersch, H., Ameloot, N., Cockx, L.: A pedotransfer function to evaluate soil profile heterogeneity using proximally sensed apparent electrical conductivity. Geoderma **150**, 389–395 (2009)

Saey, T., Van Meirvenne, M., Dewilde, M., Wyffels, F., De Smedt, P., Meerschman, E., Islam, M.M., Meeuws, F., Cockx, L.: Combining multiple signals of an electromagnetic induction sensor to prospect land for metal objects. Near Surf. Geophys. **9**, 309–317 (2011)

Saey, T., De Smedt, P., Islam, M.M., Meerschman, E., Van De Vijver, E., And, Lehouck A., Van Meirvenne, M.: Depth slicing of multi-receiver EMI measurements to enhance the delineation of contrasting subsoil features. Geoderma **189–190**, 514–521 (2012a)

Saey, T., Islam, M.M., De Smedt, P., Meerschman, E., Van De Vijver, E., Lehouck, A., Van Meirvenne, M.: Using a multi-receiver survey of apparent electrical conductivity to reconstruct a Holocene tidal channel in a polder area. Catena **95**, 104–111 (2012b)

Simpson, D.: Geoarchaeological prospection with multi-coil electromagnetic induction sensors. Ph.D. UGent (2009). ISBN 978-90-5989-344-3

Van Meirvenne, M.: Is the soil variability within the small fields of Flanders structured enough to allow precision agriculture? Precis. Agric. **4**, 193–201 (2003)

Viscarra Rossel, R.A., McBratney, A.B., Minasny, B. (eds.) Proximal Soil Sensing. Progress in Soil Science Series. Springer, New York (2010)

Viscarra Rossel, R.A., Adamchuk, V.I., Sudduth, K.A., McKenzie, N.J., Lobsey, C.: Proximal soil sensing: an effective approach for soil measurements in space and time. In: Sparks, D.L. (ed.) Advances in Agronomy, vol. 113, pp. 237–282. Academic Press, Burlington (2011)

Applications of Radar Systems in Planetary Sciences: An Overview

Fabio Tosti and Lara Pajewski

Abstract This chapter aims at reviewing remarkable results and sophistication of radar systems achieved over the history in several planetary explorations by dividing the treatment according to different planets and celestial bodies investigated. Both established and novel radar-based techniques for space exploration are described within an overall top-down approach being consolidated over years. As a result of the review, future perspectives of the research are highlighted and some benefits and limitations of different techniques are described. In line with this, increasingly reliable surveys are expected in the next few years, which can provide important information in the understanding of past and present natural phenomena as well as to sustain future human explorers and look for clues of habitable zones.

1 Introduction

This chapter includes some of the work that has been carried out on radar systems for both remote sensing and on-site inspection of the planets, and describes the main results achieved in space explorations over years. Basically, radar systems for planetary explorations can be mounted on aircraft or on satellites. In the last few years and according to ongoing and future space mission purposes already planned, typical ground-penetrating (GPR) systems installed on robotic rover vehicles have been and will be employed to detect the subsurface characteristics of planets and celestial bodies in general. Overall, terrestrial planets and satellites including Mars, Venus, Mercury and Moon have undoubtedly the Earth as

F. Tosti (✉) · L. Pajewski
Department of Engineering, Roma Tre University,
Via Vito Volterra 62, 00146 Rome, Italy
e-mail: fabio.tosti@uniroma3.it

L. Pajewski
e-mail: lara.pajewski@uniroma3.it

A. Benedetto and L. Pajewski (eds.), *Civil Engineering Applications of Ground Penetrating Radar*, Springer Transactions in Civil and Environmental Engineering,
DOI 10.1007/978-3-319-04813-0_15

the only ground-truth reference, such that surface features and processes observed on these planetary bodies can be therefore interpreted by correlation with those observed on the Earth. Based on such undeniable milestone, bright research activities and continuous technological advances have led the radar technology as a reference and essential mean in planetary exploration since the beginning of research activities in this field and throughout the years for the near future. The knowledge gathered by means of missions occurred over this relatively short time lapse for mankind history, has therefore allowed to reconstruct very complex scenarios and it has paved the way for future research purposes to be pursued.

This Chapter will endeavour to provide an insight into remarkable results and sophistication of radar systems over the history in several planetary explorations, together with the most recent findings achieved through the application of radar technology and future challenges.

The topics covered in this Chapter are the following:

- Planetary radar systems—main radar categories and investigations on Venus and Titan.
- Radar missions and explorations over history.
  - Exploration of the Moon.
  - Mars exploration—historical missions and future goals.
  - Recent advances and perspectives in GPR-based planetary missions.

2 Planetary Radar Systems: Investigations on Venus and Titan

Basically, two main categories including most of the radars flown as payload in planetary missions can be broadly outlined, namely, (i) imaging and (ii) sounding radars.

Amongst the first group, it is now established how heavy cloud cover around planets can be only penetrated by synthetic aperture radars (SAR) for high-resolution imaging (Fig. 1a, b) (Curlander and McDonough 1991). Historically speaking, SAR and radar altimeters (RA) (Fu and Cazenave 2001) were two radar systems used in two consecutive Soviet spacecrafts, i.e., Venera 15 and Venera 16 (Barsukov et al. 1986; Basilevsky et al. 1986), for imaging the planet of Venus and mapping its surface.

Still on Venus surveys, the robotic space probe Magellan, also referred to as the Venus Radar Mapper (Saunders et al. 1990), was launched by the National Aeronautics and Space Administration (NASA) later on in 1989 to map the surface of this planet by using a SAR system, and for measuring its planetary gravitational field. Still with regard to imaging radars, Titan satellite, also known in Astronomy as the largest moon of Saturn, is being currently investigated by the radar of the Cassini-Huygens probe (Porco et al. 2005), in order to penetrate its opaque atmosphere. In particular, this is a more sophisticated multimode radar system, capable to operate as a SAR, RA, scatterometer (Jones et al. 1982) and

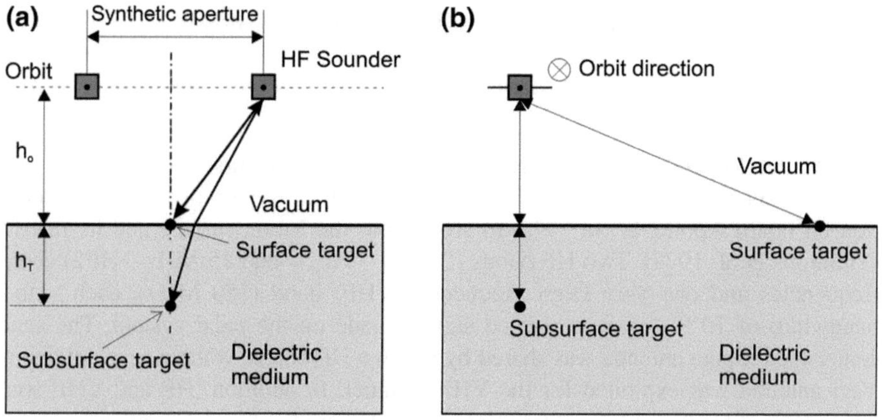

Fig. 1 Sketch of the SAR observation model, being h_o the height of the orbit, and h_T the depth of the subsurface target **a** front view, **b** lateral view

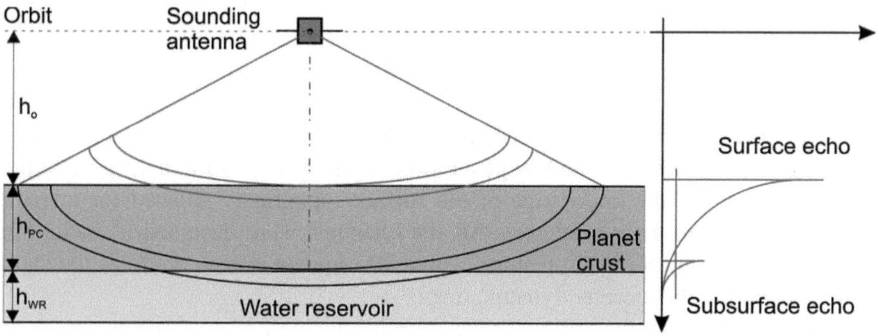

Fig. 2 Radar sounding geometry: sketch of water reservoir detection in planetary explorations being h_o the height of the orbit, h_{PC} the depth of the planet crust layer, and h_{WR} the depth of the water reservoir

radiometer (Draper 1976). Amongst the main results achieved to date, evidence for a large underground ocean of liquid water on Enceladus, a moon of Saturn, had been found and announced by NASA, thereby indicating that Enceladus is one of the most likely places in the Solar System to host microbial life.

Sounding radars (Fig. 2) are generally grouped within the low-frequency class (normally, High Frequency—HF—in the $3 \div 30$ MHz range, or lower) of GPR systems, from those used to acquire data about the sub-surface structure of planets. As a point of strength, such systems enable penetration depths below the surface up to hundreds of meters, or even kilometres, due to their low operating frequency. To overcome the receiving of unwanted echoes from other surfaces, due to the intrinsic width of the beam, which is related in turn to the low operating frequency and to the small allowable antenna dimensions, synthetic aperture techniques are normally exploited.

3 Radar Missions and Explorations Over History

3.1 Exploration of the Moon

The Apollo Lunar Sounder Experiment (ALSE) was the first radar sounder ever flown in space missions (Porcello et al. 1974). It was part of the instrumentation on board Apollo 17, in 1972, to study both the Moon surface and its interior (Simmons et al. 1974). Two HF bands (5 MHz—HF1—and 15 MHz—HF2) center frequencies and one Very High Frequency (VHF) band (150 MHz), each with a bandwidth of 10 % (using a chirped signal) made up the radar system. The same center-feed dipole antenna was shared by the two HF bands, while a seven-elements Yagi antenna was exploited for the VHF channel. In addition, HF and VHF were characterized by two different transceivers, and shared a common optical recorder. The ALSE system was designed by pursuing the best trade-off between penetration depth and resolution. According to the primary purpose of the space mission, namely, detecting information on the upper 2 km of the Moon crust, very long radar wavelengths were used, thereby exploiting the widespread dryness conditions of the Moon. Such situations enabled a much deeper penetration of the radar waves into the surface than in case of wet lunar rocks conditions. A minimum peak-to-sidelobes ratio of 45 dB after the third lobe was also provided to the system to limit any masking of weak subsurface echoes. To reduce the along-track clutter, the effective antenna footprint was narrowed by generating a synthetic aperture in the ground processing. Moreover, the knowledge of the surface topography allowed for inferring clutter from across-track scatterers. All the channels were designed to include an automatic gain control (AGC) feature, with a 30 s update rate, to optimize the signal allocation within the receiver dynamic range.

Structures beneath the surface were revealed in both Mare Crisium and Mare Serenitatis lunar basins. In particular, such layers were observed in several different parts of these basins, thereby proving to be reasonably a widespread feature. Basically, the basalt filling both of these mare basins is supposed to include these layering structures. More in details, depths of 0.9 and 1.6 km below the surface for Mare Serenitatis and 1.4 km for Mare Crisium were detected. In addition, an estimated total basalt thickness of between 2.4 and 3.4 km was found in Mare Crisium. Important information about wrinkle ridges on the Moon were also provided, to support the idea that these have formed primarily by motions along faults.

Overall, despite a renewed interest in lunar exploration has brought to the attention of the scientific community a variety of remote sensing instruments, as partly mentioned above, it is worth citing how the interpretation of these data usually depends upon geophysical models, and sometimes the obtained results can be affected by important ambiguities (e.g., Campbell 2002). Therefore, it should be crucial to validate any interpretation by in situ measurements before extending globally confident conclusions. This is the reason why the unmanned lunar exploration mission Chang'e 3 (Zou et al. 2014) operated by the China National Space Administration (CNSA) was launched in December 2013 as part of the second phase of the Chinese Lunar

Exploration Program becoming the first spacecraft to soft-land on the Moon since the Soviet Union's Luna 24 in 1976 (Akhmanova et al. 1978). Basically, it incorporates a robotic lander and the China's first lunar rover, called Yutu. With the aim to provide useful recommendations for the manufacturing of the system, Fa (2013) has attempted to analyse the optimal configuration for this radar by simulating A-scan and B-scan profiles, and the effect of lunar subsurface properties on GPR echoes was also discussed. The installed system implemented for the mission to the Moon is characterized by a dual-frequency GPR operating at frequencies of 500 and 60 MHz to provide a combination of deep penetrations depths and high subsurface resolution. In more details, such system is capable to reach a penetration depth ranging from 2 to 50 m for maria and greater than 100 m for highlands. The HF band provides data with a range resolution from 20 to 25 cm, while the low frequency band is limited to a resolution ranging from 2 up to 2.5 m (Sun and Zhang 2013). With regard to these characteristics, Chang'e 3 GPR analyses the shallow subsurface to characterize comprehensively surface geology, structure and dielectric properties.

3.2 Mars Exploration

In line with the interest aroused by Lunar explorations, planet Mars has been another very important target since the early flyby missions of the 1970s, mostly focused on understanding the role of water (liquid or solid) in the geological evolution of the planet as well as its potential in supporting biotic activity. There is a remarkable history of other sounding radar instruments flown here, which is however mainly concentrated within the very last decade. Indeed, many paths of research have dealt with the study of Mars, although they have been mostly based on the study of terrestrial analogues and on the development of more powerful remote sensing instruments, as it happened in the case of Lunar investigations. Nevertheless, the direct study of the Martian surface features and characteristics was taken as a primary key point for achieving a considerable advance in technology. Several research institutions consider the detection of water and its probable location a fundamental topic for identifying Mars as a potential source for past or even contemporary biotic activity. As a result, the use of geophysical methods such as GPR is being attracting an increasing interest among planetary scientists. More particularly concerning future goals of Mars explorations, the discovery of ground ice is certainly one of the most relevant goals to be reached, although it has been already largely discussed for years in the past (Barlow and Bradley 1990; Costard and Kargel 1995; Carr 1996), with many implications on the understanding of past and present natural phenomena.

The first approach taken for designing and testing of geophysical instruments for Mars was about focusing on individual techniques including seismic sounding (Nieto and Stewart 2003), Time Domain Electromagnetic (TDE), Surface Nuclear Magnetic Resonance (SNMR) sounding (Grimm 2003), and GPR (Olhoeft 1998; Arcone et al. 2002; Berthelier 2003). More recently, multi-frequency GPR systems (Stillman and Olhoeft 2006) were also combined with other geophysical methods in analog studies (Clifford 2006; Wainstein et al. 2008) with the perspective to maximize the capabilities of multiple tools in such very complex and remote investigations.

NASA's Mars Exploration Program includes long records of successful landers since the 1970s, with part of them also equipped with rover automated motor vehicles (NASA/JPL 2004). Amongst all, it is worth to mention the two Viking program landers in 1976 (Holmberg et al. 1980), the Mars Pathfinder probe in 1997 (Mishkin et al. 1998) equipped with the Sojourner rover, the two rovers sent in 2003 within the robotic space mission Mars Exploration Rover Mission (MER), called, respectively, MER-A Spirit (ended in May 2011) and MER-B Opportunity, and the Curiosity car-sized rover related to the NASA's Mars Science Laboratory mission (MSL). Neither from the aforementioned rover-based missions were equipped with GPR technology, although NASA has planned to launch the Mars 2020 GPR-equipped Rover within the homonymous mission in 2020 (Tahu and Shulte 2014).

Low-frequency ground penetrating radars are also currently used in other two space missions. Chronologically speaking, the Mars Advanced Radar Subsurface and Ionosphere Sounding (MARSIS) experiment on board the Mars Express Orbiter, namely, a European Space Agency (ESA) mission to Mars, was first launched in June 2003 and uses a high-gain antenna to sound from orbit (Picardi 2005). MARSIS has the twofold role to study the surface-atmosphere interaction, and to use the subsurface radar sounder for investigating possible water presence up to a depth of about 5 km.

Subsequently, the Shallow Subsurface Radar (SHARAD) was launched in August 2005. It was developed under the responsibility of the Italian Space Agency (ASI, Agenzia Spaziale Italiana) and provided to the Jet Propulsory Laboratory (JPL) for use on board the Mars Reconnaissance Orbiter Spacecraft in the frame of a NASA/ASI agreement, which foresees exploitation of the data by a joint Italian/US team. With the main goal to complement the lower frequency (maximum frequency of 5 MHz) and bandwidth (1 MHz) capabilities of the MARSIS instrument, it has allowed to probe the Martian subsurface to a depth of about 1 km through a 20 MHz center frequency and 10 MHz bandwidths (Seu et al. 2007). Strong reflections at a depth consistent with the elevation of the surrounding plains have been shown by MARSIS over the north polar cap, thereby suggesting minimal electrical attenuation through about 2 km of polar deposits and indicating possible presence of pure cold ice (Milkovich et al. 2007). The base of water-ice-rich polar layered deposits (PLD) has been reached by MARSIS at a maximum of 3 km depth at both poles. In addition, SHARAD detected icy layered sediments overlain by a layer made of carbon dioxide ice and dust mixture protecting the water ice from decay (Seu et al. 2007). Amongst the most significant findings, SHARAD has also identified layers within the South Polar Layered Deposits (SPLD), reflections originated by the contact between the SPLD and the underlying substrate at 1.4 km depth, such as also observed by MARSIS, and erosional unconformities (Putzig et al. 2007).

In addition, radio-transparent deposits have been detected by SHARAD in southern Elysium Planitia and, fairly similarly, in Amazonis Planitia, although a reflector within a depth from 40 and 90 m may suggest lava flows (Simon et al. 2012).

3.3 Future Challenges in Planned Mars Missions

After cancelling the NetLander mission (under the joint responsibility of the Centre National d'Etudes Spatiales—CNES—and ESA) in 2003, that should have been sent to Mars in 2009, the ExoMars (Exobiology on Mars) mission is currently under development by the ESA in collaboration with the Russian Federal Space Agency (Roscosmos). ExoMars is a two-mission project that is considered as a single programme at ESA. According to current plans, the ExoMars project will include four spacecrafts: two stationary landers, one orbiter and one rover. All mission elements will be sent in two launches, scheduled for 2016 and 2018. Amongst them, the Water Ice Subsurface Deposit Observation on Mars (WISDOM) GPR is one of the rover instruments that have been selected as part of the Pasteur payload (Ciarletti et al. 2011), planned to be sent in 2018. Basically, the WISDOM ultrahigh-frequency (UHF) GPR is a stepped-frequency radar with nominal step duration Δt lower than 1 μs and number of transmitted frequencies equals to 501, so that the bandwidth amounts to 2.5 GHz (from 0.5 to 3.0 GHz). It is also characterized by an output power ranging from +10 to +20 dBm, a noise factor lower than 5 dB, an instantaneous dynamics greater than 80 dB, and a power consumption lower than 11 W. The rover is equipped with a drill that can sample the subsurface down to a depth of approximately 2 m for ground-truth evidence of the radar data collected. Indeed, the WISDOM GPR main objective will be to explore the first 3 m of the soil at the centimeter-scale vertical resolution, according to the objectives and expected capabilities of the drill. The location of sedimentary layers wherein organic molecules are the most likely to be found and well preserved will be the specific target of WISDOM in order to search for subsurface evidence of past and present life at the landing site. In particular, WISDOM will help to understand the 3-D geology and geologic evolution of the landing site, to identify potential transient or persistent occurrences of liquid water and massive ground ice, identify the most promising locations for drilling as a compromise between the high scientific level and the safer conditions for drilling operations.

Later in time, the GPR equipment designed for Mars 2020 by NASA will consist of a radar called RIFMAX (Radar Imager for Mars's Subsurface Exploration), that is expected to survey the Mars surface within a 150–1440 MHz range of frequencies with a penetration depth from 10 up to 500 m. Its main role will be to gather information on the geological processes that shaped the sedimentary surface of the planet. NASA is also expecting to find evidence in the GPR surveys that Mars was a habitable environment in the past.

3.4 Recent Advances and Perspectives in GPR-Based Planetary Missions

Much more recently, the new challenge of reaching a comet nucleus has been overtaken in November 2014 by the ESA agency with the sending of the robotic space probe Rosetta (Glassmeier et al. 2007) to the comet 67P/Churyumov–Gerasimenko (67P), equipped with both an orbiter and a lander module called Philae. Within the instruments provided, the Comet Nucleus Sounding Experiment by Radiowave Transmission (CONSERT) (Kofman et al. 1998) will exploit the radar tomography technique to image the cometary interior. In this regard, the electromagnetic signal transmitted between the lander, located on the comet surface, and the spacecraft orbiting around the comet, will be measured as a function of time and as a function of the relative position of the orbiter and the lander. This operation will be repeated for a number of orbits. Concerning the electronics of the system, the transmitted signal comes from a 90 MHz center frequency antenna, modulated PSK code with a 100 ns code step. A set of 1024 codes is transmitted for each way, while the propagated signal is coherently accumulated by the receiver for increasing the Signal to Noise Ratio (SNR). Two cross half-wave dipoles, giving a circular polarization and a well-known antenna pattern, constitute the Orbiter antenna. On the other hand, the Lander antenna is made of two monopoles oriented 76° to each other, close to the nucleus surface: such pattern depends on the near subsurface dielectric properties which are currently unknown (Herique and Kolfman 2001). Basically, both the change in velocity of the electromagnetic wave-front propagated through the cometary nucleus (with respect to the velocity propagation in free space) and the energy loss during this process are dependent on the complex permittivity of the nucleus materials as well as on the ratio of the wavelength used to the size of the encountered inhomogeneities. By exploiting the above principles for a sufficient number of orbits, one will be able to obtain many cuts of the interior of the comet and therefore to build up a tomographic image of the interior.

4 Conclusions

Several radar techniques (GPR, sounders, radar tomography) which can be applied to explore surfaces and sub-surfaces of the Solar System bodies have been described and discussed in this Chapter. Such techniques have largely proved to be very promising in planetary explorations. According to a top-down approach undertaken over a timespan of few decades for investigating several planetary bodies of the Solar System, their topography is firstly reconstructed remotely along with their main physical properties at the large-scale, in order to understand the best sites to be explored. In a second step, unmanned spacecrafts are sent into the orbit of such planets to deepen those aspects emerged from the first set of surveys.

In this regard, the most widespread strategy from space agencies has been to send landers directly on the planet surface to provide finer-scale measurements. During the past few years and over the near future, the use of instrumented robotic rover vehicles is finally proving to be the last step and best approach for investigating relatively longer distances with the highest possible resolution of data. Such detailed scale of inspection, will enable to better assess the potential for biotic activity, to sustain future human explorers and look for clues of a habitable zone.

Acknowledgments The authors acknowledge the COST Action TU1208 "Civil Engineering Applications of Ground Penetrating Radar", supporting this work.

References

Akhmanova, M., Dement'ev, B., Markov, M.: Possible water in Luna 24 Regolith from the Sea of Crises. Geochem Int. **15**, 166 (1978)

Arcone, S.E., Prentice, M.L., Delaney, A.J.: Stratigraphic profiling with ground-penetrating radar in permafrost. A review of possible analogs for Mars. J. Geophys. Res. **107**(E11), 5108 (2002)

Barlow, N.G., Bradley, T.L.: Martian impact craters: correlations of ejecta and interior morphologies with diameter, latitude and terrain. Icarus **87**(1), 156–179 (1990)

Barsukov, V.L., Basilevsky, A.T., Burbal, G.A., Bobinna, N.N., Kryuchkovt, V.P., Kuzmin, R.O., Nikolaeva, O.V., Pronin, A.A., Ronca, L.B., Chernaya, L.M., Shashkina, V.P., Garanin, A.V., Kushky, E.R., Markov, M.S., Sukhanov, A.L., Kofelnikoy, V.A., Rzhiga, N., Petrov, G.M., Alexandrov, Y.N., Sidorenko, A.I., Bogomolov, A.F., Skrypnik, G.L., Bergman, M.Y., Kudrins, L.V., Bokshtein, L.M., Kronrod, M.A., Chochia, P.A., Tyuflin, Y.S., Kadnichansky, S.A., Akim, E.L.: The geology and geomorphology of the Venus surface as revealed by the radar images obtained by Veneras 15 and 16. Paper presented at the sixteenth lunar and planetary science conference, part 2, J. Geophys. Res., 91, B4, D399-D411 (1986)

Basilevsky, A.T., Pronin, A.A., Ronca, L.B., Kryuchkov, V.P., Sukhanov, A.L., Markov, M.S.: Styles of tectonic deformations on Venus: analysis of Venera 15 and 16 data. Paper presented at the sixteenth lunar and planetary science conference, part 2, J. Geophys. Res. 91, B4, D399-D411 (1986)

Berthelier, J.J.: GPR, a ground penetrating radar for the Netlander mission. J. Geophys. Res. **108**(E4), 8027 (2003)

Campbell, B.A.: Comment on "Regolith layer thickness mapping of the Moon by radar and optical data" by Y G. Shkuratov and N. V. Bondarenko. Icarus **158**, 560–561 (2002)

Carr, M.H.: Water on Mars. Oxford University Press, New York (1996)

Ciarletti, V., Corbel, C., Plettemeier, D., Caïs, P., Clifford, S.M., Hamran, S.-E.: WISDOM GPR designed for shallow and high-resolution sounding of the martian subsurface. Proc. IEEE **99**(5), 824–836 (2011)

Clifford, S.M.: Mars analog investigations of the West Egyptian desert utilising multi-frequency GPR and other electromagnetic sounding techniques. Paper presented at the 37th lunar and planetary science conference, p. 2442 (2006)

Costard, F., Kargel, J.S.: Outwash plains and thermokarst on Mars. Icarus **114**, 93–112 (1995)

Curlander, J.C., McDonough, R.N.: Synthetic Aperture Radar. Wiley Interscience, New York (1991)

Draper, C.W.: The Crookes radiometer revisited. A centennial celebration. J. Chem. Educ. **53**(6), 356 (1976)

Fa, W.: Simulation for ground penetrating radar (GPR) study of the subsurface structure of the Moon. J. Appl. Geophys. **99**, 98–108 (2013)

Fu, L.-L., Cazenave, A.: Satellite altimetry and earth sciences: a handbook of techniques and applications. Fu and Cazenave Eds., Academic Press (2001)

Glassmeier, K.-H., Boehnhardt, H., Koschny, D., Kührt, E., Richter, I.: The Rosetta mission: Flying towards the origin of the solar system. Space Sci. Rev. 128(1-4), 1–21 (2007)

Grimm, R.E.: A comparison of time domain electromagnetic and surface nuclear magnetic resonance sounding for subsurface water on Mars. J. Geophys. Res. **108**(E4), 8037 (2003)

Herique, A., Kofman, W.: Definition of the CONSERT/Rosetta radar performances. Committee on earth observation satellites (CEOS) SAR workshop, Tokyo, 2–5 Apr 2001, CEOS–SAR01–006: 275

Holmberg, N.A., Faust, R.P., Holt, H.M.: Viking'75 spacecraft design and test, Summary Volume 1, NASA Ref Pub. 1027, 174–180 (1980)

Jones, W.L., Schroeder, L.C., Bracalente, E.M., Boggs, D.H., Brown, R.A., Dome, G.J., Pierson, W.J., Wentz, F.J.: The Seasat-A satellite scatterometer—the geophysical evaluation of remotely sensed wind vectors over the ocean. J. Geophys. Res. **87**(C5), 3297–3317 (1982)

Kofman, W., Barbin, Y., Klinger, J., Levasseur-Regourd, A.-C., Barriot, J.-P., Herique, A., Hagfors, T., Nielsen, E., Grün, E., Edenhofer, P., Kochan, H., Picardi, G., Seu, R., van Zyl, J., Elachi, C., Melosh, J., Veverka, J., Weissman, P., Svedhem, L.H., Hamran, S.E., Williams, I.P.: Comet nucleus sounding experiment by radiowave transmission. Adv. Space Res. Ser. **21**(11), 1589–1598 (1998)

Milkovich, S.M., Plaut, J.J., Phillips, R.J., Picardi, G., Seu, R.: MARSIS and SHARAD radar reflections within Promethei Lingula, South Polar Layered Deposits, Mars. American Geophysical Union, Fall Meeting 2007, Abstract #P11B-0545 (2007)

Mishkin, A., Morrison, J., Nguyen, T., Stone, H., Cooper, B., Wilcox, B.: Experiences with operations and autonomy of the mars pathfinder microrover. Paper presented at the IEEE aerospace conference, vol. 2, pp. 337–351, 21–28 Mar 1998

NASA/JPL (National Aeronautics and Space Administration Jet Propulsory Laboratory).: Spacecraft: Surface Operations: Rover. Available online at http://mars.nasa.gov/mer/mission/spacecraft_rover_energy.html (2004)

Nieto, C.E., Stewart, R.R.: Geophysical investigations at a mars analog site: Devon Island, Nunavut. Paper presented at the third mars polar science conference (2003)

Olhoeft, G.R.: Ground penetrating radar on Mars. Paper presented at the 7th international conference on ground penetrating radar, pp. 387–392, 27–30 May 1998

Picardi, G.: Subsurface sounding in mars advanced radar for subsurface and ionosphere sounding (MARSIS) Geochim. Cosmochim. Acta **69**(10), 531 (2005)

Porcello, L.J., Jordan, R.L., Zelenka, J.S., Adams, G.F., Phillips, R.J., Brown, W.E., Ward, S.H., Jackson, P.L.: The Apollo lunar sounder radar system. Proc. IEEE **62**, 769–783 (1974)

Porco, C.C., Baker, E., Barbara, J., Beurle, K., Brahic, A., Burns, J.A., Charnoz, S., Cooper, N., Dawson, D.D., Del Genio, A.D., Denk, T., Dones, L., Dyudina, U., Evans, M.W., Giese, B., Grazier, K., Helfenstein, P., Ingersoll, A.P., Jacobson, R.A., Johnson, T.V., McEwen, A., Murray, C.D., Neukum, G., Owen, W.M., Perry, J., Roatsch, T., Spitale, J., Squyres, S., Thomas, P.C., Tiscareno, M., Turtle, E., Vasavada, A.R., Veverka, J., Wagner, R., West, R.: Cassini imaging science: initial results on phoebe and iapetus. Science **307**(5713), 1237–1242 (2005)

Putzig, N.E., Holt, J.W., Phillips, R.J., Seu, R., Biccari, D., Campbell, B.A., Carter, L.M., Safaeinili, A., Egan, A.F.: Internal structure of the north polar layered deposits on mars from SHARAD observations. American Geophysical Union, Fall Meeting 2007, Abstract #P11B-0544

Saunders, R.S., Pettengill, G.H., Arvidson, R.E., Sjogren, W.L., Johnson, W.T.K., Pieri, L.: The Magellan venus radar mapping mission. J. Geophys. Res. **95**(B6), 8339–8355 (1990)

Seu, R., Phillips, R.J., Biccari, D., Orosei, R., Masdea, A., Picardi, G., Safaeinili, A., Campbell, B.A., Plaut, J.J., Marinangeli, L., Smrekar, S.E., Nunes, D.C.: SHARAD sounding radar on the mars reconnaissance orbiter. J. Geophys. Res. **112**, E05S05 (2007)

Simmons, G., Strangway, D.W., Annan, A.P., Baker, R., Bannister, L., Brown, R., Cubley, D., DeBettencourt, J., England, A.W., Groener, J., Kong, J., LaTorraca, G., Meyer, J., Nanda, V., Redman, J.D., Rossiter, J., Tsang, L., Urner, J., Watts, R.: The surface electrical properties experiment, Apollo 17, Preliminary Science Report, NASA SP-330, pp. 15-1–15-14 (1974)

Simon, M.N., Carter, L.M., Campbell, B.A., Phillips, R.J., Mattei, S.: Studies of lava flows in the Tharsis region of Mars using SHARAD. Paper presented at the 43rd lunar and planetary science conference, p. 1595 (2012)

Stillman, D.E., Olhoeft, G.R.: Electromagnetic properties of Martian analog minerals at radar frequencies and Martian temperatures. Paper presented at the 37th lunar and planetary science conference, p. 2002 (2006)

Sun, Z., Jia, Y., Zhang, H.: Technological advancements and promotion roles of Chang'e-3 lunar probe mission. Sci. China **56**(11), 2702–2708 (2013)

Tahu, G., Shulte, M.: Mars 2020 Project Update. NASA Planetary Science Subcommittee (PSS) Committee on Astrobiology and Planetary Science (CAPS) Report, 3 Sept 2014

Wainstein, P.A., Wan Bun Tseung, J.-M., Moorman, B.J., Stevens, C.W.: Integrating GPR and CCRI techniques: Implications for the identification and mapping of ground ice on Mars. Int. J. Mars Sci. Explor. **4**, 1–13 (2008)

Zou, X.D., Li, C.L., Liu, J.J., Mu, L.L., Ren, X., Gao, X.Y., Zhang, X.X.: The preliminary analysis of the Crater X Near Chang'e-3 landing site. Paper presented at the 45th lunar and planetary science conference, p. 2403 (2014)

MIX
Papier aus verantwortungsvollen Quellen
Paper from responsible sources
FSC® C105338

If you have any concerns about our products,
you can contact us on
ProductSafety@springernature.com

In case Publisher is established outside the EU,
the EU authorized representative is:
Springer Nature Customer Service Center GmbH
Europaplatz 3, 69115 Heidelberg, Germany

Printed by Libri Plureos GmbH
in Hamburg, Germany